Master Handbook of
1001
MORE
Practical Electronic Circuits

Master Handbook of 1001 MORE Practical Electronic Circuits

Edited by Michael L. Fair

TAB BOOKS Inc.
BLUE RIDGE SUMMIT, PA. 17214

FIRST EDITION

FIFTH PRINTING

Printed in the United States of America

Reproduction or publication of the content in any manner, without express permission of the publisher, is prohibited. No liability is assumed with respect to the use of the information herein.

Copyright © 1979 by TAB BOOKS Inc.

Library of Congress Cataloging in Publication Data

Main entry under title:

Master handbook of 1001 more practical electronic circuits.

    Includes index.
    1. Electronic circuits—Handbooks, manuals, etc.   I. 73 magazine for radio amateurs.
TK7867.M359   621.3815'43'0202   78-10709
ISBN 0-8306-8804-8
ISBN 0-8306-7804-2 pbk.

Preface

Ever since James Watt discovered that work could be obtained from electrons, men have been designing circuits to do some specific job for them. In 1975 TAB BOOKS published a collection of designs, *Master Handbook of 1001 Practical Electronic Circuits*, that is highly popular with engineers, technicians, hobbyists, hams and others. In view of this overwhelming response TAB BOOKS sought out an additional 1001 circuits from the electronics industry.

As you might be well aware the discrete component is beginning to lose its identity with the onset of the integrated circuit. Scores of components once itemized in a parts list now are lumped into a single entry entitled hybrid module or 16-pin DIP, for example, with some unfamiliar part number. Many of the old timers hesitate to let go of this easily recognizable identification, but it is through this integrated packaging that we are able to design much more complex circuits. Integrated circuits contain years and years of proven circuits engineered and tested by people like yourself. As time goes by we'll all see more and more circuitry packaged in these chips as is evident with microprocessors; one day you are hearing of 4-bit microprocessors and the next they're talking about the 32-bit types.

Back on earth...we're still designing gadgets to do little everyday jobs. We'd like to have our garage lights come on when our headlights hit a sensor, or we've got it under our skin to build a transmitter that spans the globe. Whatever it is the job will be easier with the collection of proven circuits, maybe integrated circuits, contained on the following pages.

I'd like to be able to say that I designed all of these circuits myself, but in fact the circuits and information contained here were supplied by many of our leading semiconductor manufacturers. A list of these companies can be found on page 4. Many thanks to these companies for their kind cooperation and assistance in the preparation of this publication.

Classifying each of these 1001 circuits was a difficult task. Labels are hard to assign to circuits that can be used for many purposes. I highly recommend referring to the Index at the rear of this book when you have difficulty searching out a particular circuit.

Michael L. Fair

Acknowledgments

Analog Devices, Incorporated
Burr-Brown Research Corporation
Datel Systems, Incorporated
Fairchild Camera & Instrument Corporation
General Electric Company
GTE Sylvania Incorporated
Intel Corporation
Intersil, Incorporated
Motorola Semiconductor Products Incorporated
National Semiconductor Corporation
Texas Instruments Incorporated

Contents

Bridge Circuits 9
Solid-State Switches 13
Logarithmic Amplifiers 16
Integrators 18
FSK Circuits 21
Detectors 24
Adders 27
Servo Motor Circuits 30
Battery Chargers 36
Data Transfer Circuits 41
Smoke Detectors 43
Multivibrators 49
Frequency Doublers 53
Biomedical 55
Video Amplifiers 60
Gadgets 64
Photo-Activated Circuits 71
Filters 77
Sample & Hold Circuits 83
Test Gear & Metering Circuits 87

- Op Amp Circuits 105
- AM & FM Broadcast Receivers 124
- Converters 158
- Power Supplies, Regulators 238
- Readouts 251
- Interface Circuits 252
- Chopper Circuits 262
- Indicator Circuits 265
- Audio Amplifiers 275
- Waveform Generators 342
- Oscillators 351
- Math Function Circuits 371
- Power-Controlling Circuits 397
- Computer-Related Circuits 435
- Timers & Counters 458
- Sensing Circuits 500
- Multiplexers 510
- Transmitter & Receiver Circuits 516
- Miscellaneous Circuits 586
- TV Circuits 612
- Index 689

Bridge Circuits

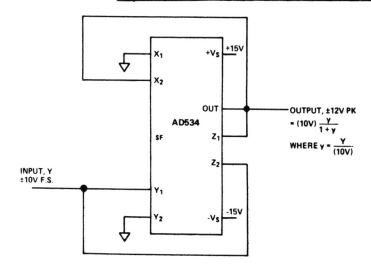

Bridge linearization function circuit using an AD534 multiplier/divider chip (courtesy Analog Devices, Inc.).

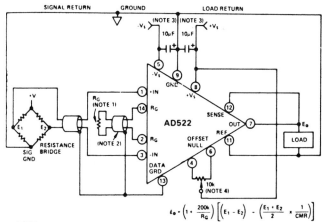

NOTES
1. GAIN RESISTOR R_G SHOULD BE 5ppm/°C (VISHAY TYPE RECOMMENDED).
2. SHIELDED CONNECTIONS TO R_G RECOMMENDED WHEN MAXIMUM SYSTEM BANDWIDTH AND AC CMR IS REQUIRED, AND WHEN R_G IS LOCATED MORE THAN SIX INCHES FROM AD522. NO INSTABILITIES ARE CAUSED BY REMOTE R_G LOCATIONS. WHEN NOT USED, THE DATA GUARD PIN CAN BE LEFT UNCONNECTED.
3. POWER SUPPLY FILTERS ARE RECOMMENDED FOR MINIMUM NOISE IN NOISY ENVIRONMENTS.
4. NO TRIM REQUIRED FOR MOST APPLICATIONS. IF REQUIRED, A 10kΩ, 25ppm/°C, 25 TURN TRIM POT (SUCH AS VISHAY 1202 Y 10k) IS RECOMMENDED.

Bridge amplifier. Typical supply voltage is ±15 volts (courtesy Analog Devices, Inc.).

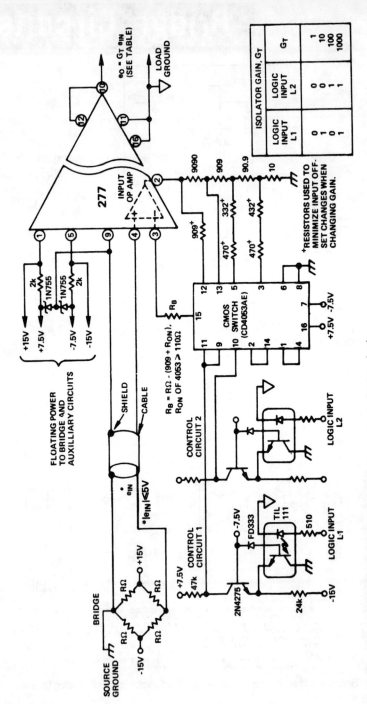

Programmable gain bridge transducer amplifier (courtesy Analog Devices, Inc.).

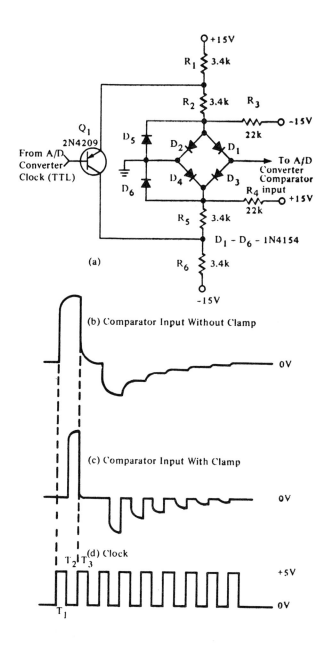

Diode-bridge clamping circuit to improve A/D converter performance (courtesy Burr-Brown Corporation).

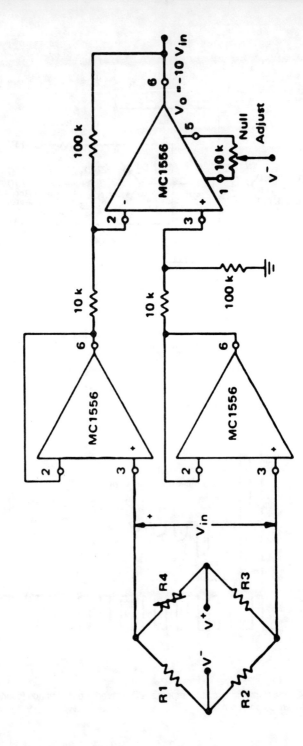

High-impedance bridge amplifier using three MC1556 op amps (courtesy Motorola Semiconductor Products Inc.).

Solid-State Switches

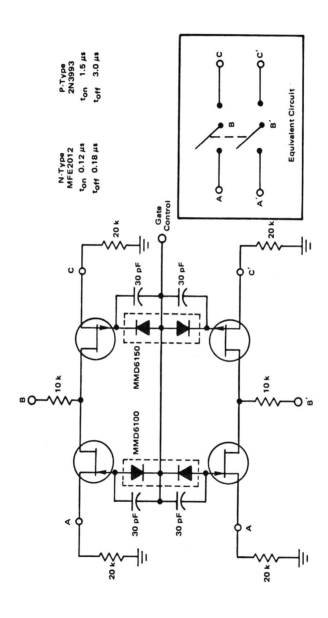

DPDT FET switch (courtesy Motorola Semiconductor Products Inc.).

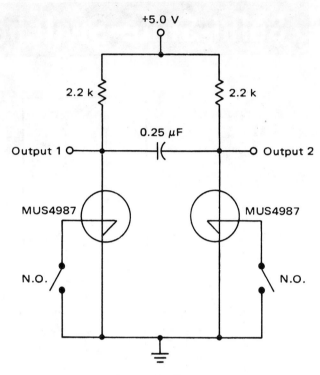

Bistable (memory) switch using two SUS devices. The mechanical switches can be replaced with other solid-state devices or negative-going pulses referenced to the supply line (courtesy Motorola Semiconductor Products Inc.).

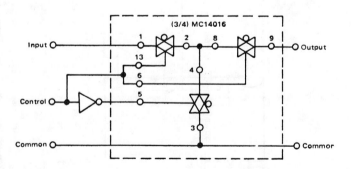

Series-shunt analog switch, useful in applications requiring extremely low input-to-output signal feedthrough (courtesy Motorola Semiconductor Products Inc.).

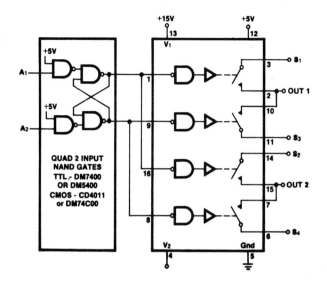

Truth Table (IH5052)

COMMAND		STATE OF SWITCHES AFTER COMMAND	
A_2	A_1	S_3 & S_4	S_1 & S_2
0	0	same	same
0	1	on	off
1	0	off	on
1	1	INDETERMINATE	

Latching DPDT switch. The A_1 and A_2 inputs are normally low. A high input to A_2 turns S_1 and S_2 on. A high to A_1 turns S_3 and S_4 on. This feature is desirable with limit detectors, peak detectors or mechanical closures (courtesy Intersil, Inc.).

Logarithmic Amplifiers

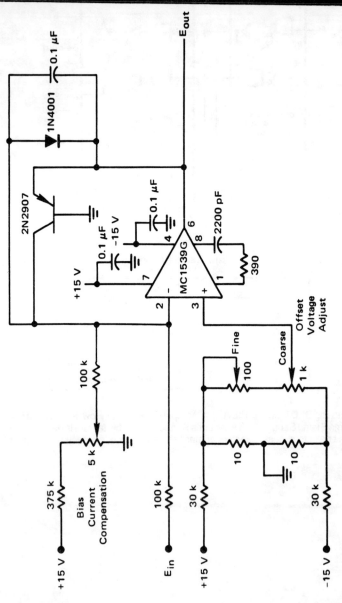

Logarithmic amplifier using an MC1539G op amp (courtesy Motorola Semiconductor Products Inc.).

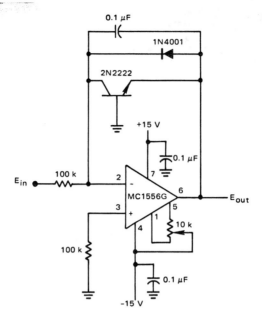

Logarithmic amplifier using an MC1556 op amp. The 10K pot is an offset adjustment (courtesy Motorola Semiconductor Products Inc.).

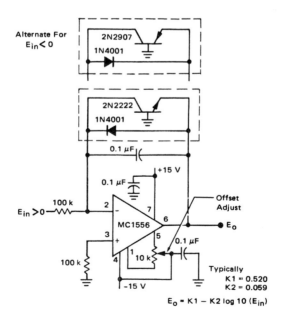

Typically
K1 = 0.520
K2 = 0.059

$E_o = K1 - K2 \log 10\, (E_{in})$

Logarithmic amplifier using an MC1556 op amp (courtesy Motorola Semiconductor Products Inc.).

Integrators

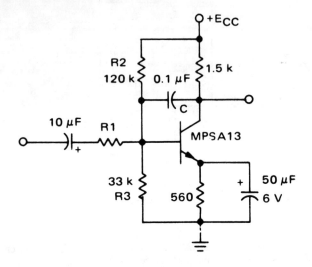

Operational integrator. Two cascaded 2N3904s can replace the MPSA 13 (courtesy Motorola Semiconductor Products Inc.).

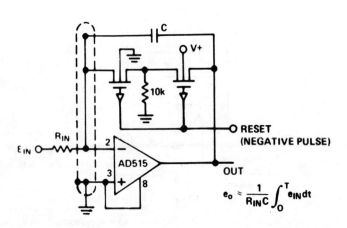

$$e_o = \frac{1}{R_{IN}C} \int_0^T e_{IN} dt$$

Low-drift integrator with low-leakage-guarded reset (courtesy Analog Devices, Inc.).

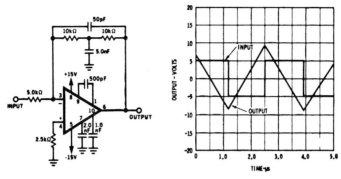

High-speed integrator using an ECG915 operational amplifier (courtesy GTE Sylvania Incorporated).

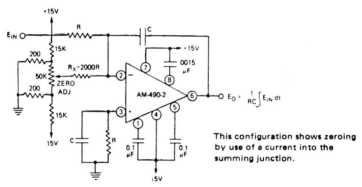

Precision integrator using the Datel AM-490-2 8-pin TO-99 (courtesy Datel Systems, Inc.).

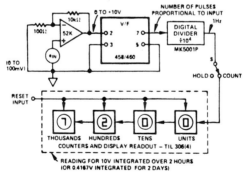

Long-term precision integrator for such things as pollution monitoring. The analog signal is applied to precision input amplifier model 52K, then to the V/F converter input. Model 458 is a 100 kHz V/F converter while model 460 is a 1 MHz V/F converter. The V/F output is connected to a large capacity counter and display, operating as a totalizer. The total pulse count is equal to the time integral of the analog input signal (courtesy Analog Devices, Inc.).

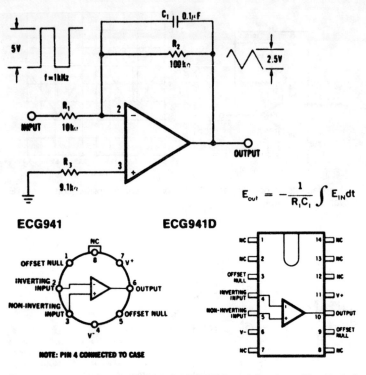

Simple integrator using an ECG941/941D/941M operational amplifier. Typical supply voltage is ±15 volts (courtesy GTE Sylvania Incorporated).

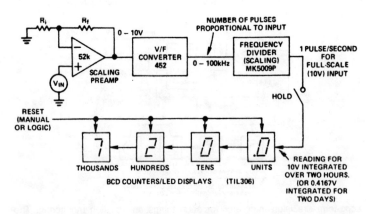

Long-term precision integrator for such applications as pollution control. The analog signal is applied to precision amplifier model 52K, then to the V/F converter input. Model 452's output is connected to a high capacity counter and display operating as a totalizer. The total pulse is equal to the time integral of the analog input (courtesy Analog Devices, Inc.).

FSK Circuits

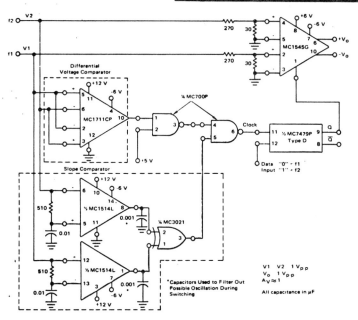

FSK with slope and voltage detection (courtesy Motorola Semiconductor Products Inc.).

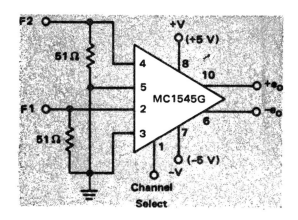

Frequency shift keyer using the MC1545G wide-band amplifier (courtesy Motorola Semiconductor Products Inc.).

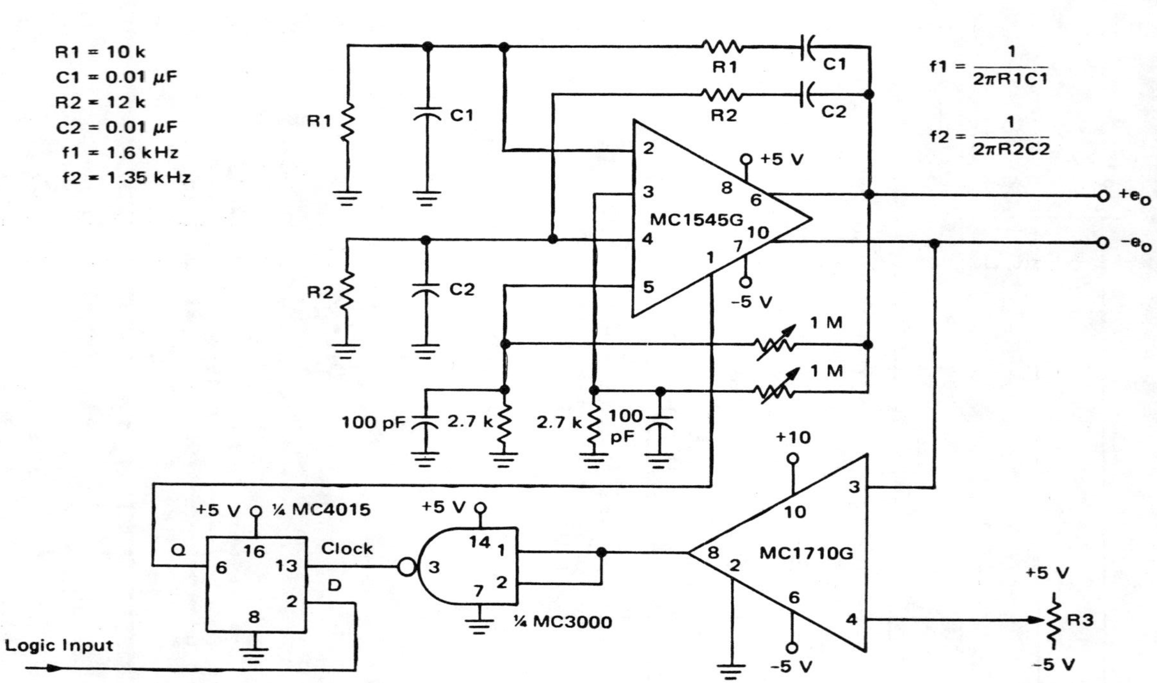

$$f_1 = \frac{1}{2\pi R_1 C_1}$$

$$f_2 = \frac{1}{2\pi R_2 C_2}$$

R1 = 10 k
C1 = 0.01 µF
R2 = 12 k
C2 = 0.01 µF
f1 = 1.6 kHz
f2 = 1.35 kHz

Self-generating FSK (courtesy Motorola Semiconductor Products Inc.).

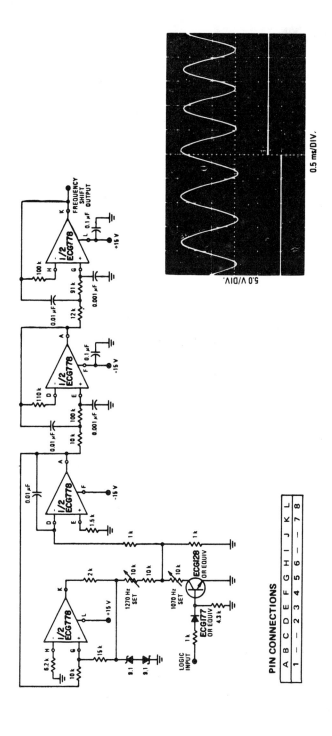

Frequency-shift keyer tone generator using two 8-pin DIPs and one transistor (courtesy GTE Sylvania Incorporated).

Detectors

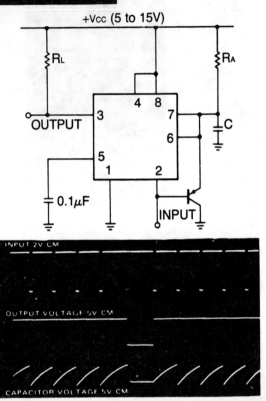

Missing pulse detector using an ECG955M timer/oscillator chip. The timing cycle is continuously reset by the input pulse train. A change in frequency or missing pulse allows completion of the timing cycle, which causes a change in the output level. The time delay should be set a little longer than normal between pulse for this reason (courtesy GTE Sylvania Incorporated).

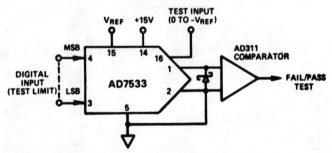

Digitally programmable limit detector (courtesy Analog Devices, Inc.).

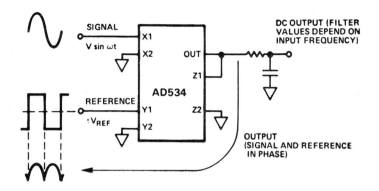

+V $V_R \sin \omega t$ DURING POSITIVE HALF-CYCLE
(−V)(−V_R) $\sin \omega t$ DURING NEGATIVE HALF-CYCLE

POSITIVE − SIGNAL AND REFERENCE IN PHASE
NEGATIVE − SIGNAL AND REFERENCE 180° OUT OF PHASE

Phase-sensitive detector with square-wave reference. If the input and reference are in phase the output is positive. If they are 180° out of phase the output is negative (courtesy Analog Devices, Inc.).

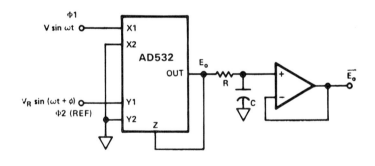

$$E_o = \frac{V V_R}{10} \sin \omega t (\sin \omega t \cos \phi + \cos \omega t \sin \phi)$$

$$E_o = \frac{V V_R}{10} (\sin^2 \omega t \cos \phi + \sin \omega t \cos \omega t \sin \phi)$$

$$E_o = \frac{V V_R}{20} ([1 - \cos 2\omega t]\cos \phi + \sin 2\omega t \sin \phi)$$

$$\overline{E_o} = \frac{V V_R}{20} \cos \phi = \frac{V V_R}{20} \text{ (IN PHASE)}$$

$$-\frac{V V_R}{20} \text{ (180° OUT OF PHASE)}$$

Phase-sensitive detector for sinusoidal signals. This circuit measures the magnitude of in-phase or 180°-out-of-phase inputs with the proper polarity, depending on the relationship to the reference with less than 1% error. The op amp shown is a AD741J (courtesy Analog Devices, Inc.).

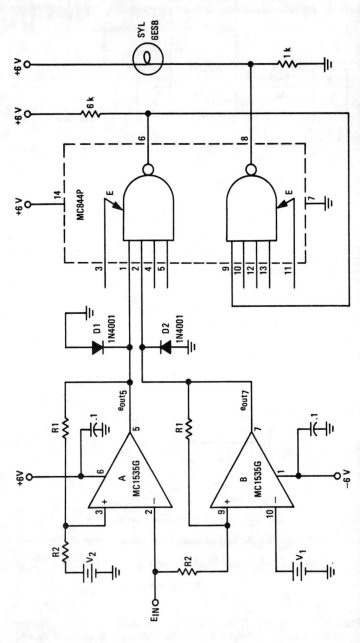

Signal-level envelope detector. The MC1535G is a dual op amp and the MC844P a dual power gate. This circuit indicates by way of the lamp when the input signal is out of range (courtesy Motorola Semiconductor Products Inc.).

Adders

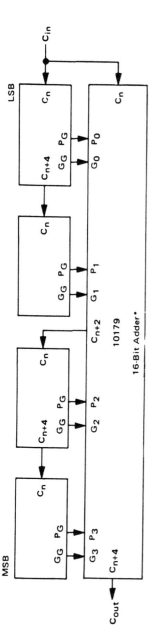

16-bit look-ahead carry adder using an MC10179 and four MC10181s (courtesy Motorola Semiconductor Products Inc.).

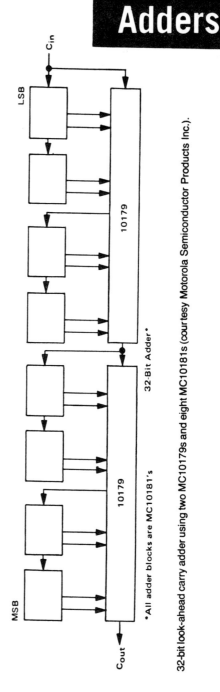

*All adder blocks are MC10181's

32-bit look-ahead carry adder using two MC10179s and eight MC10181s (courtesy Motorola Semiconductor Products Inc.).

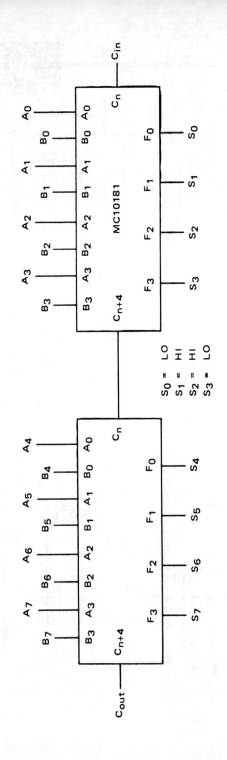

8-bit parallel adder using the MC10181 (courtesy Motorola Semiconductor Products Inc.).

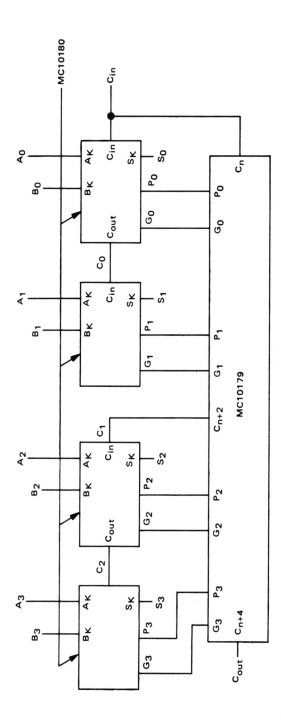

4-bit look-ahead carry adder using an MC10179 and four MC10180s (courtesy Motorola Semiconductor Products Inc.).

Servo Motor Circuits

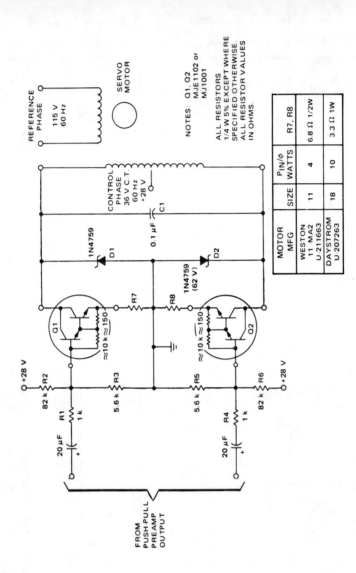

Servo motor power amplifier (courtesy Motorola Semiconductor Products Inc.).

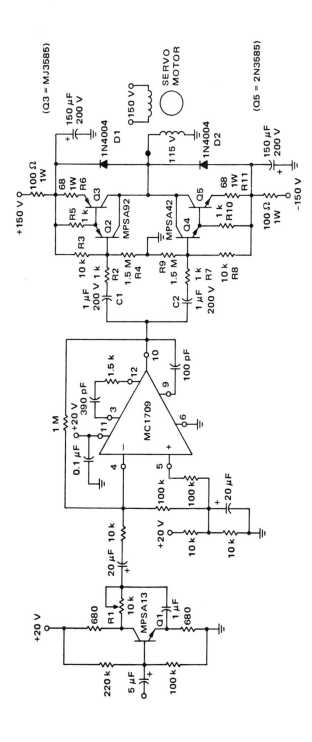

Line-operated servo motor amplifier (courtesy Motorola Semiconductor Products Inc.).

Servo motor preamplifier (courtesy Motorola Semiconductor Products Inc.).

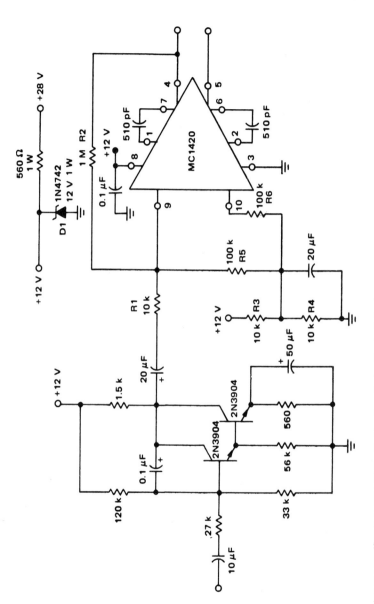

Servo motor preamplifier (courtesy Motorola Semiconductor Products Inc.).

Servo motor preamplifier (courtesy Motorola Semiconductor Products Inc.).

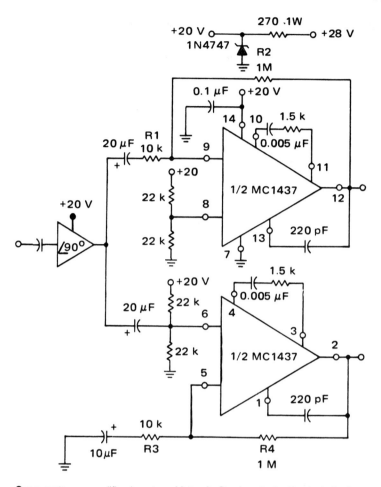

Servo motor preamplifier (courtesy Motorola Semiconductor Products Inc.).

Battery Chargers

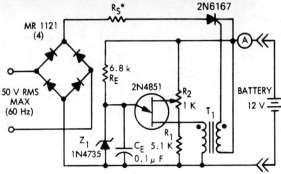

T_1 - PRIMARY = 30 TURNS #22
SECONDARY = 45 TURNS #22
CORE = FERROXCUBE 203 F 181-3C3
* R_S - SERIES RESISTANCE TO LIMIT CURRENT THROUGH SCR.
2N6167 IS RATED AT 20 AMPS RMS.

12-volt battery charger with 20-ampere RMS maximum. This circuit is a simple relaxation oscillator using a UJT. The circuit will not work unless the polarity of the battery is observed when connected (courtesy Motorola Semiconductor Products Inc.).

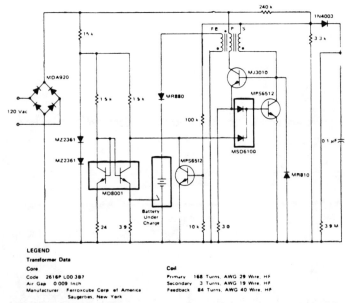

25 kHz nicad battery charger with third-electrode sensing (courtesy Motorola Semiconductor Products Inc.).

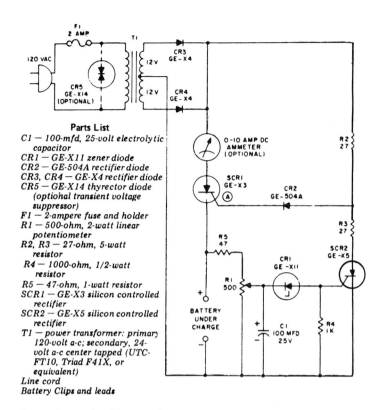

Parts List

C1 — 100-mfd, 25-volt electrolytic capacitor
CR1 — GE-X11 zener diode
CR2 — GE-504A rectifier diode
CR3, CR4 — GE-X4 rectifier diode
CR5 — GE-X14 thyrector diode (optional transient voltage suppressor)
F1 — 2-ampere fuse and holder
R1 — 500-ohm, 2-watt linear potentiometer
R2, R3 — 27-ohm, 5-watt resistor
R4 — 1000-ohm, 1/2-watt resistor
R5 — 47-ohm, 1-watt resistor
SCR1 — GE-X3 silicon controlled rectifier
SCR2 — GE-X5 silicon controlled rectifier
T1 — power transformer: primary 120-volt a-c; secondary, 24-volt a-c center tapped (UTC-FT10, Triad F41X, or equivalent)
Line cord
Battery Clips and leads

Automotive regulated battery charger. Adjust R1 for a fully charged battery voltage. This inexpensive unit can charge any 12-volt lead-acid battery. Should the battery become discharged while the charger is connected the unit will come back on again to recharge the battery (courtesy General Electric Company).

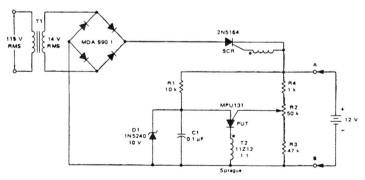

12-volt battery charger with an SCR and a PUT (courtesy Motorola Semiconductor Products Inc.).

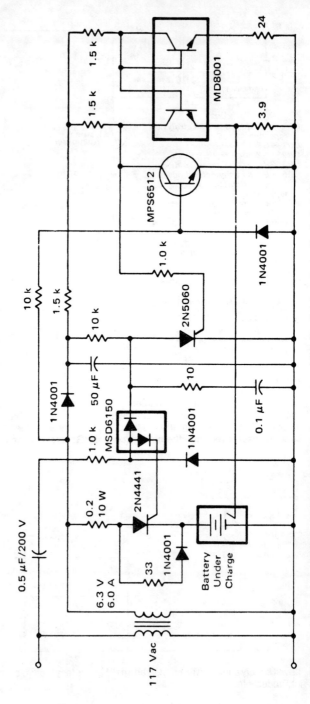

60-hertz nicad battery charger with third-electrode sensing (courtesy Motorola Semiconductor Products Inc.).

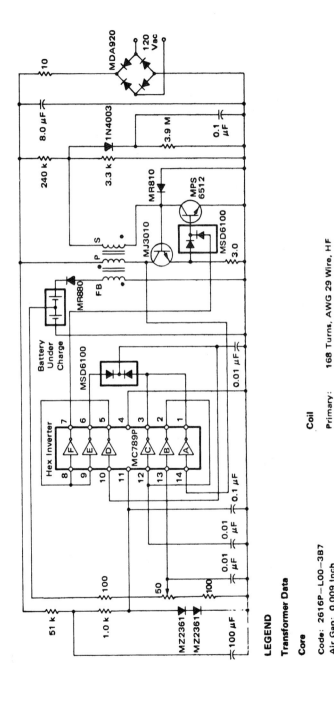

20 kHz nicad battery charger with voltage sensing (courtesy Motorola Semiconductor Products Inc.).

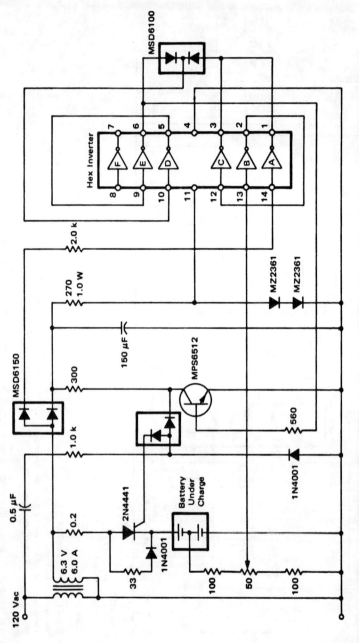

60-hertz nicad battery charger with voltage sensing. Hex inverter is an MC789P (courtesy Motorola Semiconductor Products Inc.).

Data Transfer Circuits

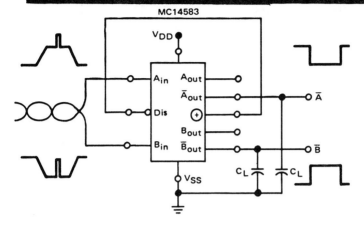

Dual Schmitt trigger used as a 2-wire differential line receiver (courtesy Motorola Semiconductor Products Inc.).

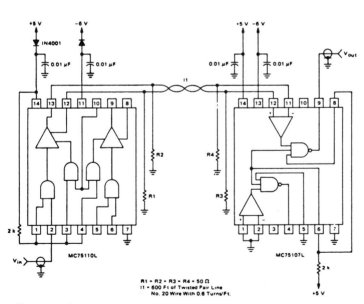

Two-wire differential data transmission system using an MC75110 driver and an MC75107 receiver (courtesy Motorola Semiconductor Products Inc.).

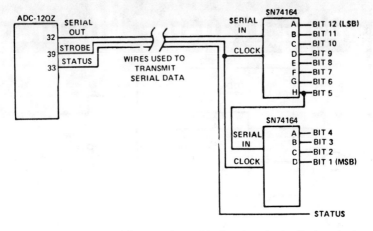

3-wire plus digital ground data transfer system (courtesy Analog Devices, Inc.).

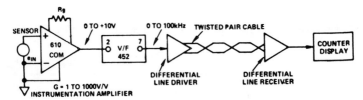

High-noise immunity data transmission system using a V/F converter. Model 610, an instrumentation amplifier, amplifies the low-level differential transducer signal up to the 10-volt full scale input level of the 452 V/F converter. A differential line driver is used to drive a twisted pair. A differential line receiver is used to drive the digital counter and display (courtesy Analog Devices, Inc.).

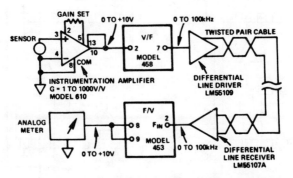

High-performance high-noise-rejection two-wire data transmission system. Instrumentation amplifier 610 amplifies the low-level transducer signal to apply to the 458 V/F converter. A differential line driver is used to drive the twisted pair. A differential line receiver is used to drive the 453 F/V converter, which in turn powers the analog meter (courtesy Analog Devices, Inc.).

Smoke Detectors

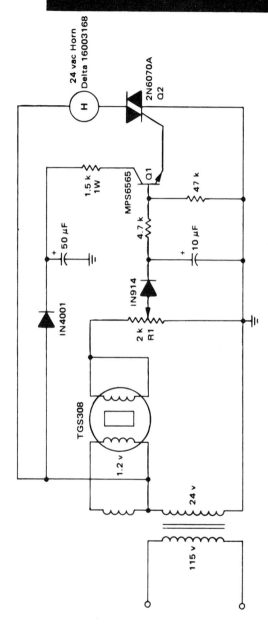

TGS gas/smoke detector with triac control (courtesy Motorola Semiconductor Products Inc.).

TGS gas/smoke detector using a McMOS gated oscillator for triac control (courtesy Motorola Semiconductor Products Inc.).

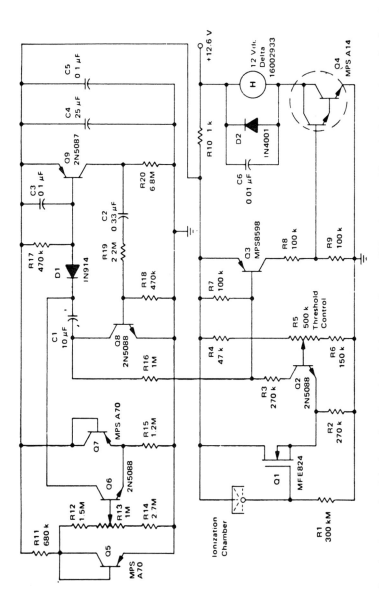

Ionization-chamber smoke detector using discrete transistors for alarm oscillators (courtesy Motorola Semiconductor Products Inc.).

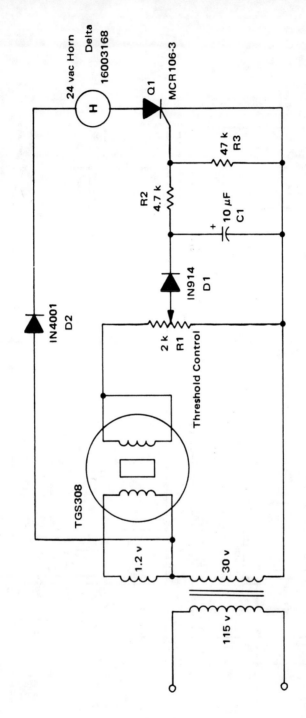

SCR gas/smoke detector with half-wave control of 24-volt AC horn. The Taguchi gas sensor (TGS) consists of an N-type semiconductor of tin dioxide encased in a noble metal wire heater, which also serves as an electrode (courtesy Motorola Semiconductor Products Inc.).

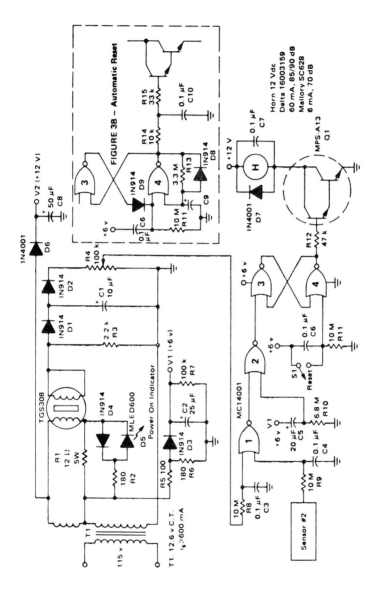

TGS gas/smoke detector with 2-minute time delay using a McMOS latch (courtesy Motorola Semiconductor Products Inc.).

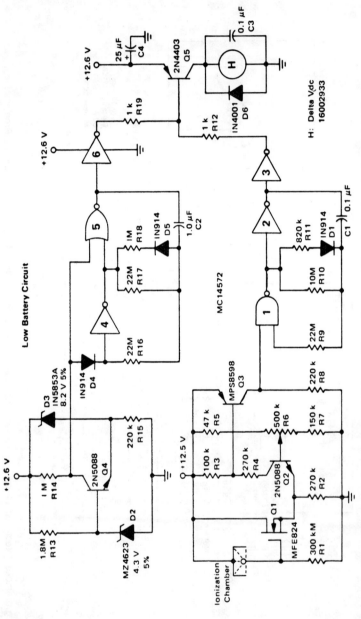

Ionization-chamber smoke detector using McMOS alarm oscillators (courtesy Motorola Semiconductor Products Inc.).

Multivibrators

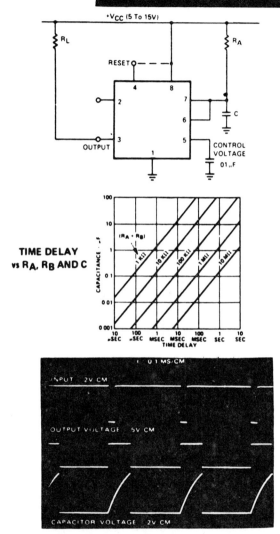

TIME DELAY vs R_A, R_B AND C

One-shot multivibrator using an ECG955M timer/oscillator chip. Upon application of a negative trigger to pin 2 the flip-flop is set, which releases the short circuit across the external capacitor. This drives the output high. The photo shows the actual waveforms generated in this mode. The time that the output is high is given by $t = 1.1 R_A C$ and can be easily determined by the graph showing time delay. Applying a negative pulse to reset terminal 4 during the timing cycle discharges the capacitor and causes the cycle to start again. When the reset function is not used it is recommended that it be connected to supply voltage to avoid false triggering (courtesy GTE Sylvania Incorporated).

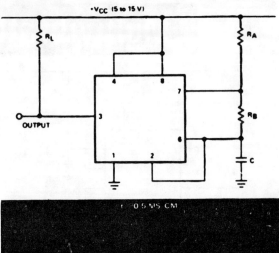
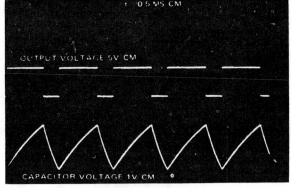

FREE RUNNING FREQUENCY
vs R_A, R_B AND C

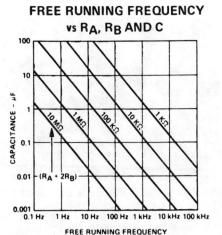

Free-running multivibrator using an ECG955M timer/oscillator chip. The external capacitor charges through R_A and R_B, but discharges through R_B only. The photo shows the actual waveforms generated. The free-running frequency can be easily found by referring to the graph (courtesy GTE Sylvania Incorporated).

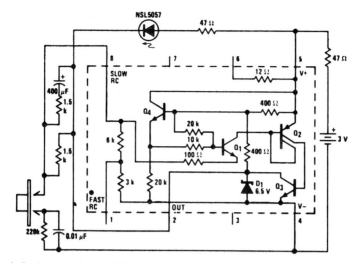

Indicating one-shot multivibrator. The circuit delivers a 0.5-second flash from the LED every time the pushbutton makes contact (courtesy National Semiconductor Corporation).

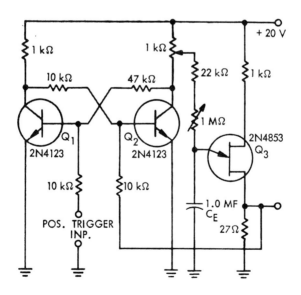

One-shot multivibrator using a UJT. This circuit is insensitive to bias voltage changes (courtesy Motorola Semiconductor Products Inc.).

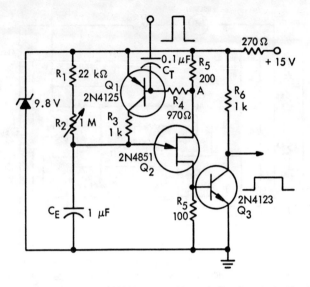

One-shot multivibrator using a UJT (courtesy Motorola Semiconductor Products Inc.).

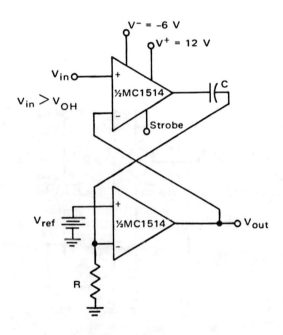

Monostable multivibrator (courtesy Motorola Semiconductor Products Inc.).

Frequency Doublers

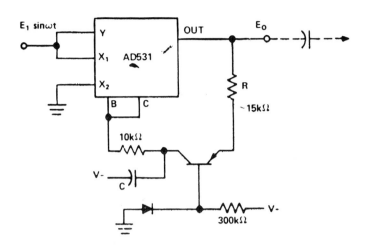

$$E_O = \frac{E_1^2 - E_1^2 \cos 2\omega t}{2 \overline{E_O}}$$

$$E_O = KE_1 - KE_1 \cos 2\omega t$$

Frequency doubler with linear amplitude response (courtesy Analog Devices, Inc.).

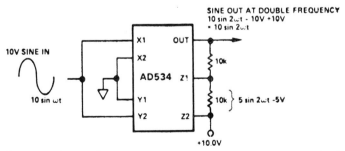

Frequency doubler. This circuit accepts a sinusoidal signal with a 10-volt amplitude and produces a double-frequency signal also having a 10-volt amplitude with no DC offset (courtesy Analog Devices, Inc.).

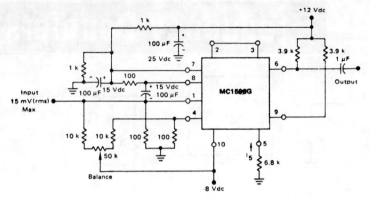

Low-frequency doubler using an MC1596G. This circuit works well in the low-frequency and audio range below 1 MHz (courtesy Motorola Semiconductor Products Inc.).

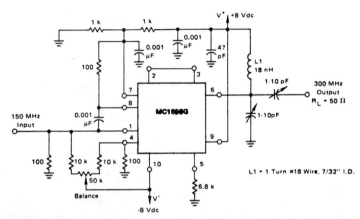

150 MHz to 300 MHz frequency doubler using an MC1596G. Spurious outputs are 20 dB below the desired output (courtesy Motorola Semiconductor Products Inc.).

Biomedical

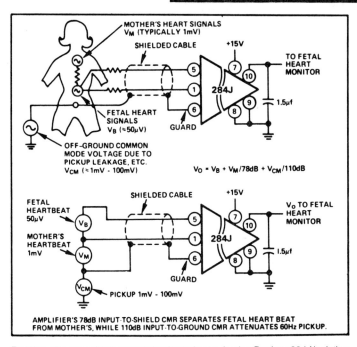

Fetal heartbeat monitoring input circuitry using an Analog Devices 284J isolation amplifier (courtesy Analog Devices, Inc.).

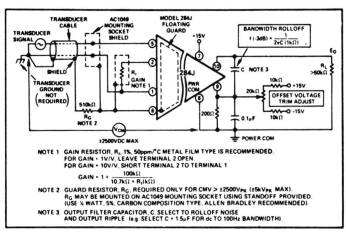

Isolation amplifier for biomedical and industrial applications. (courtesy Analog Devices, Inc.).

Multilead EKG recorder input circuitry using a 284-J isolation amplifier (courtesy Analog Devices, Inc.).

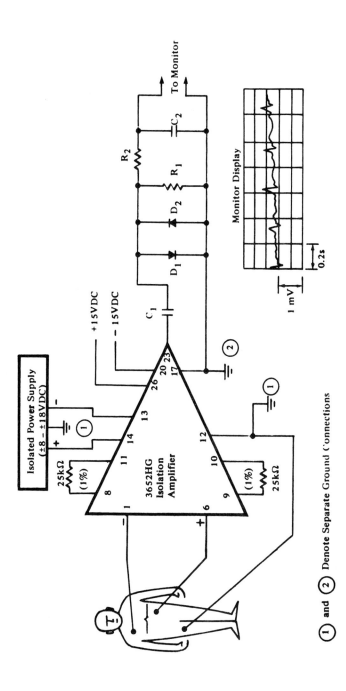

EKG input amplifier using an optically coupled 3652 HG isolation amplifier to protect the patient from possible lethal potentials (courtesy Burr-Brown Research Corporation).

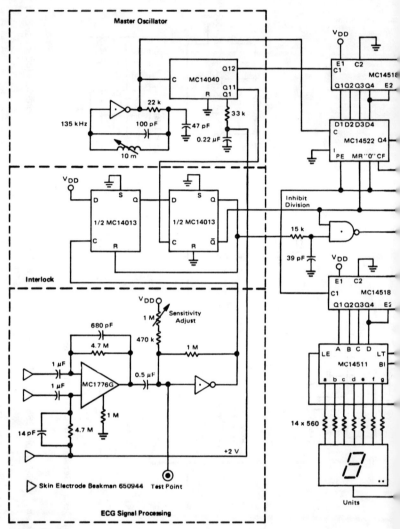

*MC14572 Hex Functional Gate   **Common Cathode Display   HP5082-7740

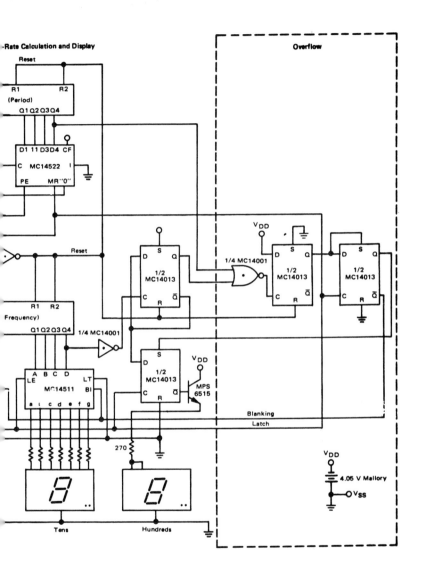

Heart rate monitor (courtesy Motorola Semiconductor Products Inc.).

Video Amplifiers

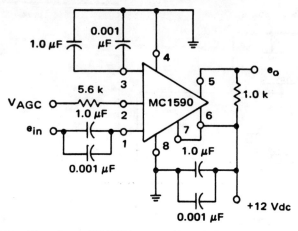

Video amplifier using an MC1590 (courtesy Motorola Semiconductor Products Inc.).

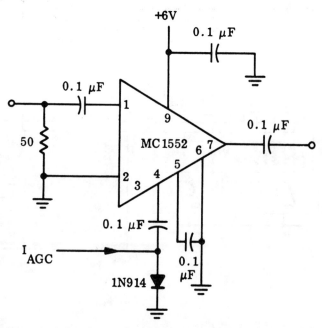

Video amplifier with AGC using an MC1552 (courtesy Motorola Semiconductor Products Inc.).

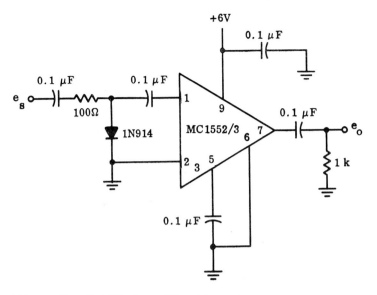

Video amplifier with AGC using an MC1552/1553 wide-band amplifier (courtesy Motorola Semiconductor Products Inc.).

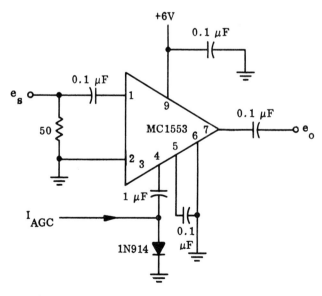

Video amplifier with AGC using an MC1553 wide-band amplifier (courtesy Motorola Semiconductor Products Inc.).

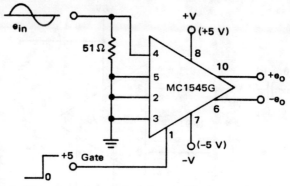

Video switch. With a logic one at pin 1 the amplifier is turned on (courtesy Motorola Semiconductor Products Inc.).

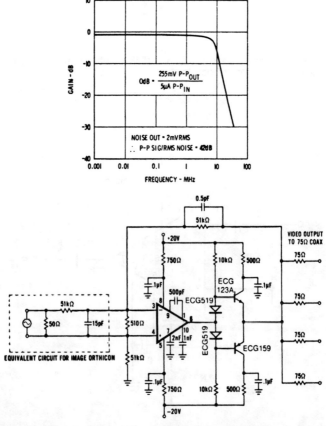

Wide-band video amplifier using an ECG915 operational amplifier with 75-ohm coax drive capability (courtesy GTE Sylvania Incorporated).

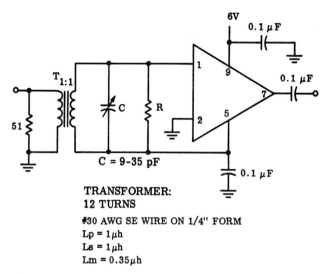

30 MHz video amplifier using an MC1552G. Set R to 1K for a voltage gain of 55 (courtesy Motorola Semiconductor Products Inc.).

Gadgets

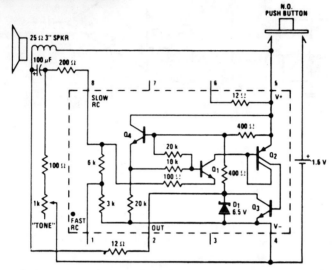

Electronic trombone using an LM3909 chip. Circuitry inside dashed lines is the LM3909. The location and surroundings of the speaker determine the frequency output. Construct a cubical box roughly 64 cubic inches with one end that is able to slide in and out like a piston. The box should be stiffened with thin layers of pressed wood. Minimum volume should be 10 cubic inches. Speaker, circuit and battery should be mounted in the sliding end. A tube 2½ inches long with an inside diameter of 5/16 inch should be provided to bleed air in and out as the piston moves (courtesy National Semiconductor Corporation).

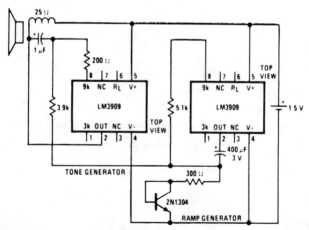

Whooper siren using two LM3909 chips (courtesy National Semiconductor Corporation).

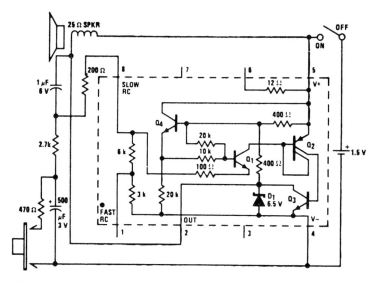

Fire siren using an LM3909 chip. Circuitry inside dashed lines is the LM3909. This circuit produces a rapidly rising sound when pressing the pushbutton and a coasting sound upon release (courtesy National Semiconductor Corporation).

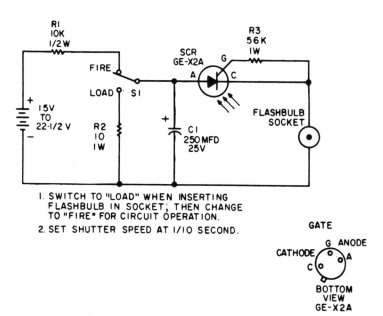

1. SWITCH TO "LOAD" WHEN INSERTING FLASHBULB IN SOCKET; THEN CHANGE TO "FIRE" FOR CIRCUIT OPERATION.
2. SET SHUTTER SPEED AT 1/10 SECOND.

Light-triggered photoflash slave. This circuit is activated when you operate your camera's flash. If a 22.5-volt battery is not available use two 9-volt batteries. To increase the LASCR sensitivity mount it behind a lens or reflector (courtesy General Electric Company).

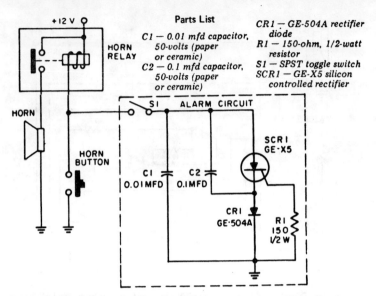

Automotive burglar alarm for 12-volt negative ground systems. The entire alarm circuit (inside dashed lines) can be mounted on the SPST toggle switch. The circuit is triggered whenever any electrical device in the car is turned on. To turn off the alarm it is necessary to turn off S1 or hit the horn button. Mount S1 in the truck (courtesy General Electric Company).

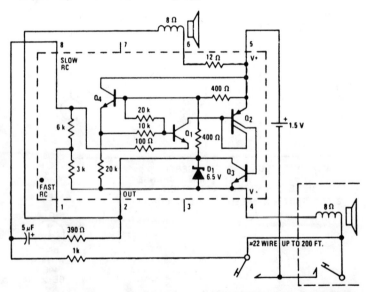

Morse code set using an LM3909 chip. Circuitry inside dashed lines is the LM3909. The three-wire system and parallel telegraph keys allow the user to practice with send/receive switches (courtesy National Semiconductor Corporation).

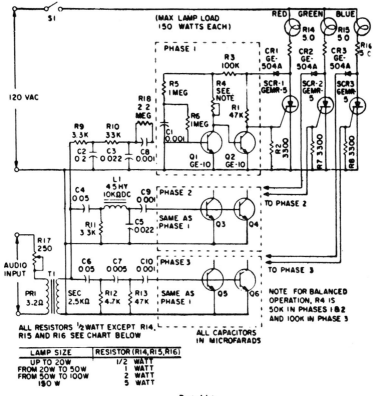

Parts List

C1, C8, C9, C10 — 0.001-mfd, 100-volt capacitor, paper (total 6)
C2 — 0.2-mfd, 50-volt capacitor, paper
C3, C5 — 0.022-mfd, 50-volt capacitor, paper
C4, C6 — 0.05-mfd, 50-volt capacitor, paper
C7 — 0.005-mfd, 50-volt capacitor, paper
CR1 thru CR3 — GE-504A rectifier diode (total 3)
L1 — 4.5 henry, 10,000-ohm DC choke (Triad S-12X, or equivalent)
Q1 thru Q6 — GE-10 transistor (total 6)
R1, R13 — 47K, 1/2-watt, 10% resistor (total 4)
R2, R7, R8, R9, R11 — 3.3K, 1/2-watt, 10% resistor
R3 — 100K, 1/2-watt, 10% resistor (total 3)
R4 — Two 50K potentiometers (Phase 1 and 2) and one 100K potentiometer (Phase 3)

R5, R6 — 1-megohm, 1/2-watt, 10% resistor (total 6)
R10 — 33K, 1/2-watt, 10% resistor
R12 — 4.7K, 1/2-watt, 10% resistor
R13 — 47K, 1/2-watt, 10% resistor
R14 thru R16 — 5 ohms (see text for rating)
R17 — 250-ohm potentiometer
R18 — 2.2-megohms, 1/2-watt, 10% resistor
S1 — SPST toggle switch, 5-amps (minimum)
SCR1 thru SCR3 — GEMR-5 silicon controlled rectifier
T1 — 2.5K/3.2-ohms audio output transformer (Triad A-3332, or equivalent)
Minibox — 6" x 5" x 4" (Bud CU-3007-A, or equivalent)
Line cord and grommet
Sockets, AC (total 3)
Vectorboard and pins

Audio dancing lights with three channels. This circuit uses the audio output from your stereo and separates the high, low and medium frequencies to activate three colored lights. Input sensitivity is controlled by R17 (courtesy General Electric Company).

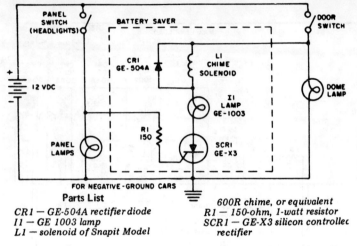

Parts List

CR1 — GE-504A rectifier diode
I1 — GE 1003 lamp
L1 — solenoid of Snapit Model
600R chime, or equivalent
R1 — 150-ohm, 1-watt resistor
SCR1 — GE-X3 silicon controlled rectifier

Automotive battery saver. This circuit uses an SCR and doorbell chime. This complete circuit can be built inside the doorbell chime housing. It is only operational when the dome light and panel light are on. The only time this occurs is when the door is open and the parking or head lights are on. Note that the SCR gate is connected to the car's panel lights (courtesy General Electric Company).

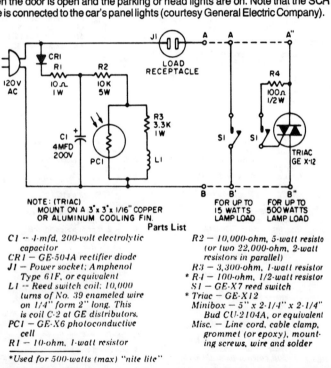

Parts List

C1 — 4-mfd, 200-volt electrolytic capacitor
CR1 — GE-504A rectifier diode
J1 — Power socket; Amphenol Type 61F, or equivalent
L1 — Reed switch coil; 10,000 turns of No. 39 enameled wire on 1/4" form 2" long. This is coil C-2 at GE distributors.
PC1 — GE-X6 photoconductive cell
R1 — 10-ohm, 1-watt resistor

R2 — 10,000-ohm, 5-watt resistor (or two 22,000-ohm, 2-watt resistors in parallel)
R3 — 3,300-ohm, 1-watt resistor
* R4 — 100-ohm, 1/2-watt resistor
S1 — GE-X7 reed switch
* Triac — GE-X12
Minibox — 5" x 2-1/4" x 2-1/4" Bud CU-2104A, or equivalent
Misc. — Line cord, cable clamp, grommet (or epoxy), mounting screws, wire and solder

*Used for 500-watts (max) "nite lite"

Automatic night light to discourage prowlers or burglars. One of two circuits can be built as shown in the schematic, depending on the requirements. For small loads, 15 watts, the reed switch can directly control the light (courtesy General Electric Company).

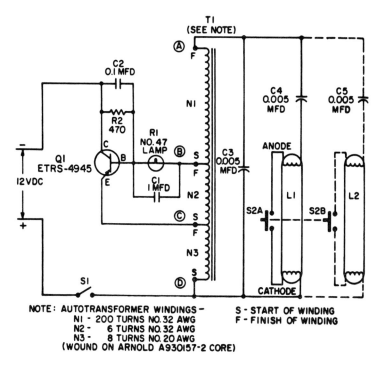

```
NOTE: AUTOTRANSFORMER WINDINGS—
      N1 - 200 TURNS NO. 32 AWG
      N2 -   6 TURNS NO. 32 AWG
      N3 -   8 TURNS NO. 20 AWG
   (WOUND ON ARNOLD A930157-2 CORE)
```

S - START OF WINDING
F - FINISH OF WINDING

Parts List

C1 — 1-mfd, 50-V paper capacitor
C2 — 0.1-mfd, 50-V paper capacitor
C3, C4, C5 — 0.005-mfd, 1000-V disc-ceramic capacitors (delete C5 for single lamp)
Q1 — ETRS-4945 transistor*
R1 — GE No. 47 bulb
R2 — 470-ohm, 1/2-watt resistor
L1, L2 — GE F8T5-CW fluorescent lamps
S1 — SPST toggle switch
S2 — DPST momentary push button switch (if only one lamp is needed, a SPST push button is used)

T1 — Autotransformer — core available from GE distributors as ETRS-4891, or from General Electric Co., Dept. B, 3800 N. Milwaukee Ave., Chicago, Ill. 60641. See text for winding details.
Minibox — 12" x 2-1/2" x 2-1/4" (Bud CU-2114-A), or equivalent
Pin Sockets — GE ALF141-33, or equivalent

*Available from General Electric Co., Dept. B, 3800 N. Milwaukee Ave., Chicago, Ill. 60641

Battery operated fluorescent light for one or two 8-watt units. The circuit is a simple transistor inverter operated by a 12-volt battery. T1 is 200 turns of AWG #32 magnet wire in two layers of 100 turns each. Cover the first layer with electrical tape before starting the second. Then cover the second with tape. Mark the two ends as start (S) and finish (F); mark this winding as N1. Next wind N2 as six turns of AWG #32 magnet wire over N1 in the same direction. Mark each end of N2 with start (S) and finish (F). Then wind N3 as eight turns of AWG #20 magnet wire over N2 in the same direction. Mark N3 with S and F leads. Each layer should have electrical tape over it (courtesy General Electric Company).

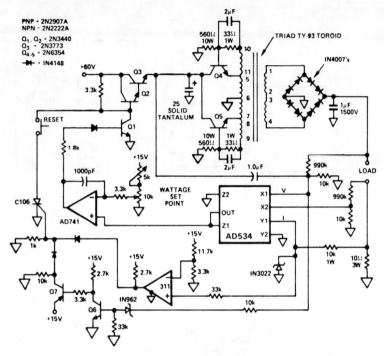

High-voltage booster. Care should be taken with this circuit because it takes 300 mW and turns it into 1000 volts at 300 watts. This is many times greater than is necessary to kill and is absolutely lethal. The AD534 is used to control the output of toroidal DC-to-DC converter. Both the AD534 multiplier and the DC-to-DC converter are inside the AD741's feedback loop. The inverter frequency is 4 kHz (courtesy Analog Devices, Inc.).

Photo-Activated Circuits

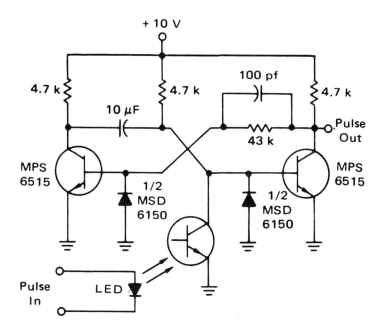

Photo-driven pulse stretcher (courtesy Motorola Semiconductor Products Inc.).

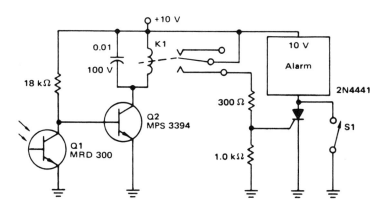

Light-relay-operated SCR alarm circuit (courtesy Motorola Semiconductor Products Inc.).

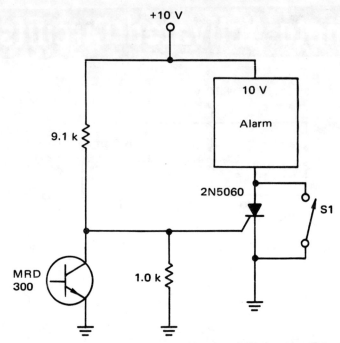

Light-operated SCR alarm using a sensitive gate SCR (courtesy Motorola Semiconductor Products Inc.).

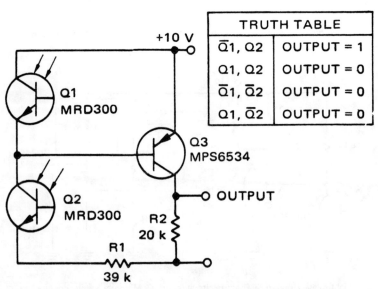

Photo-activated logic driver (courtesy Motorola Semiconductor Products Inc.).

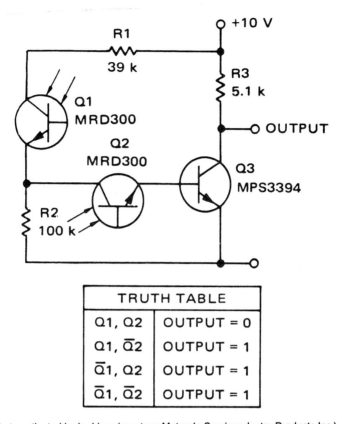

Photo-activated logic driver (courtesy Motorola Semiconductor Products Inc.).

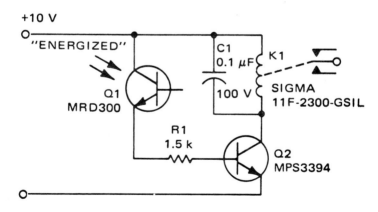

Light-operated relay. The phototransistor can be activated by a flashlight (courtesy Motorola Semiconductor Products Inc.).

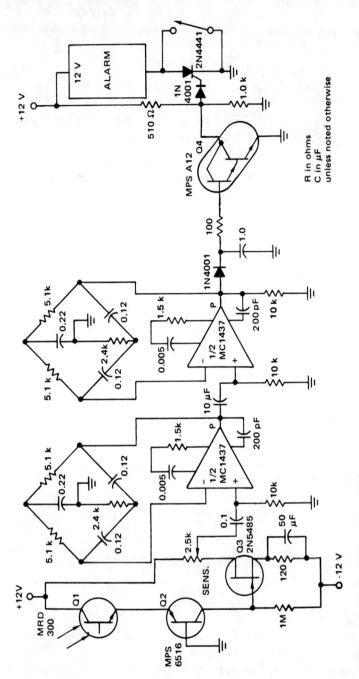

Frequency-sensitive photo-activated alarm (courtesy Motorola Semiconductor Products Inc.).

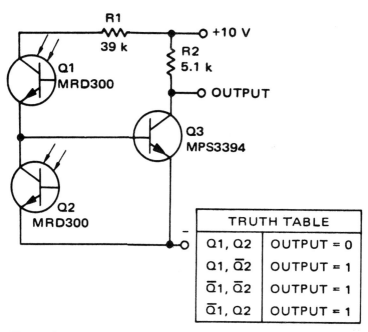

Photo-activated logic driver (courtesy Motorola Semiconductor Products Inc.).

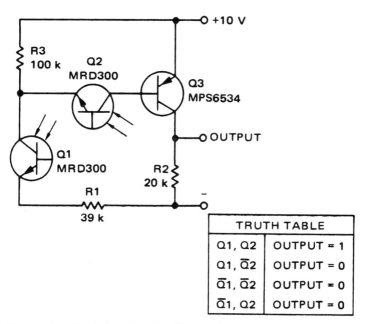

Photo-activated logic driver (courtesy Motorola Semiconductor Products Inc.).

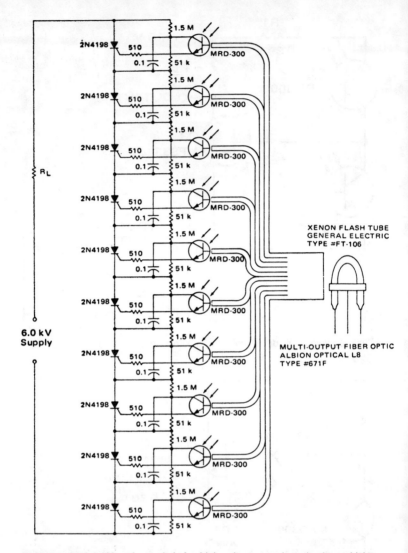

Light-operated 6 kV series switch for high-voltage crowbar circuits or high-voltage pulse-forming networks (courtesy Motorola Semiconductor Products Inc.).

Filters

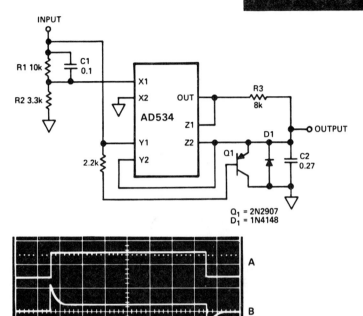

Derivative-controlled low-pass filter. This circuit settles rapidly in response to step changes, then assumes a long time constant for filtering noise. In the oscilloscope photo trace A is the input step, trace B is the control input and trace C is the signal output (courtesy Analog Devices, Inc.).

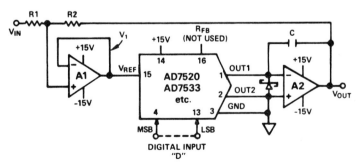

Digitally programmable low-pass filter. The two op amps are a AD747 (courtesy Analog Devices, Inc.).

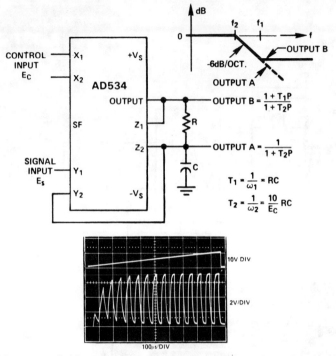

Voltage-controlled low-pass filter. The oscilloscope photo shows the circuit's response to a square wave with a ramp input. As an example if R is 8K and C is 0.002 μF output A has a pole frequency from 100 Hz to 10 kHz for E_C ranging from 100 mV to 10V. Output B has an additional zero at 10 kHz. The circuit can be converted to a high-pass filter by interchanging C and R (courtesy Analog Devices, Inc.).

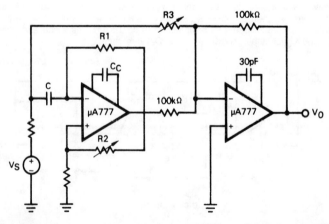

Notch filter for up to 50 kHz using two μA777 op amps (courtesy Fairchild Semiconductor).

1 MHz bandpass filter insertion-loss tester. The test is performed by sweeping the amplitude of the input signal and comparing the envelopes of the input and output signals. The test signal is produced by modulating a 1 MHz sinusoidal carrier with a 10-volt 1 kHz ramp using a 429 multiplier. The test signal is demodulated by the diode-connected AD812A and network R1-R2-C1. The signal coming out of the filter under test is passed through an identical network consisting of AD812B, R3, R4 and C2. The AD521 compares the two demodulated outputs (courtesy Analog Devices, Inc.).

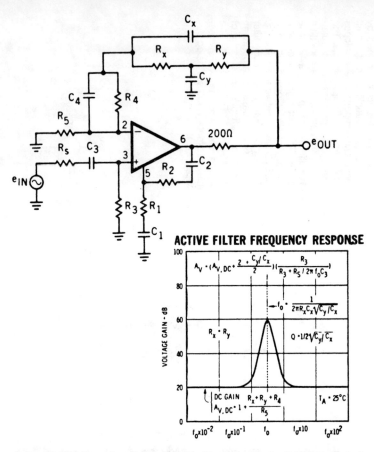

ACTIVE FILTER FREQUENCY RESPONSE

$$A_V = (A_{V,DC} + \frac{2 + C_y/C_x}{2})(\frac{R_3}{R_3 + R_5 / 2\pi f_0 C_3})$$

$$f_0 = \frac{1}{2\pi R_x C_x \sqrt{C_y/C_x}}$$

$$R_x = R_y$$

$$Q = 1/2\sqrt{C_y/C_x}$$

DC GAIN $A_{V,DC} = 1 + \frac{R_x + R_y + R_4}{R_5}$

$T_A = 25°C$

Active filter having a bandpass with 60 dB gain using the ECG925. Typical supply voltage is ±15 volts, but it can be powered with supplies ranging from ±3 volts to ±22 volts.

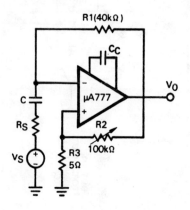

Bandpass filter using a µA 777 op amp for center frequencies up to 50 kHz. Set capacitor Cc to 30 pF for frequencies below 10 kHz and 3 pF for frequencies above 10 kHz. Set capacitor C for desired frequency (courtesy Fairchild Semiconductor).

NOTCH FREQUENCY AS A FUNCTION OF C_1

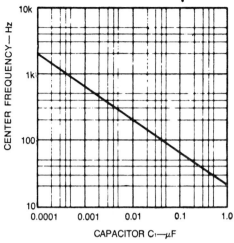

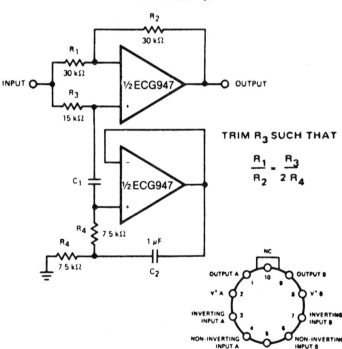

TRIM R_3 SUCH THAT

$$\frac{R_1}{R_2} = \frac{R_3}{2R_4}$$

Notch filter using an ECG947 dual operational amplifier as a gyrator. The ECG947 is short-circuit protected and requires no external components for frequency compensation (courtesy GTE Sylvania Incorporated).

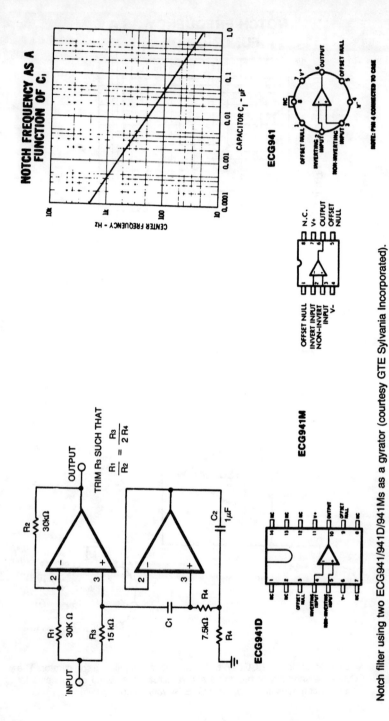

Notch filter using two ECG941/941D/941Ms as a gyrator (courtesy GTE Sylvania Incorporated).

Sample & Hold Circuits

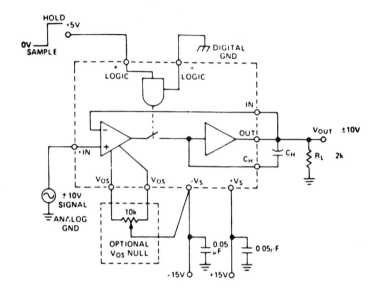

Sample-and-hold circuit with $A_V = +1$. The chip is an AD582 sample-and-hold amplifier (courtesy Analog Devices, Inc.).

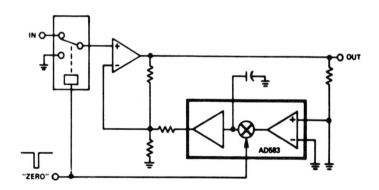

Sample-and-hold circuit used to automatically zero a high-gain amplifier (courtesy Analog Devices, Inc.).

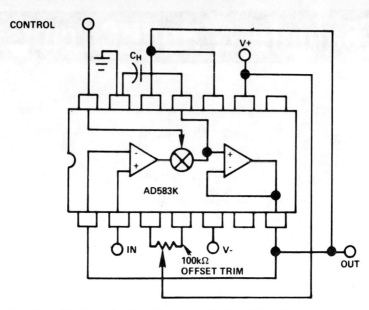

Sample-and-hold circuit with unity gain and offset nulling (courtesy Analog Devices, Inc.).

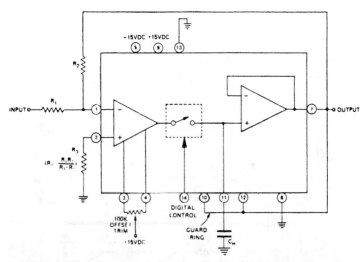

Inverting sample-and-hold circuit with gain equal to $-R2/R1$. The chip used is a Datel SHM-IC-1 14-pin DIP. For a gain of -1 the bandwidth is one-half of that for the noninverting mode. R3 is equal to the parallel combination of R1 and R2 and is used to compensate for voltage offset caused by input bias current. R1 and R2 should be 100 PPM/°C metal film resistors for a gain of -1. For high gains the ratio should be matched closely or trimmed with a small value carbon composition type resistor (courtesy Datel Systems, Inc.).

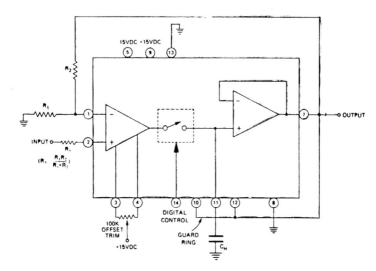

Noninverting sample-and-hold circuit with gain equal to 1 + (R2/R1). The chip is a Datel SHM-IC-1 14-pin DIP. Bandwidth decreases proportionally with gain. R3 is equal to the parallel combination of R1 and R2, and is used to compensate for voltage offset caused by input bias current. R1 and R2 should be 100 PPM/°C metal film resistors (courtesy Datel Systems, Inc.).

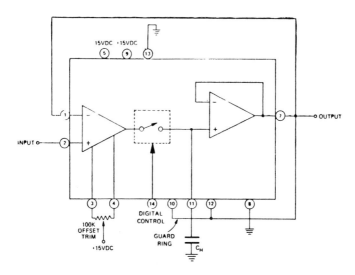

Noninverting unity-gain sample-and-hold circuit using a Datel SHM-IC-1. The SHM-IC-1 is a 14-pin DIP with a 2 MHz bandwidth. The 100K offset pot should be a 100 PPM/°C 15-turn type, available from Datel. To zero ground pin 2 and pin 14, then adjust offset for zero output at pin 7 (courtesy Datel Systems, Inc.).

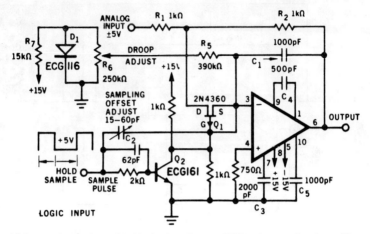

High-speed sample-and-hold circuit using an ECG915 operational amplifier (courtesy GTE Sylvania Incorporated).

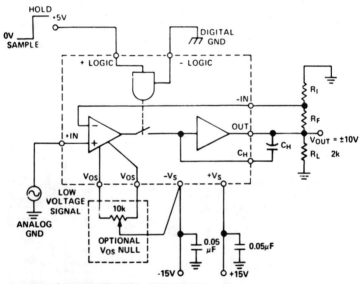

Sample-and-hold circuit with $A_V = (1 + R_F/R_L)$. Circuitry inside dashed lines is the AD582 sample-and-hold amplifier (courtesy Analog Devices, Inc.).

Test Equipment & Metering Circuits

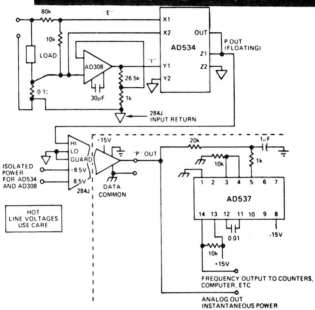

Watt-hour meter. The output of the 284J represents the instantaneous power delivered to the load, which is a 40-watt light bulb. This signal is averaged and converted to a frequency by the AD537 V/F converter. The pulse repetition rate of the AD537 varies in direct proportion to the average power consumed by the 40-watt light bulb. Different sensitivities (e.g., watt-minutes, watt-seconds, etc.) can be obtained by altering the scale factor of the AD537, the gain of the 284J or the count ratio on the AD537 output. If an analog output is desired, an analog integrator can be successfully employed within the limitations shown (courtesy Analog Devices, Inc.).

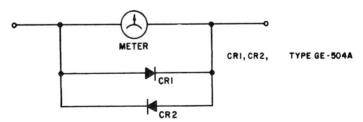

DC meter protection circuit. The diodes will not conduct until the voltage across the meter reaches 0.5 to 0.7 volts. For a typical meter movement with an internal resistance of 1200 ohms and a full-scale current rating of 50 μA the rectifiers will introduce less than 1% error (courtesy General Electric Company).

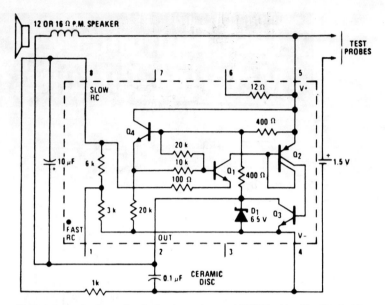

Buzz box continuity and coil checker using an LM3909 chip. Circuitry inside dashes is the LM3909. A short up to 100 ohms will cause a tone to be generated. By probing two values, such as a direct short and 5 ohms, a difference can be detected if done in quick succession (courtesy National Semiconductor Corporation).

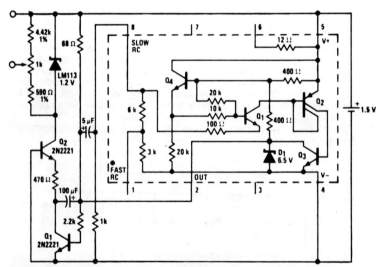

Oscilloscope calibrator using an LM3909 chip. The output is a clean rectangular wave that is exactly 1 volt peak to peak. The rectangular wave is approximately 1.5 ms on and 5.5 ms off. Battery life from a 1.5-volt D-cell is approximately 500 hours (courtesy National Semiconductor Corporation).

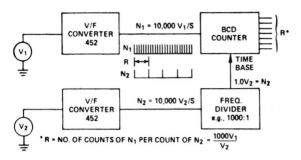

Wide-range ratiometer using two 100 kHz converters with less than a 0.1% error over a dynamic range of 10,000 to 1 (courtesy Analog Devices, Inc.).

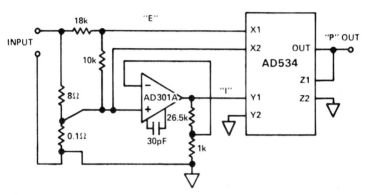

Wattmeter for an audio amplifier to 100 watts. To use the circuit connect a nonreactive 8-ohm 10-watt load to the speaker terminals with the 0.1-ohm shunt in series with it. The output of the amplifier should be connected to the circuit with AWG #16 wire (courtesy Analog Devices, Inc.).

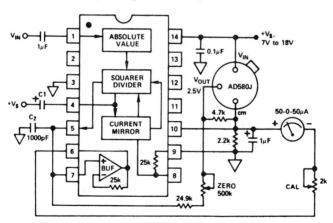

Decibel measurement circuit using an AD536 true RMS-to-DC converter chip with power output to a linear meter display (courtesy Analog Devices, Inc.).

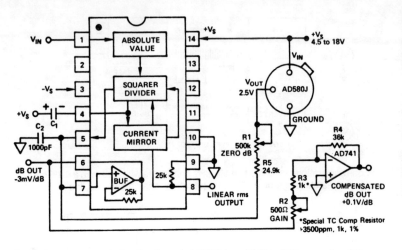

Decibel measurement circuit using an AD536 true RMS-to DC converter chip (courtesy Analog Devices, Inc.).

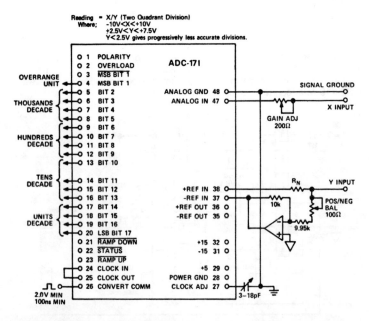

Ratiometric measurement circuit using an ADC171 dual-slope integrating A/D converter (courtesy Analog Devices, Inc.).

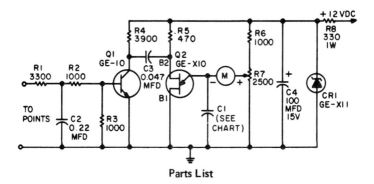

Parts List

C1 — See chart
C2 — 0.22-mfd, 400-volt capacitor
C3 — 0.047-mfd, 200-volt capacitor
C4 — 100-mfd, 15-volt electrolytic capacitor
CR1 — GE-X11 zener diode
Q1 — GE-10 transistor
Q2 — GE-X10 unijunction transistor
R1 — 3300-ohm, 1/2-watt resistor
R2, R3, R6 — 1000-ohm, 1/2-watt resistor
R4 — 3900-ohm, 1/2-watt resistor
R5 — 470-ohm, 1/2-watt resistor
R7 — 2500-ohm, 2-watt potentiometer
R8 — 330-ohm, 1-watt resistor
M — Meter (GE Type D092, Cat. No. 50 171 111EMEM) rated 0-500 microamperes; 220-ohms terminal resistance. New faceplate calibrated 0-6000 rpm. See text.

High-precision tachometer. Capacitor C1 is rated at 200 volts. The meter shown is a standard panel meter with a 0 to 500 µA range. It is available from GE with a modified faceplate. For information about the modified meter write to General Electric Company, Tube Dept., Attention R.G. Kempton, 316 East Ninth St., Owenboro, KY 42301. R7 is used to calibrate the tachometer before it is installed. Feed a 60-hertz signal to the input (to point). On four stroke engines with four cylinders adjust for 1800 RPM, with six cylinders adjust for 1200 RPM and with eight cylinders adjust for 900 RPM. On two stroke engines with three cylinders adjust for 1200 RPM and with four cylinders adjust for 900 RPM. The 60-hertz signal should be no greater than 9 volts AC (courtesy General Electric Company).

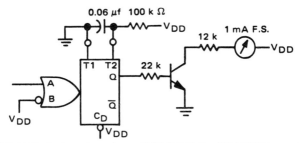

Precision analog tachometer using an MC14538B. The MC14538B operates as a monostable multivibrator with an output pulse of 6 ms. This pulse drives the transistor which in turn drives the meter. Full-scale calibration should be made at 167 hertz, which corresponds to 10,020 RPM (courtesy Motorola Semiconductor Products Inc.).

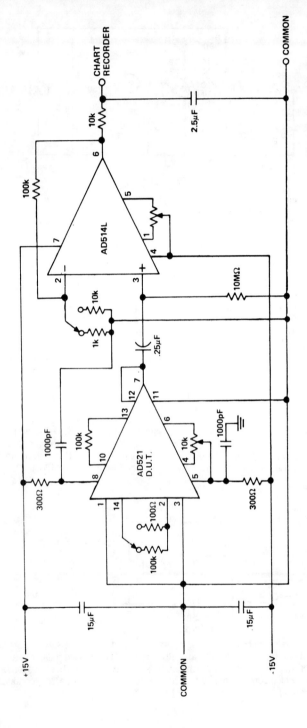

Peak-to-peak noise measuring circuit for frequencies between 0.1 hertz and 10 hertz. Typical measurements are made by reading the maximum peak-to-peak voltage noise of the device under test (DUT) for three periods for 10 seconds each (courtesy Analog Devices, Inc.).

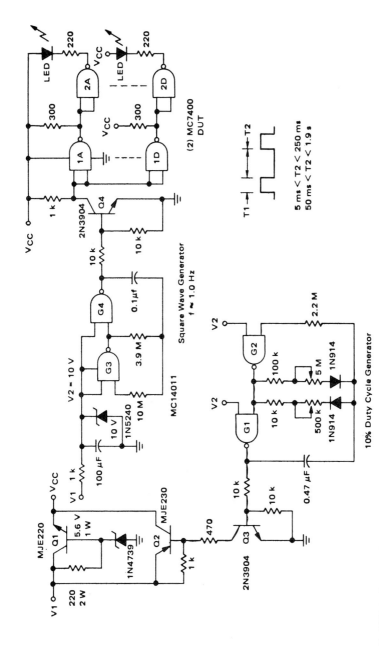

TTL SOA (safe operating area) test circuit (courtesy Motorola Semiconductor Products Inc.).

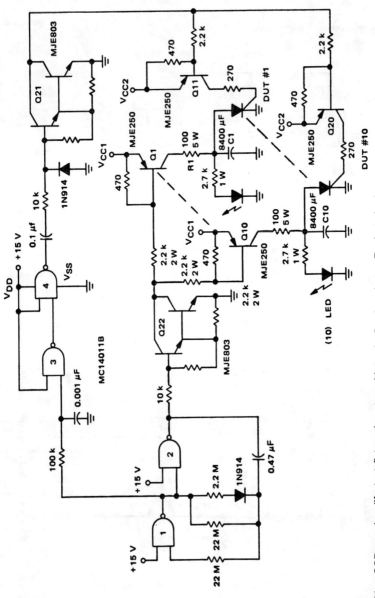

10-position SCR crowbar life test fixture (courtesy Motorola Semiconductor Products Inc.).

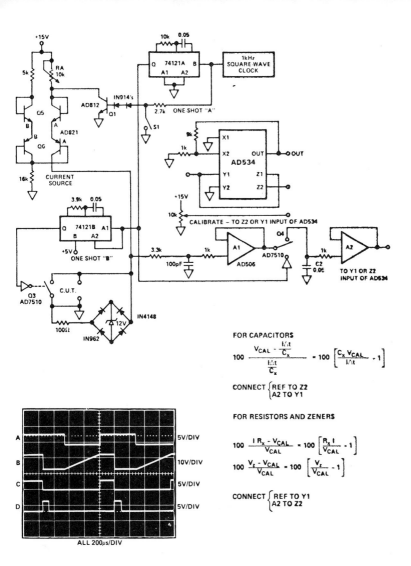

FOR CAPACITORS

$$100 \,\frac{V_{CAL} - \frac{I\triangle t}{C_x}}{\frac{I\triangle t}{C_x}} = 100 \left[\frac{C_x V_{CAL}}{I\triangle t} - 1 \right]$$

CONNECT $\begin{cases} \text{REF TO Z2} \\ \text{A2 TO Y1} \end{cases}$

FOR RESISTORS AND ZENERS

$$100 \,\frac{I R_x - V_{CAL}}{V_{CAL}} = 100 \left[\frac{R_x I}{V_{CAL}} - 1 \right]$$

$$100 \,\frac{V_z - V_{CAL}}{V_{CAL}} = 100 \left[\frac{V_z}{V_{CAL}} - 1 \right]$$

CONNECT $\begin{cases} \text{REF TO Y1} \\ \text{A2 TO Z2} \end{cases}$

Component sorter of resistors, capacitors, and zeners. Trace A is a 1 kHz square-wave clock that is applied to Q1. The component under test (CUT) in this case is a 0.01 μF capacitor. It is allowed to charge until the clock goes high, turning off the current source (trace B). The voltage the capacitor sits at is inversely proportional to its absolute value. The AD506 follows this potential and feeds the sample-and-hold circuit Q4-C2-A2. The sample-and-hold circuit is enabled by one-shot A for 200 μs when the clock goes high (trace C). After this time one-shot A goes low triggering one-shot B on for 100 μs. This pulse drives Q3 on (trace D) and discharges the CUT. This same fixture can check resistors and zeners by closing S1. This allows the current source to run all of the time. This is necessary since resistors and zeners have no memory (courtesy Analog Devices, Inc.).

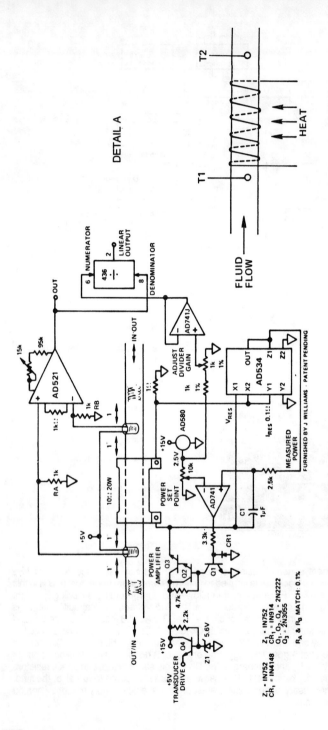

Flowmeter circuit for measuring the flow rate of liquids flowing at slow speeds. To understand the principle of operation refer to Detail A. T1 and T2 are temperature sensors. With no flow through the pipe power is dissipated into the medium symmetrically and there is no difference in temperature at T1 and T2. As flow begins T1 will take on the temperature upstream, but T2 will be influenced by the power dissipated into the moving stream. The time response of the flowmeter is in the order of 10 to 15 seconds (courtesy Analog Devices, Inc.).

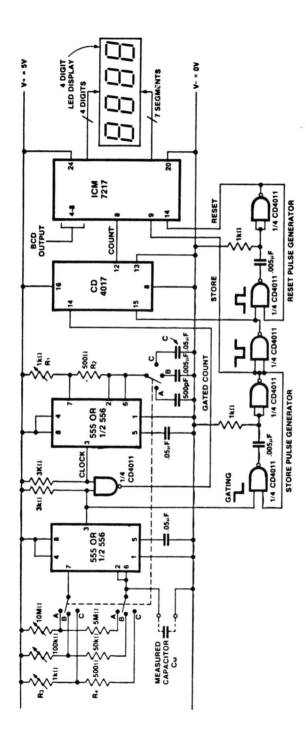

Inexpensive capacitance meter. This circuit uses two 555 timers, or one 556, to generate a gated count for an ICM7217, dependent on the value of capacitance. The clock timer operates as a fixed oscillator whose output period is determined by R1, R2 and C. The gating timer also operates as an oscillator, but its output high time is determined by the value of measured capacitance in combination with R3 and R4. Range A reads 1 to 9999 pF, range B reads 1 to 9999 nF and range C reads 1 to 9999 µF (courtesy Intersil, Inc.).

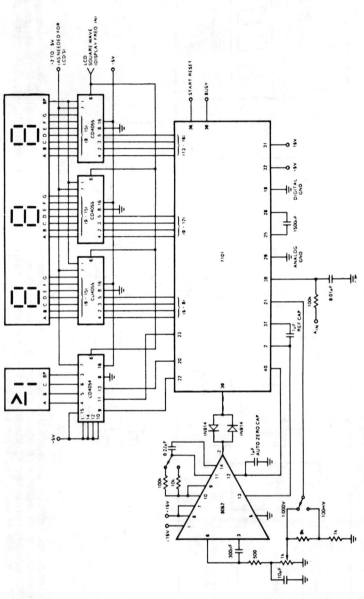

3½-digit LCD DPM/DVM using the 8052/7101 A/D pair (courtesy Intersil, Inc.).

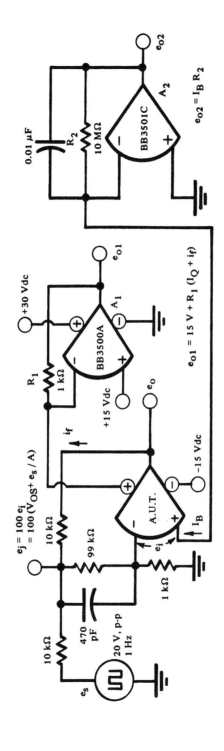

Five-in-one op amp specs test circuit. Tests can be made for open-loop gain, offset voltage, input bias current, quiescent current and output voltage swing (courtesy Burr-Brown Research Corporation).

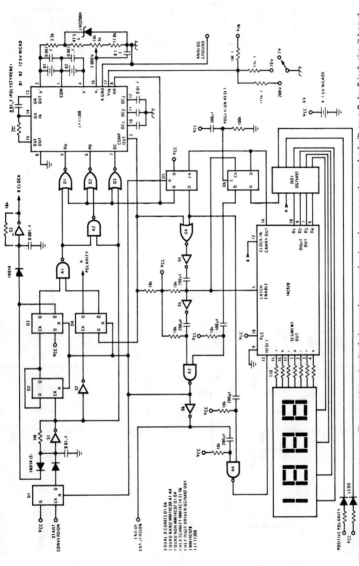

Dual-polarity 3½-digit voltmeter. The complete circuit of nine packages and external components can be built on a 3- by 5-inch circuit board (courtesy National Semiconductor Corporation).

2½-digit DVM using a binary D/A converter to accomplish conversion of the BCD input signal (courtesy Motorola Semiconductor Products Inc.).

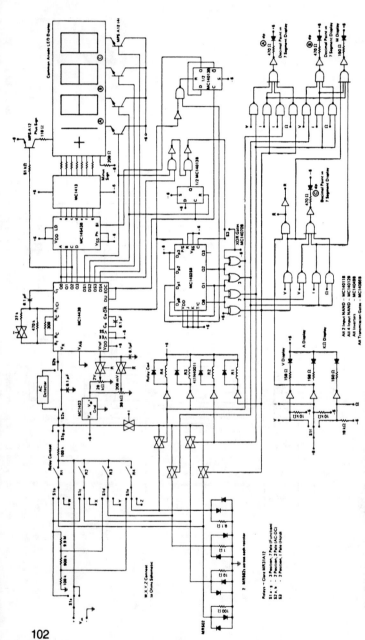

3½-digit autoranging multimeter using an MC14433. The multimeter includes ranges from 200 millivolts to 200 volts, AC and DC ampere ranges from 2 milliamperes to 2 amperes full scale and resistance ranges from 2K to 2M full scale (courtesy Motorola Semiconductor Products Inc.).

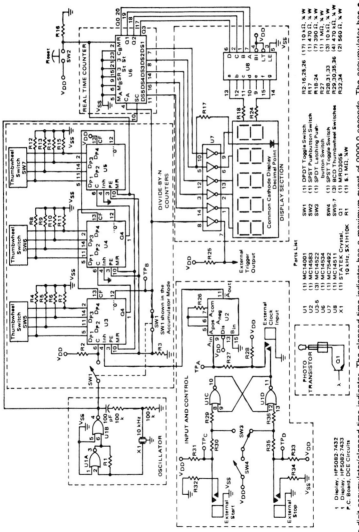

Five-digit accumulator/elapsed time indicator. The elapsed time indicator has a maximum event of 9999.9 seconds. The accumulator has a maximum count of 99,999 times N, where N is a prescaler number from 1 to 999 (courtesy Motorola Semiconductor Products Inc.).

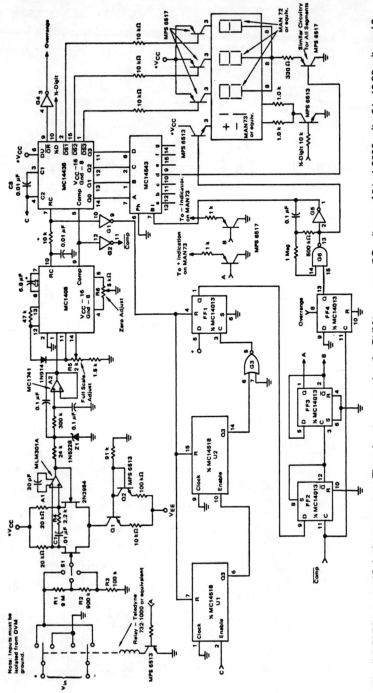

3½-digit DVM using an IC dual-ramp system. Three input ranges allow the DVM to measure DC voltages of 0 to 1.9999 volts, 0 to 19.99 volts and 0 to 199.9 volts (courtesy Motorola Semiconductor Products Inc.).

Op Amp Circuits

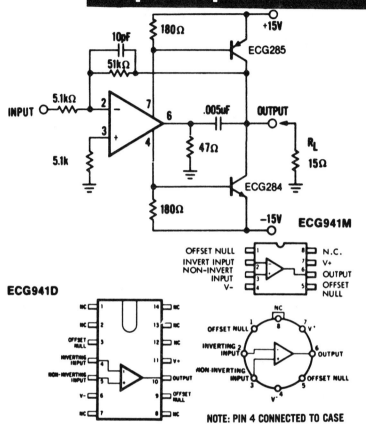

High-slew-rate power amplifier using an ECG941/941D/941M operational amplifier (courtesy GTE Sylvania Incorporated).

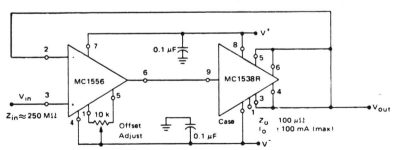

High-impedance input, high-current output voltage follower (courtesy Motorola Semiconductor Products Inc.).

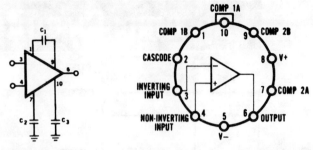

Frequency compensation circuit using an ECG915 operational amplifier. See table for component values. Supply voltage is ±15 volts (courtesy GTE Sylvania Incorporated).

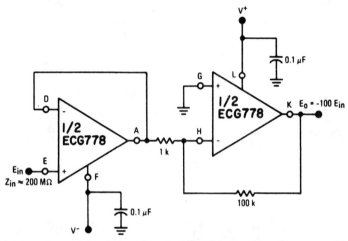

High-impedance high-gain inverting amplifier. Typical supply voltages are +15 volts and −15 volts (courtesy GTE Sylvania Incorporated).

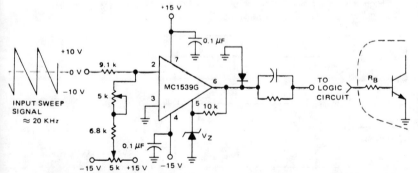

Voltage comparator using an MC1539G op amp (courtesy Motorola Semiconductor Products Inc.).

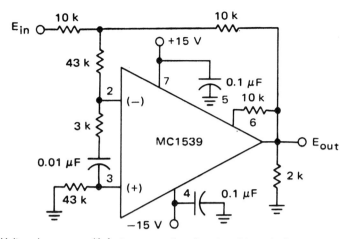

Unity-gain op amp with fast response time (courtesy Motorola Semiconductor Products Inc.).

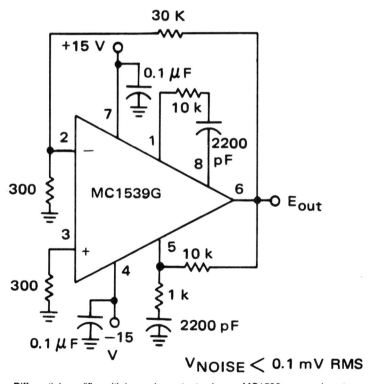

$V_{NOISE} < 0.1$ mV RMS

Differential amplifier with low-noise output using an MC1539 op amp (courtesy Motorola Semiconductor Inc.).

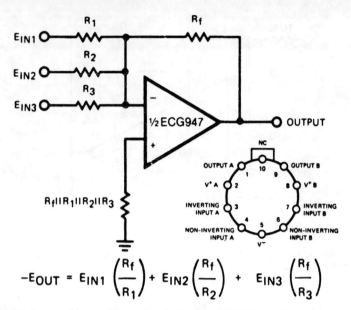

Weighted averaging amplifier using half of an ECG947 dual operational amplifier. The ECG947 is short-circuit protected and requires no external components for frequency compensation (courtesy GTE Sylvania Incorporated).

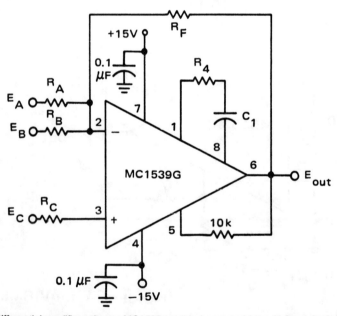

Differential amplifier using an MC1539 op amp (courtesy Motorola Semiconductor Products Inc.).

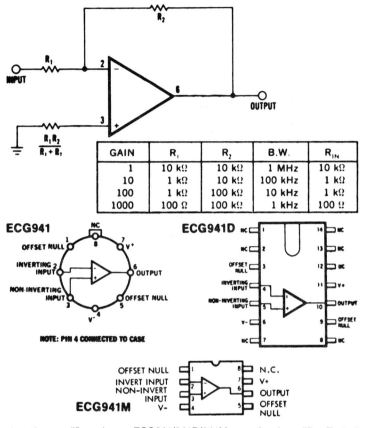

Inverting amplifier using an ECG941/941D/941M operational amplifier. Typical supply voltage is ±15 volts (courtesy GTE Sylvania Incorporated).

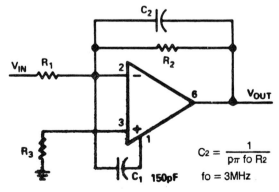

$$C_2 = \frac{1}{p\pi\, f_o\, R_2}$$

$f_o = 3\text{MHz}$

Feedforward frequency compensation circuit using an AD101A/201A/301A op amp. Typical supply voltage is ±15 volts (courtesy Analog Devices, Inc.).

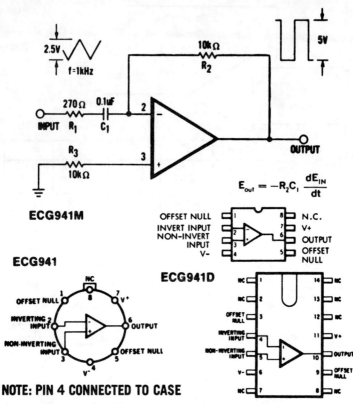

Simple differentiator using an ECG941/941D/941M operational amplifier. Typical supply voltage is ±15 volts (courtesy GTE Sylvania Incorporated).

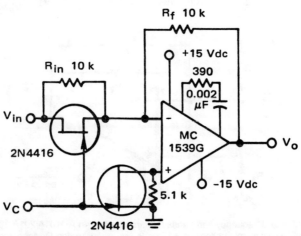

Op amp with FET AGC circuit (courtesy Motorola Semiconductor Products Inc.).

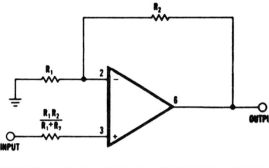

GAIN	R_1	R_2	B.W.	R_{IN}
10	1 kΩ	9 kΩ	100 kHz	400 MΩ
100	100 Ω	9.9 kΩ	10 kHz	280 MΩ
1000	100 Ω	99.9 kΩ	1 kHz	80 MΩ

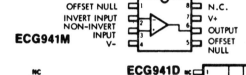

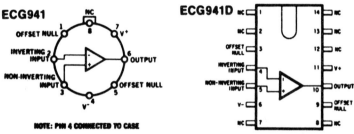

Noninverting amplifier using an ECG941/941D/941M operational amplifier. Typical supply voltage is ±15 volts (courtesy GTE Sylvania Incorporated).

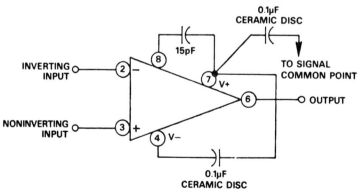

Unity-gain op amp using an AD509 8-pin TO99 (courtesy Analog Devices, Inc.).

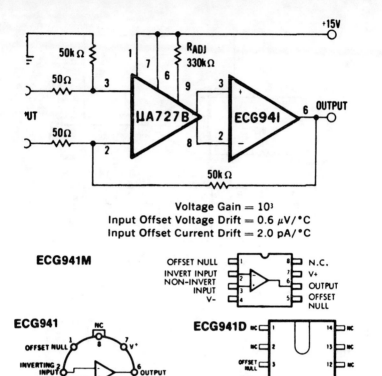

Low-drift low-noise amplifier using an ECG941/941D/941M operational amplifier and a μA727B temperature controlled differential amplifier. Typical supply voltage is ±15 volts (courtesy GTE Sylvania Incorporated).

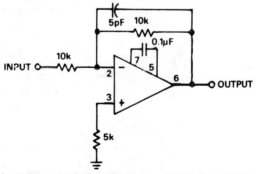

Op amp with minimum settling time using an AD518 8-pin TO99 (courtesy Analog Devices, Inc.).

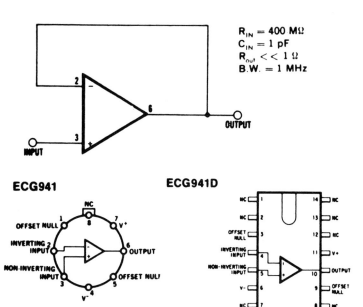

Unity-gain voltage follower using an ECG 941/941D/941M operational amplifier. Typical supply voltage is ±15 volts (courtesy GTE Sylvania Incorporated).

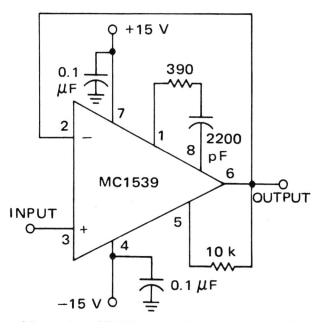

Voltage follower using an MC1539 op amp with unity-gain compensation (courtesy Motorola Semiconductor Products Inc.).

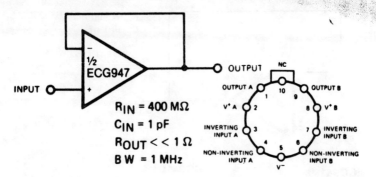

Unity-gain voltage follower using half of an ECG947 dual operational amplifier. The ECG947 is short-circuit protected and requires no external components for frequency compensation (courtesy GTE Sylvania Incorporated).

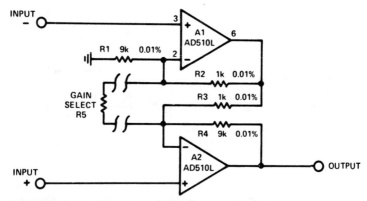

Instrumentation amplifier using two AD510 8-pin TO99 op amps. As shown the gain is 10. By adding R5 the gain will increase. For a gain of 10 the frequency response is down 3 dB at 500 kHz. Full output of ±10 volts can be attained up to 1800 hertz (courtesy Analog Devices, Inc.).

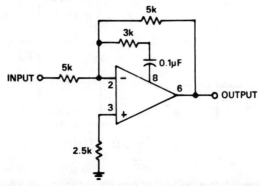

Op amp with high bandwidth using an AD518 8-pin TO99. Bandwidth is nearly 25 MHz with the feedforward technique shown (courtesy Analog Devices, Inc.).

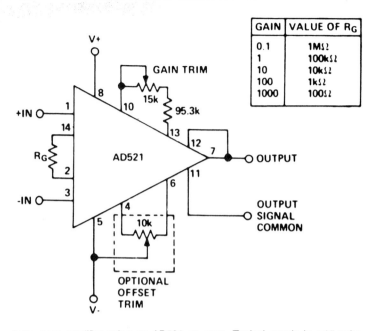

Instrument amplifier using an AD521 op amp. Typical supply is ±15 volts (courtesy Analog Devices, Inc.).

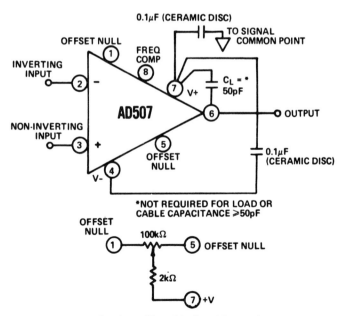

General-purpose operational amplifier with closed-loop gain greater than 10. Typical supply voltage is ±15 volts (courtesy Analog Devices, Inc.).

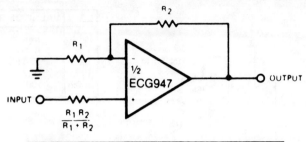

GAIN	R₁	R₂	B.W.	R_IN
10	1 kΩ	9 kΩ	100 kHz	400 MΩ
100	100 Ω	9.9 kΩ	10 kHz	280 MΩ
1000	100 Ω	99.9 kΩ	1 kHz	80 MΩ

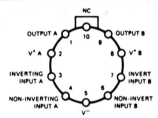

Noninverting amplifier using half of an ECG947 dual operational amplifier. The ECG947 is short-circuit protected and requires no external components for frequency compensation (courtesy GTE Sylvania Incorporated).

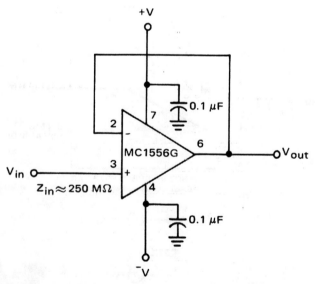

Voltage follower using an MC1556 op amp (courtesy Motorola Semiconductor Products Inc.).

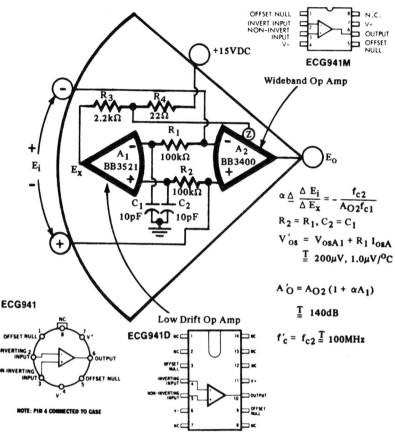

Differential-input composite op amp (courtesy Burr-Brown Research Corporation).

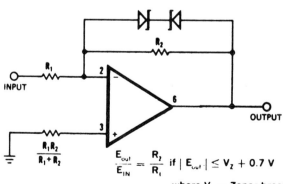

Clipping amplifier using an ECG941/941D/941M operational amplifier. Typical supply voltage is ±15 volts (courtesy GTE Sylvania Incorporated).

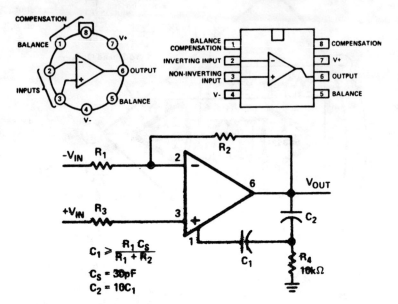

Two pole frequency compensation circuit using an AD101A/201A/301A op amp. Typical supply voltage is ±15 volts (courtesy Analog Devices, Inc.).

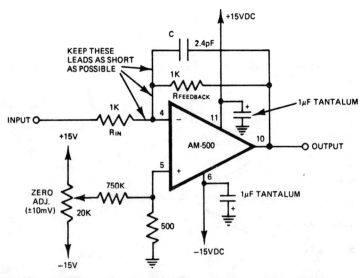

Fast-settling op amp with gain of −1. For gains larger than −1 use an input resistor valued at 500 ohms or less and pick a feedback resistor for the required gain, i.e., 1K for −2, 1.5K for −3, etc. (courtesy Datel Systems, Inc.).

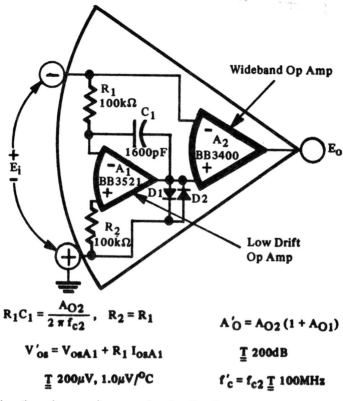

$$R_1 C_1 = \frac{A_{O2}}{2\pi f_{c2}}, \quad R_2 = R_1$$

$$V'_{os} = V_{osA1} + R_1 I_{osA1}$$

$$\underline{\underline{\top}}\ 200\mu V,\ 1.0\mu V/^\circ C$$

$$A'_O = A_{O2}(1 + A_{O1})$$

$$\underline{\underline{\top}}\ 200 dB$$

$$f'_c = f_{c2} \underline{\underline{\top}} 100 MHz$$

Inverting-only composite op amp (courtesy Burr-Brown Research Corporation).

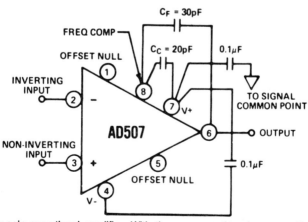

Unity-gain operational amplifier. With the compensation shown a unity-gain frequency of approximately 10 MHz to 12 MHz results (courtesy Analog Devices, Inc.).

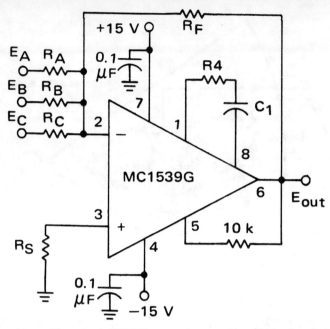

Summing amplifier using an MC1539 op amp (courtesy Motorola Semiconductor Products Inc.).

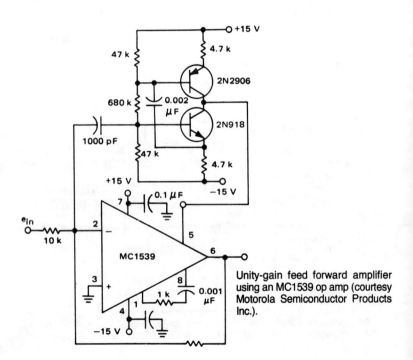

Unity-gain feed forward amplifier using an MC1539 op amp (courtesy Motorola Semiconductor Products Inc.).

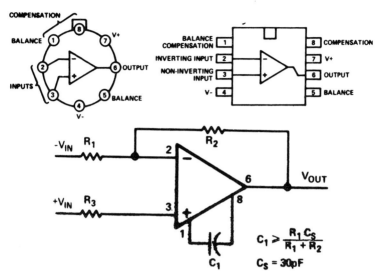

Single pole frequency compensation circuit using an AD101A/201A/301A op amp. Typical supply voltage is ±15 volts. Voltage gain is 88 dB (courtesy Analog Devices, Inc.).

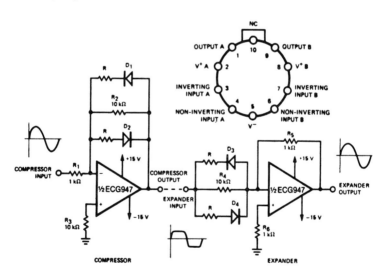

Compressor/expander amplifiers using an ECG947 dual operational amplifier. The ECG947 is short-circuit protected and requires no external components for frequency compensation (courtesy GTE Sylvania Incorporated).

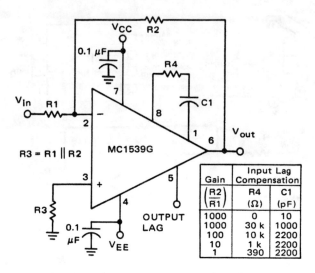

Op amp in closed-loop frequency-compensated configuration (courtesy Motorola Semiconductor Products Inc.).

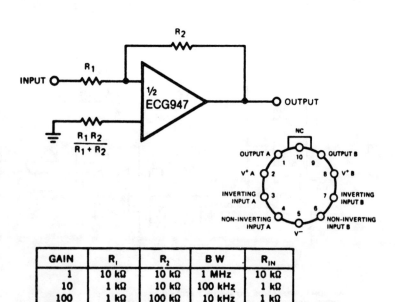

Inverting amplifier using half of an ECG947 dual operational amplifier. The ECG947 is short-circuit protected and requires no external components for frequency compensation (courtesy GTE Sylvania Incorporated).

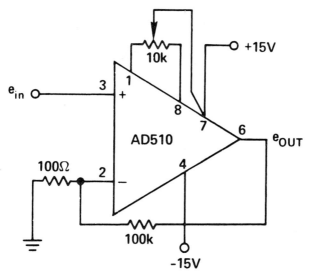

Noninverting op amp using an AD510 8-pin TO99 (courtesy Analog Devices, Inc.).

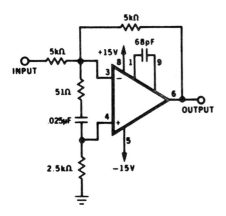

Inverting unity-gain high-slew-rate circuit using an ECG915 operational amplifier (courtesy GTE Sylvania Incorporated).

Voltage follower using an ECG915 operational amplifier. Output is taken on pin 6 (courtesy GTE Sylvania Incorporated).

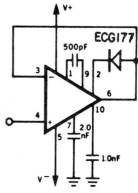

AM & FM Broadcast Receiver Circuits

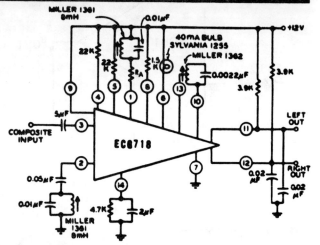

FM stereo processor. Supply voltage is typically +12 volts. The ECG718 is a 14-pin DIP with the functions of a 19 kHz amplifier, frequency doubler, stereo indicator lamp driver, audio mute, stereo/mono switch, and stereo demodulator (courtesy GTE Sylvania Incorporated).

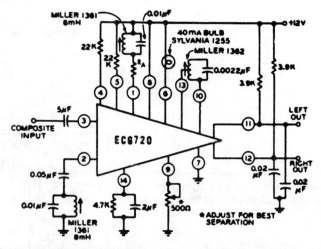

FM stereo processor. Supply voltage is typically +12 volts. The ECG720 has the functions of a 19 kHz amplifier, frequency doubler, stereo indicator lamp driver, audio mute, stereo/mono switch, and stereo demodulator. It also permits adjustable stereo channel separation (courtesy GTE Sylvania Incorporated).

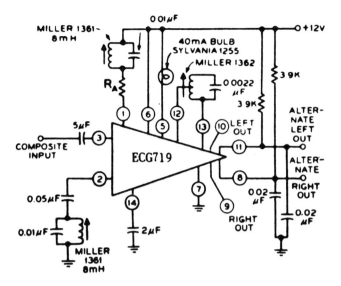

FM stereo processor. The 14-pin DIP performs the standard stereo functions of 19 kHz amplifier, frequency doubler, stereo indicator lamp driver, and stereo demodulator. The IC also includes two alternate emitter follower outputs (courtesy GTE Sylvania Incorporated).

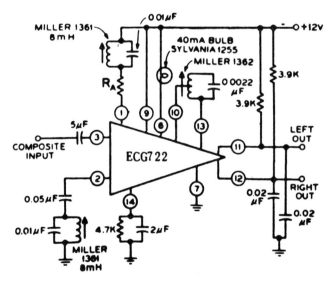

FM stereo processor. The 14-pin ECG722 perform the standard stereo functions of 19 kHz amplifier, frequency doubler, stereo indicator lamp driver, and stereo demodulator. Current drain at 12 volts is typically 14 mA (courtesy GTE Sylvania Incorporated).

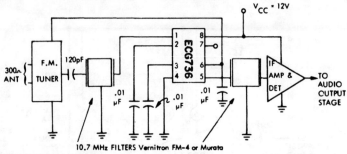

FM IF gain block shown connected to typical FM receiver circuitry. The IF amplifier and detector chip can be an ECG708, for example (courtesy GTE Sylvania Incorporated).

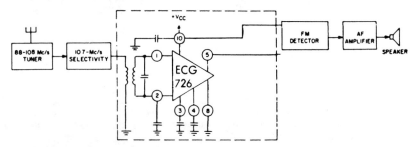

FM IF amplifier for FM broadcast receiver. Supply voltage can be between 6 and 12 volts, the typical being 7.5 volts. Voltage gain is between 55 and 67 dB, depending on the IF (courtesy GTE Sylvania Incorporated).

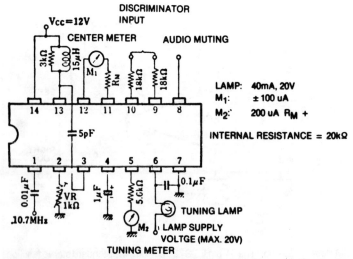

FM tuning indicator circuit using an ECG1149 14-pin DIP. Recommended supply voltage is 12 volts. Lamp turn-on voltage at 10.7 MHz is 10 mV (courtesy GTE Sylvania Incorporated).

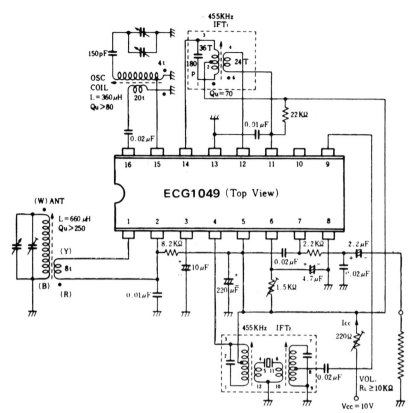

AM broadcast tuner with two IF amplifiers. Audio output with 400-hertz modulation at 30% is 80 mV. All coils, transformers and variable capacitors are standard. The tuning capacitor is a two-section type with the antenna section at 18 to 28 pF and trimmed and the oscillator section at 25 to 35 pF and trimmed. Parts can be purchased at Radio Shack (courtesy GTE Sylvania Incorporated).

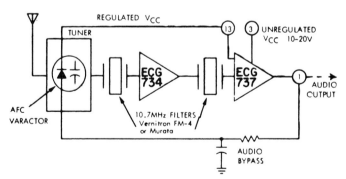

Smith AFC system for an FM receiver. The ECG737 is an FM detector and limiter. It uses quadrature detection and includes a voltage regulator at pin 3. The ECG734 is an FM gain block. This type of circuit is ideal for FM mobile operation (courtesy GTE Sylvania Incorporated).

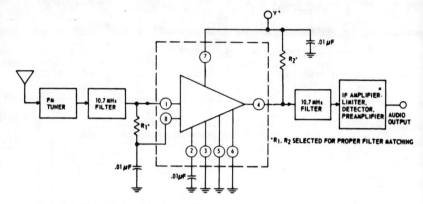

10.7 MHz FM wide-band high-gain IF amplifier-limiter using an ECG781 8-lead TO-99. Supply voltage is 8.5 volts. The circuit provides a voltage gain of 80 dB. Typical sensitivity is 50 μV at 10.7 MHz (courtesy GTE Sylvania Incorporated).

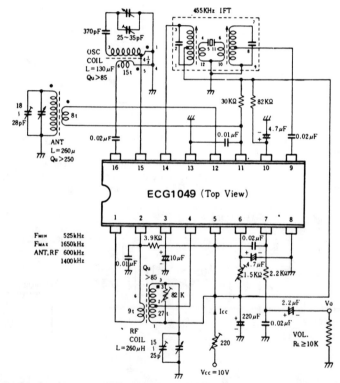

AM broadcast tuner with RF amplifier. All coils and transformers are standard and can be purchased at Radio Shack. The three-section tuning capacitor is also standard. Audio output with 400-hertz modulation at 30% is 80 mV (courtesy GTE Sylvania Incorporated).

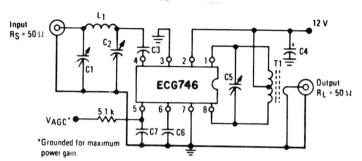

Note 1. Primary: 120 µH (center-tapped)
Q_U = 140 at 455 kHz
Primary: Secondary turns ratio ≈ 13

Note 2. Primary: 6.0 µH
Primary winding = 24 turns #36 AWG (close-wound on 1/4" dia. form)
Core = Arnold Type TH or equiv.
Secondary winding = 1-1/2 turns #36 AWG, 1/4" dia. (wound over center-tap)

	Frequency	
Component	455 kHz	10.7 MHz
C1	—	80–450 pF
C2	—	5.0–80 pF
C3	0.05 µF	0.001 µF
C4	0.05 µF	0.05 µF
C5	0.001 µF	36 pF
C6	0.05 µF	0.05 µF
C7	0.05 µF	0.05 µF
L1	—	4.6 µH
T1	Note 1	Note 2

IF amplifier for 455 kHz or 10.7 MHz. See table for component selection (courtesy GTE Sylvania Incorporated).

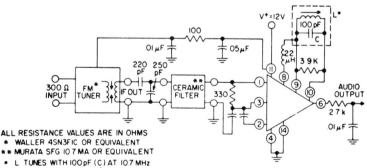

ALL RESISTANCE VALUES ARE IN OHMS
* WALLER 4SN3FIC OR EQUIVALENT
** MURATA SFG 107MA OR EQUIVALENT
• L TUNES WITH 100 pF (C) AT 10.7 MHz
  Q_O UNLOADED ≅ 75 (G.I EX22741 OR EQUIVALENT)

Complete 10.7 MHz FM IF system for FM broadcast receivers using an ECG788 16-pin DIP. The ECG788 chip also has delayed AGC at pin 15 for an RF amplifier, a tuning meter output at pin 13, a mute drive at pin 12, muting sensitivity at pin 5 and an AFC output at pin 7 (courtesy GTE Sylvania Incorporated).

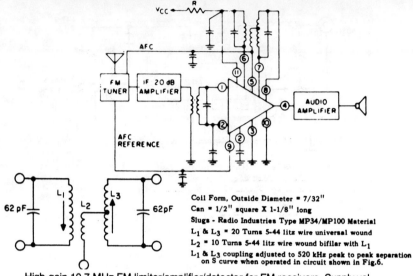

High-gain 10.7 MHz FM limiter/amplifier/detector for FM receivers. Supply voltage should be 30 volts with resistor R being 750 ohms. See lower diagram for specific discriminator information. The bypass capacitor at pin 5 is 0.001 μF. Bypass capacitors at pins 2, 9, 11, and 12 are 0.05 μF. The input transformer is a standard 10.7 MHz IF type (courtesy GTE Sylvania Incorporated).

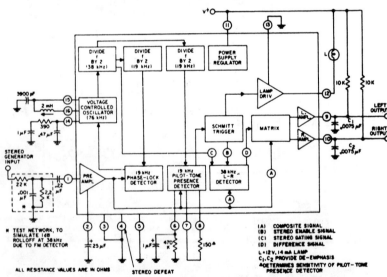

FM stereo multiplex decoder using an ECG789 16-pin QIP. Typical supply voltage is 12 volts; however, the ECG789 can operate over a wide variation of supply voltage up to 16 volts. The internal lamp driver can be made to drive a lamp of higher power than the 14 mA one shown by controlling an external NPN or PNP transistor. To drive PNP type, pin 13 is grounded and pin 12 is connected to the base. To drive an NPN type, pin 12 is connected to supply and pin 13 is connected to the base (courtesy GTE Sylvania Incorporated).

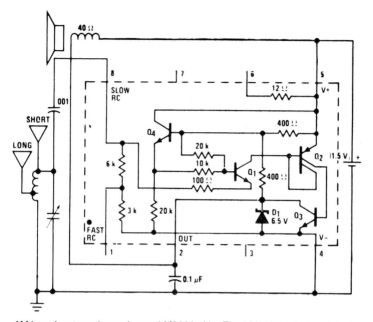

AM broadcast receiver using an LM3909 chip. The LM3909 acts as a detector amplifier. It does not oscillate because there is no positive feedback. The receiver is only good for local stations and sensitivity is similar to a crystal set. It will drive a 6-inch loudspeaker. The antenna can be the short version shown, which is 10 to 20 feet long, or the long version, which is 30 to 100 feet. The long one works better (courtesy National Semiconductor Corporation).

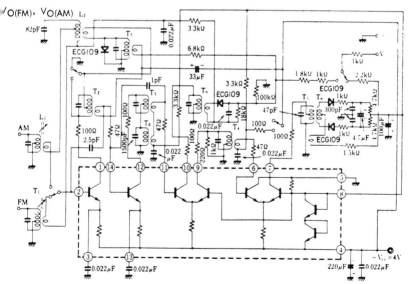

AM/FM IF amplifier for 455 kHz and 10.7 MHz using an ECG1054 14-pin DIP. IF coils and transformers are all standard and can be purchased at Radio Shack. Supply voltage is 4 volts (courtesy GTE Sylvania Incorporated).

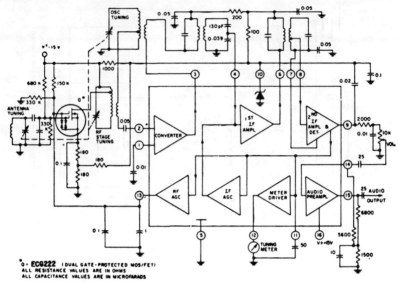

AM broadcast receiver using an ECG787 16-pin DIP. The RF stage is optional. Coils and tuning capacitor can be salvaged from a junk transistor radio or purchased at Radio Shack. The ECG787 provides 75 mV of audio output at the detector that can be fed to any of the audio amplifier circuits shown in this volume, above (courtesy GTE Sylvania Incorporated). PLL FM stereo multiplex decoder using an ECG743 16-pin DIP. The ECG743 features 45 dB channel separation, automatic stereo/mono switching, stereo indicator lamp driver with current limiting, high-impedance input with low-impedance output, 70 dB SCA rejection and one adjustment alignment. Supply voltage rang is 10 to 16 volts. This circuit is suitable for line-operated and automotive FM stereo receivers below (courtesy GTE Sylvania Incorporated).

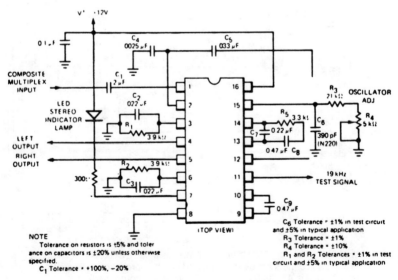

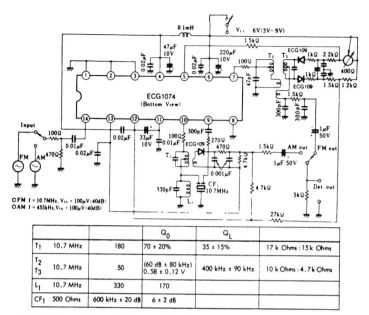

		Q_0	Q_L	
T_1	10.7 MHz	180	70 ± 20%	35 ± 15%	17 k Ohms : 15 k Ohms
T_2 T_3	10.7 MHz	50	(60 dB @ 80 kHz) 0.58 ± 0.12 V	400 kHz ± 90 kHz	10 k Ohms : 4.7 k Ohms
L_1	10.7 MHz	330	170		
CF_1	500 Ohms	600 kHz ± 20 dB	6 ± 2 dB		

AM/FM IF amplifier and detector with tuning meter. AF output is about 10 mV for 400-hertz modulation at 30%. Typical voltage gain is 90 dB for FM and 80 dB for AM (courtesy GTE Sylvania Incorporated).

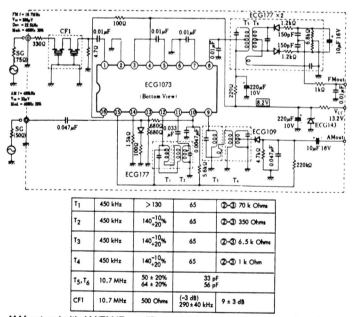

T_1	450 kHz	>130	65	②-③ 70 k Ohms
T_2	450 kHz	140^{-10}_{+20}%	65	②-③ 350 Ohms
T_3	450 kHz	140^{-10}_{+20}%	65	②-③ 6.5 k Ohms
T_4	450 kHz	140^{-10}_{+20}%	65	②-③ 1 k Ohm
T_5, T_6	10.7 MHz	50 ± 20% 64 ± 20%	33 pF 56 pF	
CF1	10.7 MHz	500 Ohms	(-3 dB) 290 ± 40 kHz	9 ± 3 dB

AM front end with AM/FM IF amplifier and detector. Since supply voltage is 13.2 volts, the circuit is ideal for automotive applications. CF1 is a 10.7 MHz ceramic filter (courtesy GTE Sylvania Incorporated).

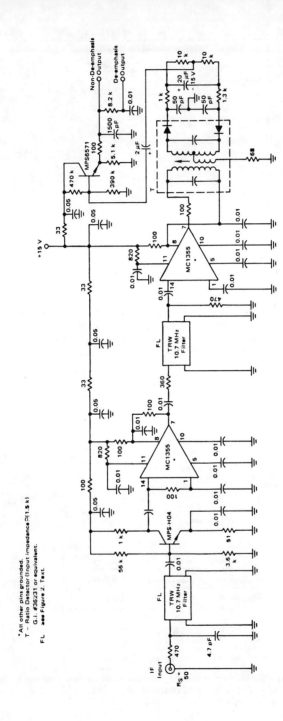

FM IF amplifier using two MC1355 chips. Two TRW phase linear five-pole 10.7 MHz filters provide a combined 240 kHz bandwidth (courtesy Motorola Semiconductor Products Inc.).

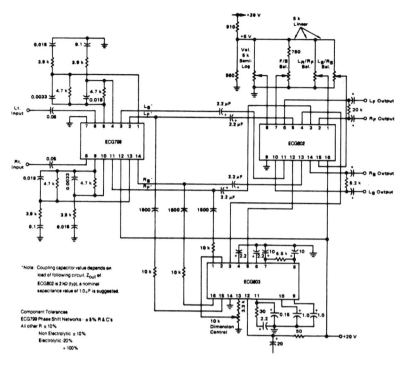

4-channel SQ logic decoder. Current drain at 20 volts is 60 mA. Input impedance is 2M. Output impedance is 2K (courtesy GTE Sylvania Incorporated).

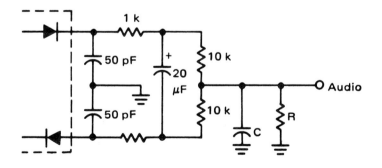

FM ratio detector circuit. With R equal to 22K and C equal to 100 pF the rolloff occurs at about 70 kHz (courtesy Motorola Semiconductor Products Inc.).

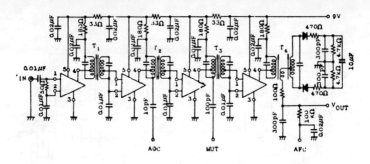

Characteristic	Symbol	Test Condition	Value	Unit
Supply Voltage	V_{cc}	--	9	V
Supply Current	I_{cc}	--	24	mA
Detector Output Voltage	V_{OD}	V_{IN} = 60 dB(μV), f = 400 Hz, ΔF = 22.5 kHz	70	mV
Input Limiting Voltage	$V_{IN(lim)}$	-3dB	21	dB(μV)
Band Width	BW	6 dB Band Width	±110	kHz
Total Harmonic Distortion	THD	V_{IN} = 60 dB(μV), f = 400 Hz, ΔF = 75 kHz	0.5	%
AM Rejection	AMR	FM f = 400 Hz, ΔF = 75 kHz, AM f = 1 kHz 30%	45	dB
Capture Ratio	--	f = 400 Hz, ΔF = 75 kHz	3	dB

IF TRANSFORMER

	C_o (pF)	f (MHz)	Q_o 1-6	Q_o 3-4	TURNS 1-6	1-2	3-4	4-5
T_1	120	10.7	65	65	13	6	13	6
T_2	120	10.7	65	65	13	6	13	6
T_3	120	10.7	65	65	13	9	13	6

	C (pF) 1-3	C (pF) 4-6	f (MHz)	Q_o 1-3	TURNS 1-3	1-2	5-CT	4-CT	6-CT
T_4	22	47	10.7	65	31-1/2	11	9-1/2	11	11

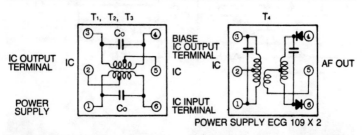

10.7 MHz FM IF amplifier and detector using four ECG1104 5-pin modules. Coil data and specifications also are shown (courtesy GTE Sylvania Incorporated).

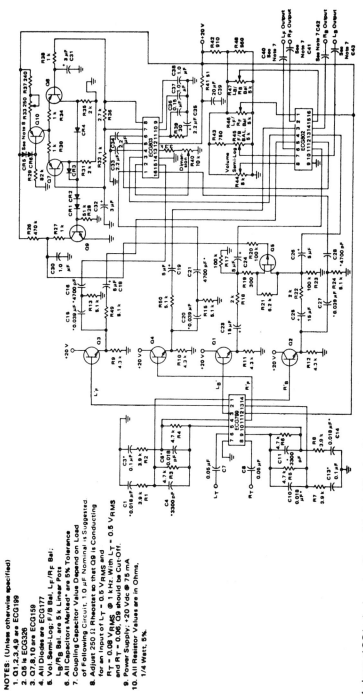

4-channel SQ logic decoder. Current drain is 75 mA at 20 volts. Input impedance is 2M, while output impedance is 2K (courtesy GTE Sylvania Incorporated).

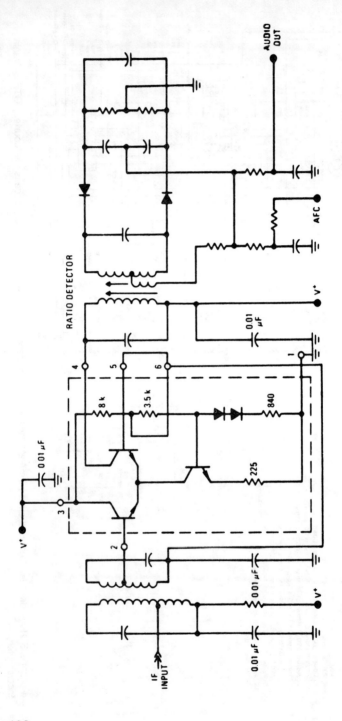

FM IF amplifier and discriminator for 10.7 MHz using an ECG760 (enclosed in dashed lines). Gain is typically 40 dB at 10.7 MHz. Power supply voltage should be 20 volts (courtesy GTE Sylvania Incorporated).

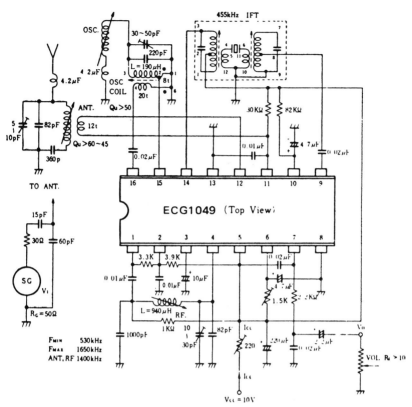

AM broadcast tuner with RF amplifier for automobile application. Audio output is 80 mV with 400-hertz modulation at 30%. The IF is 455 kHz. All coils and transformers are standard and can be purchased at Radio Shack (courtesy GTE Sylvania Incorporated).

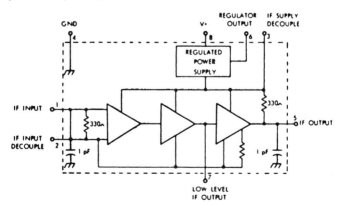

FM IF gain block with voltage regulator. The gain at 10.7 MHz is typically 35 dB. Operating voltage range is 10 to 20 volts. The internal voltage regulator regulates at 7.7 volts and is for external use (courtesy GTE Sylvania Incorporated).

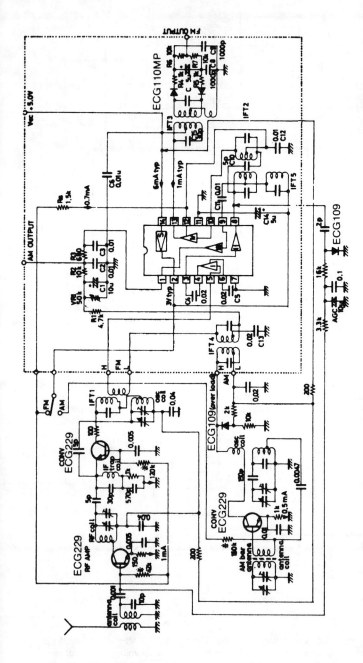

AM/FM radio receiver without AF amplifier using an ECG1003 AM/FM IF amplifier. AM specifications at 1000 kHz are: max sensitivity, 26 dB/m; available sensitivity, 44.5 dB/m; selectivity, +10 kHz 23 dB, −10 kHz 24 dB. FM specifications at 83 MHz are: max sensitivity, −1 dBP μV; available sensitivity, 4.0 dB/μV; −3 dB limiting sensitivity, 12 dB/m; HFM available sensitivity, 20 dB/μV; 3 dB bandwidth, 200 kHz; demodulated output, 60 mV (courtesy GTE Sylvania Incorporated).

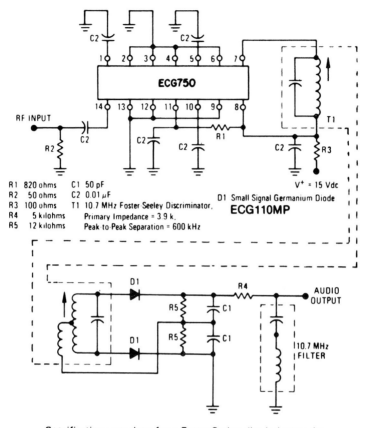

R1 820 ohms   C1 50 pF
R2 50 ohms   C2 0.01 µF
R3 100 ohms   T1 10.7 MHz Foster-Seeley Discriminator,
R4 5 kilohms   Primary Impedance = 3.9 k.
R5 12 kilohms   Peak-to-Peak Separation = 600 kHz

D1 Small Signal Germanium Diode **ECG110MP**

Specifications are given for a Foster-Seeley discriminator. Improved AM rejection at low signal levels can be obtained with a ratio detector.

For optimum circuit stability it is important to ground pins 2, 3, 4, 6, 9, 12, and 13.

FM limiting amplifier and Foster-Seeley discriminator designed for a 10.7 MHz IF (courtesy GTE Sylvania Incorporated).

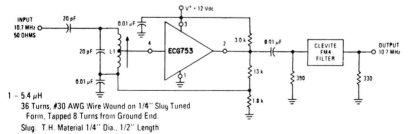

FM IF amplifier for 10.7 MHz (courtesy GTE Sylvania Incorporated).

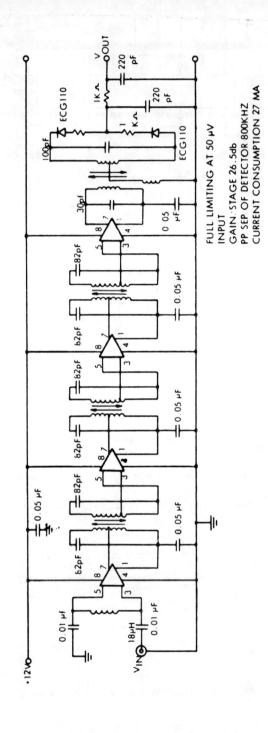

Complete 10.7 MHz IF strip for an FM tuner using four ECG703A chips (courtesy GTE Sylvania Incorporated).

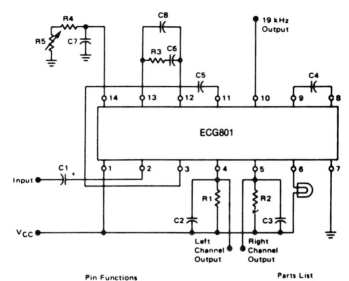

Pin Functions

Pin 1 = V_{CC}
Pin 2 = Input
Pin 3 = Amplifier Output
Pin 4 = Left Channel Output
Pin 5 = Right Channel Output
Pin 6 = Lamp Indicator
Pin 7 = Ground
Pin 8 = Switch Filter
Pin 9 = Switch Filter
Pin 10 = 19 kHz Output
Pin 11 = Modulator Input
Pin 12 = Loop Filter
Pin 13 = Loop Filter
Pin 14 = Oscillator RC Network

Parts List

C1 = 2.0 µF
C2 = 0.02 µF
C3 = 0.02 µF
C4 = 0.25 µF
C5 = 0.05 µF
C6 = 0.5 µF
C7 = 470 pF
C8 = 0.25 µF
R1 = 3.9 kΩ
R2 = 3.9 kΩ
R3 = 1.0 kΩ
R4 = 16 kΩ
R5 = 5.0 kΩ

FM stereo demodulator with integral stereo/monaural switch 75 mA lamp driving capability. Typical supply voltage is 12 volts but can be operated between 8 and 14 volts. Alignment is simple since there is only one adjustment, R5. Adjust R5 until 19 kHz is read at pin 10 on a frequency counter (courtesy GTE Sylvania Incorporated).

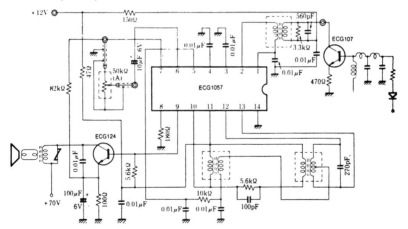

FM IF amplifier, detector and audio amplifier with 1-watt output. The ECG1057 is a 14-pin DIP. The circuit shown can be adapted for either 4.5 MHz or 10.7 MHz. Select the IF transformers accordingly. Coils and transformers are standard and can be purchased at Radio Shack (courtesy GTE Sylvania Incorporated).

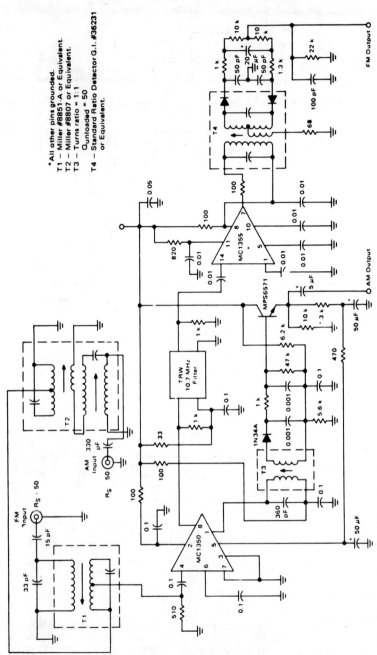

Composite AM/FM IF amplifier and detector circuitry (courtesy Motorola Semiconductor Products Inc.).

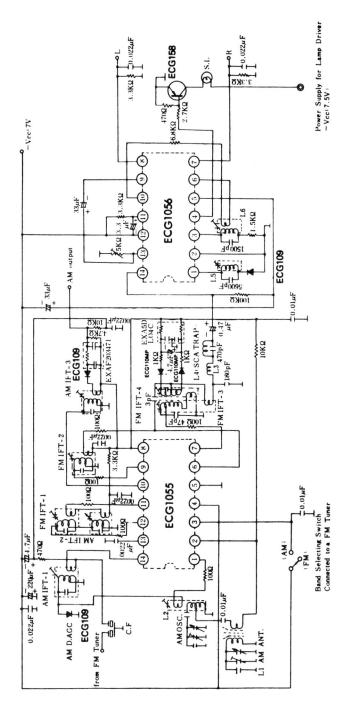

AM/FM stereo tuner. Both ICs are 14-pin DIPs. The ECG1055 is an AM/FM IF amplifier and AM converter, while the ECG1056 is a stereo multiplex demodulator with a composite amplifier, 19 kHz pilot signal filter, a 38 kHz subcarrier generator and either a matix or switching circuit for left and right stereo audio channels (courtesy GTE Sylvania Incorporated).

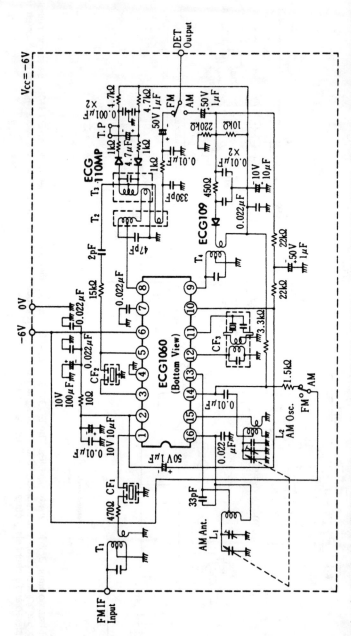

AM receiver front end with AM/FM IF amplifier-detector circuit. FM IF is 10.7 MHz, and the AM IF is 455 kHz. Detector output in FM is between 17 and 76 mV; for AM it is between 14.5 and 42 mV. All coils, transformers, variable capacitor and ceramic filters are standard and can be purchased at Radio Shack (courtesy GTE Sylvania Incorporated).

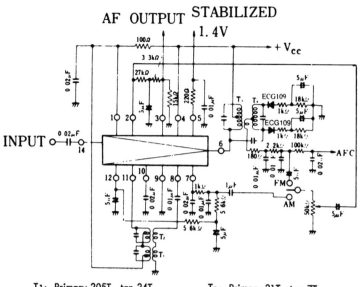

T$_1$: Primary 205T, tap 24T
Secondary 8T, Q$_O$ = 100

T$_2$: Primary 22T, tap 5T
Secondary 1T, Q$_O$ = 120

T$_3$: Primary 21T, tap 7T
Secondary 5T, Q$_O$ = 120

T$_4$: 11T×2

Performance Characteristics

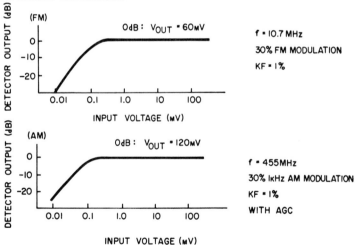

AM/FM IF amplifier and detector using an ECG1108 14-pin DIP. This chip is composed of three major sections. The input signal is fed to pin 14, then amplified by four transistors. AGC is applied to pin 12. Output from this section is at pin 11. The next section receives its input at pin 8. This section is an AM IF and detector as well as an FM IF and limiter. The last section is an audio amplifier with its input at pin 2; output is from pin 3. Typical supply voltage is 5 volts (courtesy GTE Sylvania Incorporated).

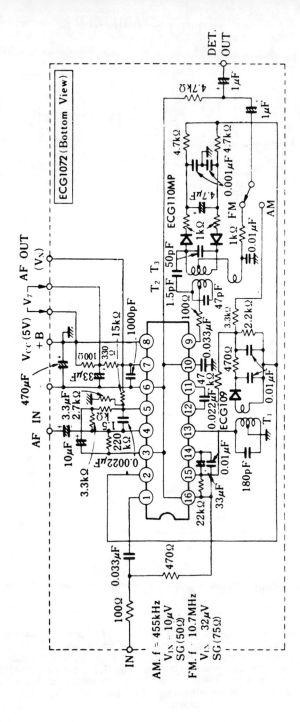

AM/FM IF amplifier with AF amplifier. Note in the diagram that the AF amplifier is not connected to the detector output. To use the AF amplifier connect the detector output to the AF input. The AF output with 1 kHz at 1 mV input is 0.47 volt typically (courtesy GTE Sylvania Incorporated).

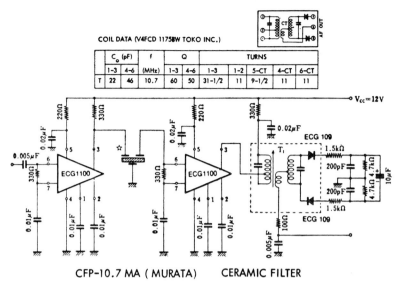

CFP-10.7 MA (MURATA)   CERAMIC FILTER

10.7 MHz FM IF amplifier with detector. Coil data for the transformer is shown in the table. The ECG1100 is a 7-pin module. Typical voltage gain per unit is 56 dB. Total supply current for the entire circuit is 21 mA. The detector output for a 60 dB input is 90 mV (courtesy GTE Sylvania Incorporated).

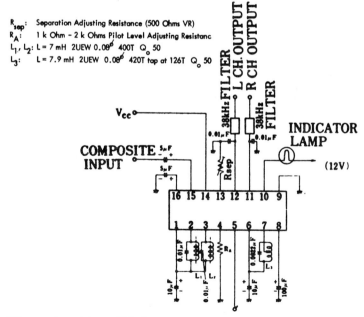

FM stereo demodulator with indicator lamp using an ECG1106 16-pin DIP. Input impedance is 20K. The indicator lamp operates at a 19 kHz input level of 5 mV RMS. Audio muting on voltage is 0.75 volt and off voltage is 1.0 volt (courtesy GTE Sylvania Incorporated).

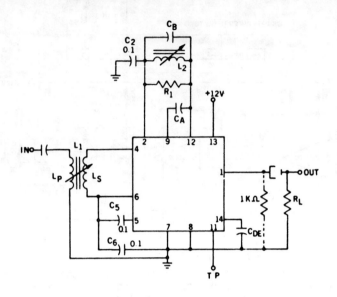

COMPONENT CHART

	Component Value		
	TV (4.5 MHz)	FM (10.7 MHz)	Notes
L_2 Inductance	7-14 µH	1.5-3 µH	1
L_2 Nom. Unloaded Q	50	50	—
L_2 Nom. D-C Resistance	50Ω	50Ω	—
C_A	3.0 pF	4.7 pF	—
C_B	120 pF	120 pF	2
R_1	20 kΩ	3.1 kΩ	—
Loaded Network Q	30	20	—
C_5 and C_6	0.1 µF	0.1 µF	—
C_2	0.1 µF	0.1 µF	—
C_{de}	0.01 µF	0.01 µF	—

NOTES:
1. Suggested coil source: 1.5 - 3µH Miller 9050, 7-14µH Miller 9052.
2. Use NPO type capacitor.

FM detector stage using one ECG708 IC. The output can be employed to directly drive a transistor power output stage. Either a 4.5 MHz IF or a 10.7 MHz IF can be detected (see component chart). An output of 0.6 volt can be obtained with less than 1% distortion. The output of the three-stage amplifier at the input is brought out to pin 10 so that the chip can be used as a wide-band 60 dB amplifier (courtesy GTE Sylvania Incorporated).

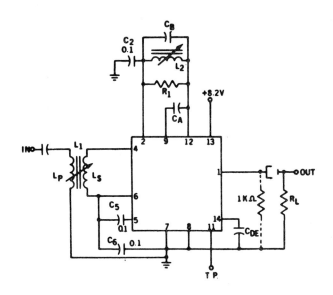

COMPONENT CHART

	Component Value		Notes
	TV (4.5 MHz)	FM (10.7 MHz)	
L_2 Inductance	7-14 μH	1.5-3 μH	1
L_2 Nom. Unloaded Q	50	50	—
L_2 Nom. D-C Resistance	<50Ω	<50Ω	—
C_A	3.0 pF	4.7 pF	—
C_B	120 pF	120 pF	2
R_1	20 kΩ	3.1 kΩ	—
Loaded Network Q	30	20	—
C_5 and C_6	0.1 μF	0.1 μF	...
C_2	0.1 μF	0.1 μF	.-
C_{de}	0.01 μF	0.01 μF	—

NOTES:
1. Suggested coil source: 1.5 - 3μH Miller 9050, 7-14μH Miller 9052.
2. Use NPO type capacitor.

FM detector and limiter stage for 4.5 MHz or 10.7 MHz, using one ECG709. See component chart for appropriate frequency. Designed for use with an 8-volt supply, the ECG709 is a three-stage limiter with a balanced product detector. There are 19 transistors, 6 diodes, and 18 resistors inside the chip. Only one screwdriver adjustment is necessary to tune the circuit (courtesy of GTE Sylvania Incorporated).

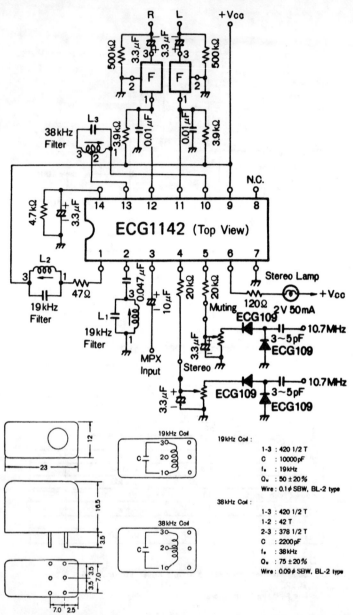

FM stereo demodulator with stereo-monaural switch and audio muting. Recommended supply voltage is 9 volts. Standby current is 10 mA. Input impedance is 20K. Audio muting is on at 0.85 volt and off at 1.08 volts. Stereo indicator is on at input voltage of 12 mV RMS and off at input voltage of 8.4 mV RMS. Stereo-monoaural switching occurs at 1.13 volts for stereo and 0.82 volt for monaural. Component F is a 38 kHz band eliminate filter (courtesy GTE Sylvania Incorporated).

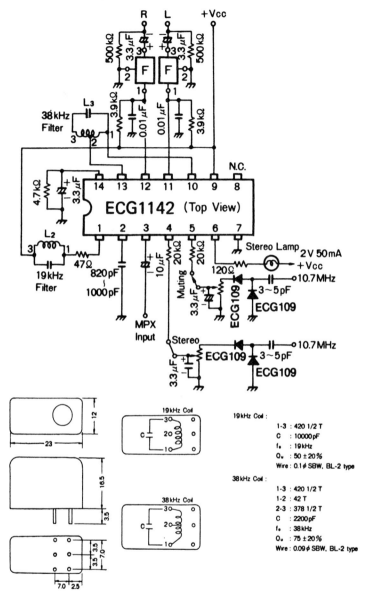

FM stereo demodulator with stereo-monaural switch and audio muting. Recommended supply voltage is 9 volts. Standby current is 10 mA. Input impedance is 20K. Audio muting is on at 0.85 volt and off at 1.08 volts. Stereo indicator is on at input voltage of 12 mV RMS and off at input voltage of 8.4 mV RMS. Stereo-monaural switching occurs at 1.13 volts for stereo and 0.82 volt for monaural. Component F is a 38 kHz band eliminate filter (courtesy GTE Sylvania Incorporated).

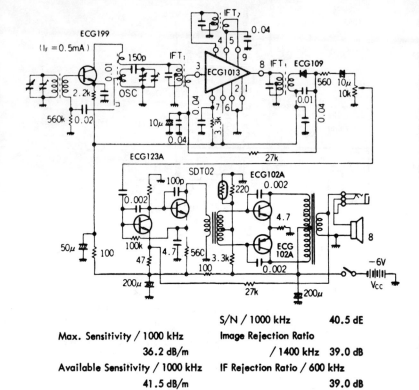

	S/N / 1000 kHz 40.5 dE
Max. Sensitivity / 1000 kHz	Image Rejection Ratio
36.2 dB/m	/ 1400 kHz 39.0 dB
Available Sensitivity / 1000 kHz	IF Rejection Ratio / 600 kHz
41.5 dB/m	39.0 dB

AM radio receiver with a maximum sensitivity of 36.2 dB/m at 1000 kHz. Available sensitivity is 41.5 dB/m at 1000 kHz. Signal to noise ratio is 40.5 dB. Image rejection ratio is 39.0 dB at 1400 kHz. IF rejection ratio at 600 kHz is 39 dB. IF transformers and audio transformers are standard and can be purchased at Radio Shack. The same is true for the tuning capacitor (courtesy GTE Sylvania Incorporated).

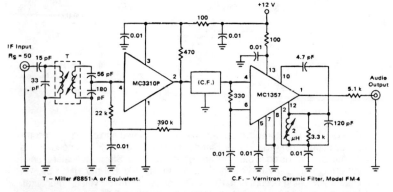

FM IF amplifier with quadrature detector (courtesy Motorola Semiconductor Products Inc.).

TYPICAL PERFORMANCE:

20 μV Sensitivity for 3 dB Limiting
680 mV(rms) Recovered Audio;
75 kHz Deviation
THD ≈ 1%

When using the device as a non-saturating limiter the load must be chosen to prevent voltage saturation of the output stage. The load impedance can be calculated from:

$$R_L \leq \frac{2(V^+ - 5.3)}{5.0} \text{ kilohms}$$

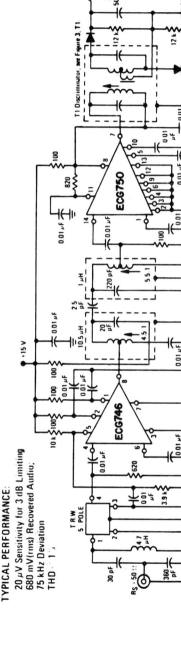

Complete limiting FM IF amplifier and detector. Typical AM rejection is 60 dB. ECG746 is an 8-pin DIP and ECG750 is a 14-pin DIP. The ECG750 is designed for use with a Foster-Seeley discriminator or ratio detector. Transformer T1 is a 10.7 MHz Foster-Seeley discriminator with a primary impedance of 3.9K and a peak-to-peak separation of 600 kHz. The discriminator diodes are ECG110 MP (courtesy GTE Sylvania Incorporated).

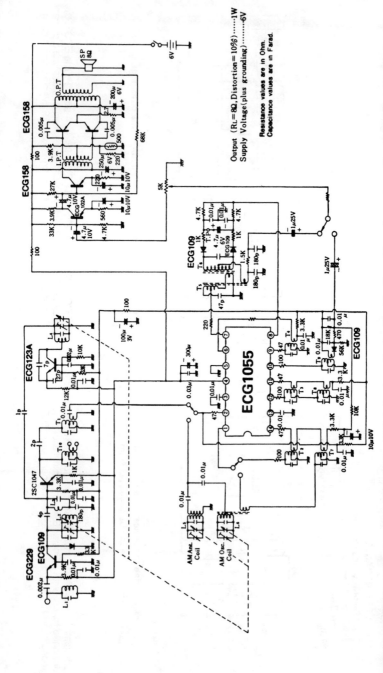

Complete AM/FM broadcast receiver with 1-watt output and powered by 6 volts. All coils and transformers are standard and can be purchased at Radio Shack. FM sensitivity is 1 μV, while AM is 30 μV (courtesy GTE Sylvania Incorporated).

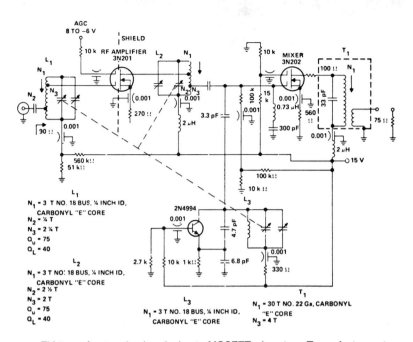

FM tuner front end using dual-gate MOSFETs (courtesy Texas Instruments Incorporated).

Converters

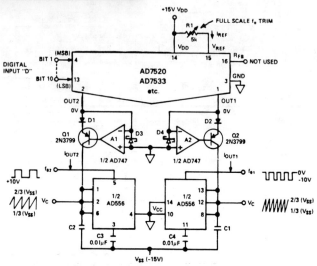

D/F converter with complementary output frequencies. This circuit provides two output frequencies. One output is proportional to the fractional binary equivalent of D, while the other output is proportional to 1 - D. Excellent linearity is obtained from 10 hertz to 10 kilohertz. The 556 timer provides either pulse or sawtooth output waveforms. D1 and D2 are required to protect the emitter-base junctions of Q1 and Q2 during powerup. D3 and D4 protect the AD7520 OUT 1 and OUT 2 terminals (courtesy Analog Devices, Inc.).

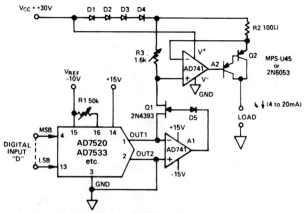

4 to 20 mA converter. The circuit drives a digitally programmed current into a grounded load, R_L, according the the relationship I_K 4mA (D) (16 mA). With a 10-bit D/A converter such as the AD7520 the circuit provides an output from 4 mA to 20 mA with a resolution of 15⅝ μA. The maximum compliance voltage of the load is +25 volts, equivalent to a resistance of 1250 ohms maximum. Higher voltages across the load can be developed if the lower end of R_L is returned to a negative voltage, but Q2 must be able to handle the additional breakdown voltage (courtesy Analog Devices, Inc.).

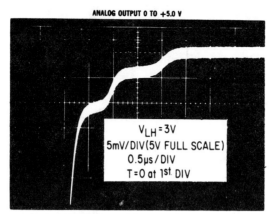

μA722/ECG915 op amp switching ON, as it should with typical logic voltage on least significant bits. Note complete absence of ringing.

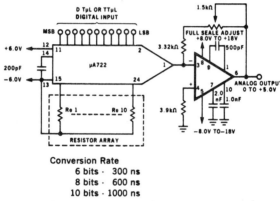

Conversion Rate
6 bits - 300 ns
8 bits - 600 ns
10 bits - 1000 ns

High-speed 10-bit A/D converter using the μA 722 and an ECG915 operational amplifier (courtesy GTE Sylvania Incorporated).

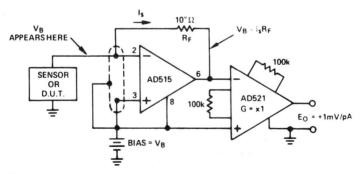

Current-to-voltage converter with grounded bias and sensor (courtesy Analog Devices, Inc.).

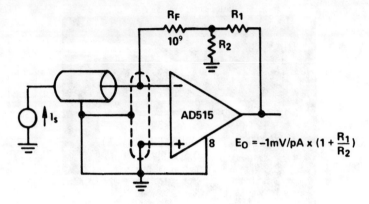

Picoampere-to-voltage converter with gain (courtesy Analog Devices, Inc.).

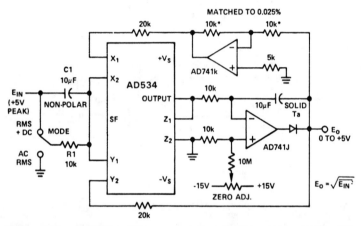

High-performance RMS-to-DC converter. To calibrate switch to RMS + DC, apply an input of say 1.00 volt DC and set the zero adjust until the output reads the same as the input. Check the inputs of ±5 volts (courtesy Analog Devices, Inc.).

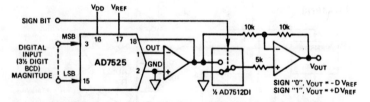

Sign-magnitude D/A converter with BCD coding. The AD7525 is used to provide a choice of positive or negative gains for the analog input signal, with sign-magnitude BVD coding. Gains range from −1.999 to +1.999. The BCD coding of the AD7525 permits it to be used with thumbwheel switches for manual gain setting without BCD-binary translation (courtesy Analog Devices, Inc.).

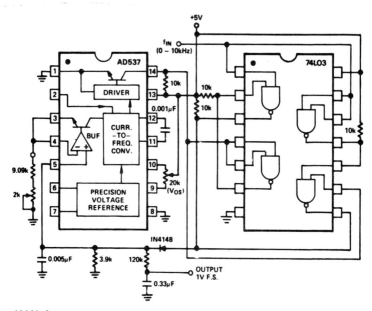

10 kHz frequency-to-voltage converter. The input signal should be a pulse train or square wave with characteristics similar to TTL or 5-volt CMOS outputs. Minimum pulse width is 40 μs. Full scale output is 1 volt for a 10 kHz input. To trim first set Vos to midrange and trim the 2K pot for 1-volt output. Then apply a 10-hertz waveform and trim Vos for a 10 mV output (courtesy Analog Devices, Inc.).

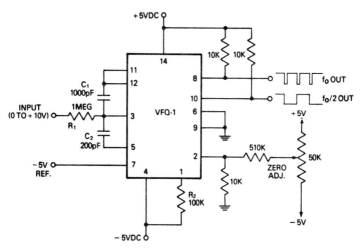

Voltage-to-frequency converter. Full scale frequency is 10 kHz as shown. The chip is the Datel VFQ-1. Not the fo output and fo/2 outputs. Full scale output can be extended to 100 kHz by changing C1 and C2 as indicated on schematic (courtesy Datel Systems, Inc.).

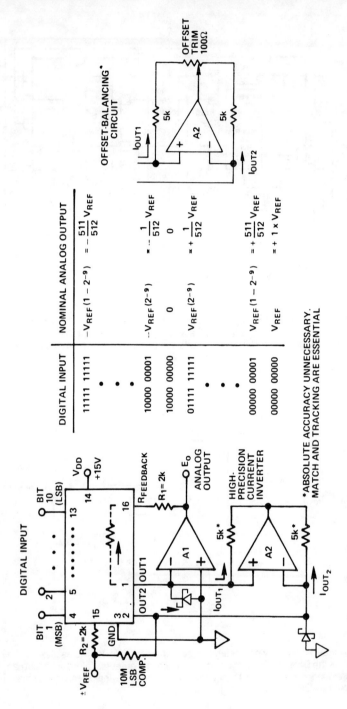

Bipolar offset-binary digital-to-voltage converter (four-quadrant multiplier) using an AD7520. R1 and R2 are used for gain adjustment (courtesy Analog Devices, Inc.).

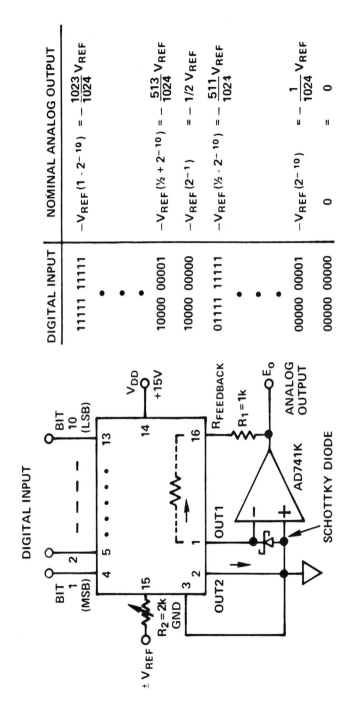

Unipolar binary digital-to-voltage converter (two-quadrant multiplier) using an AD7520 D/A converter. R1 and R2 provide gain adjustment capability (courtesy Analog Devices, Inc.).

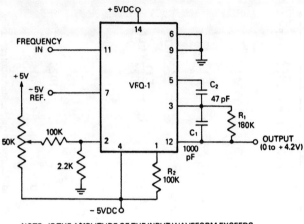

NOTE: IF THE AMPLITUDE OF THE INPUT WAVEFORM EXCEEDS THE SPECIFIED MAXIMUM, THE FOLLOWING INPUT CIRCUIT SHOULD BE ADDED

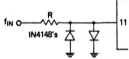

Frequency-to-voltage converter for a 0 to 100 kHz input. Chip used is a Datel VFQ-1 (courtesy Datel Systems, Inc.).

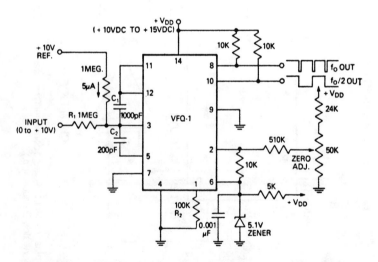

Voltage-to-frequency converter for single supply operation. Full scale output is 10 kHz. The chip used is a Datel VFQ-1 14-pin DIP (courtesy Datel Systems, Inc.).

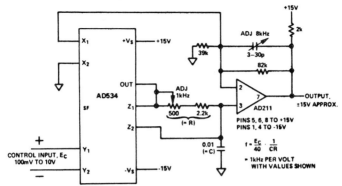

CALIBRATION PROCEDURE:

WITH E_C = 1.0V, ADJUST POT TO SET f = 1.000kHz. WITH E_C = 8.0V, ADJUST TRIMMER CAPACITOR TO SET f = 8.000kHz. LINEARITY WILL TYPICALLY BE WITHIN ±0.1% OF F.S. FOR ANY OTHER INPUT.

DUE TO DELAYS IN THE COMPARATOR, THIS TECHNIQUE IS NOT SUITABLE FOR MAXIMUM FREQUENCIES ABOVE 10kHz, BUT SCALING TO OTHER F.S. VALUES IS STRAIGHTFORWARD.

A TRIANGLE-WAVE OF ±5V PK APPEARS ACROSS THE 0.01μF CAPACITOR; IF USED AS AN OUTPUT A VOLTAGE-FOLLOWER SHOULD BE INTERPOSED.

Differential-input voltage-to-frequency converter using an AD534 multiplier/divider chip (courtesy Analog Devices, Inc.).

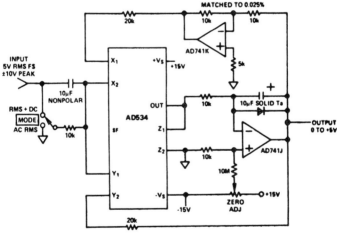

CALIBRATION PROCEDURE:

WITH 'MODE' SWITCH IN 'RMS + DC' POSITION, APPLY AN INPUT OF +1.00VDC. ADJUST ZERO UNTIL OUTPUT READS SAME AS INPUT. CHECK FOR INPUTS OF ±10V; OUTPUT SHOULD BE WITHIN ±0.05% (5mV).

ACCURACY IS MAINTAINED UP TO 100kHz, AND IS TYPICALLY HIGH BY 0.5% AT 1MHz FOR V_{IN} = 4V RMS (SINE, SQUARE OR TRIANGULAR WAVE).

PROVIDED THAT THE PEAK INPUT IS NOT EXCEEDED, CREST-FACTORS UP TO AT LEAST TEN HAVE NO APPRECIABLE EFFECT ON ACCURACY.

INPUT IMPEDANCE IS ABOUT 10kΩ; FOR HIGH (10MΩ) IMPEDANCE, REMOVE MODE SWITCH AND INPUT COUPLING COMPONENTS.

Wide-band high-crest-factor RMS-to-DC converter using an AD534 multiplier/divider chip (courtesy Analog Devices, Inc.).

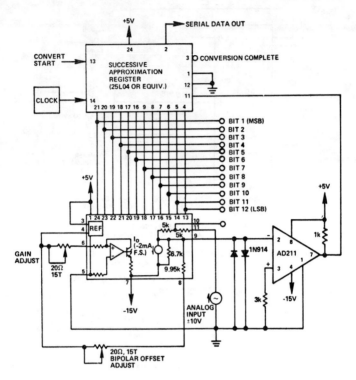

Successive approximation A/D converter (courtesy Analog Devices, Inc.).

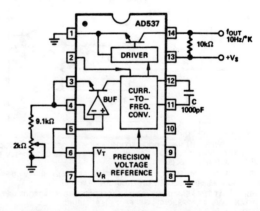

Kelvin temperature-to-frequency converter. No sensor is required since the AD537 has an internal V$_{TEMP}$ output. The output tracks the temperature at 10 hertz per degree Kelvin. To calibrate, select a 25°C standard, which corresponds to 298°K, and adjust the 2K pot for 2.98 kHz (courtesy Analog Devices, Inc.).

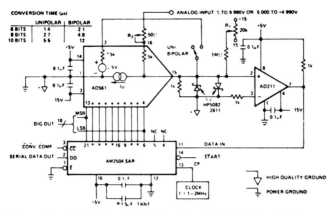

Fast precision 10-bit A/D converter (courtesy Analog Devices, Inc.).

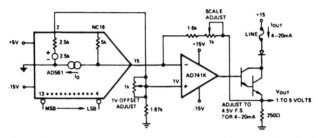

Digital 4-to-20 mA or 1-to-5 volt converter (courtesy Analog Devices, Inc.).

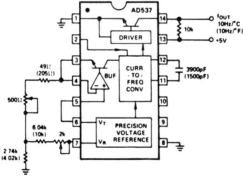

Celsius (Fahrenheit) temperature-to-frequency converter. No sensor is required since the AD537 has an internal V_{TEMP} output. The converter output tracks the temperature at a rate of 10 hertz per degree Celsius or Fahrenheit, depending on the resistor-capacitor values chosen. All values in parentheses are for Fahrenheit. To calibrate measure the room temperature in Kelvin. Measure the temperature output at pin 6 at that temperature. Calculate the offset adjustment by dividing the V_{TEMP} at pin 6 in millivolts by the room temperature in Kelvin, then multiplying by 273.2 The offset voltage will be in millivolts. Then disconnect the 49-ohm resistor and adjust the 2K pot for the offset voltage at pin 7. Connect the 49-ohm resistor. Then adjust the 500-ohm slope trimmer for room temperature, or 25°C, which results in 250 hertz (courtesy Analog Devices, Inc.).

167

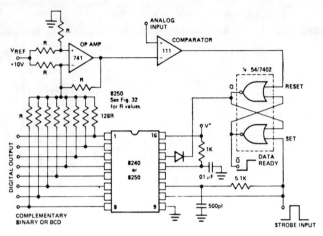

A/D converter. The circuit shown is an 8-bit binary A/D converter when using the 8240 or a 2-digit BCD A/D converter when using the 8250. The input strobe first resets then triggers the 8240/8250 and set the flip-flop, which enables the counter. The staircase from the op amp counts down until it reaches the analog input, at which time the comparator resets the flip-flop and stops the count. The digital word at the eight outputs is the complementary binary or BCD equivalent of the analog input. As shown the maximum conversion time is 2.6 ms, determined by R and C at pin 14 of the counter. The not-Q output of the flip-flop is convenient to use as a data ready flag since it is high when conversion is complete (courtesy Intersil, Inc.).

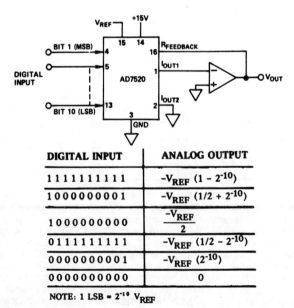

DIGITAL INPUT	ANALOG OUTPUT
1 1 1 1 1 1 1 1 1 1	$-V_{REF}(1 - 2^{-10})$
1 0 0 0 0 0 0 0 0 1	$-V_{REF}(1/2 + 2^{-10})$
1 0 0 0 0 0 0 0 0 0	$\dfrac{-V_{REF}}{2}$
0 1 1 1 1 1 1 1 1 1	$-V_{REF}(1/2 - 2^{-10})$
0 0 0 0 0 0 0 0 0 1	$-V_{REF}(2^{-10})$
0 0 0 0 0 0 0 0 0 0	0

NOTE: 1 LSB = $2^{-10} V_{REF}$

10-bit D/A converter connected for two-quadrant multiplication (courtesy Analog Devices, Inc.).

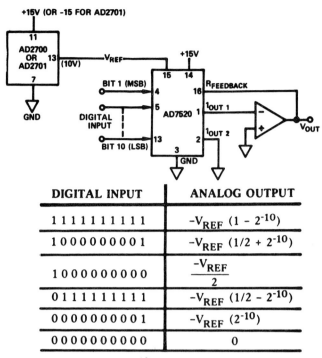

DIGITAL INPUT	ANALOG OUTPUT
1 1 1 1 1 1 1 1 1 1	$-V_{REF}(1 - 2^{-10})$
1 0 0 0 0 0 0 0 0 1	$-V_{REF}(1/2 + 2^{-10})$
1 0 0 0 0 0 0 0 0 0	$\dfrac{-V_{REF}}{2}$
0 1 1 1 1 1 1 1 1 1	$-V_{REF}(1/2 - 2^{-10})$
0 0 0 0 0 0 0 0 0 1	$-V_{REF}(2^{-10})$
0 0 0 0 0 0 0 0 0 0	0

NOTE: 1 LSB = $2^{-10} V_{REF}$

D/A converter with 10-volt reference (courtesy Analog Devices, Inc.).

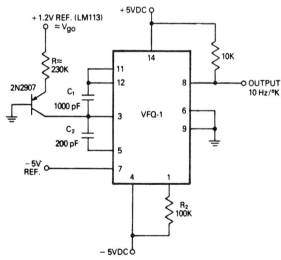

Temperature-to-frequency converter using a Datel VFQ-1 14-pin DIP (courtesy Datel Systems, Inc.).

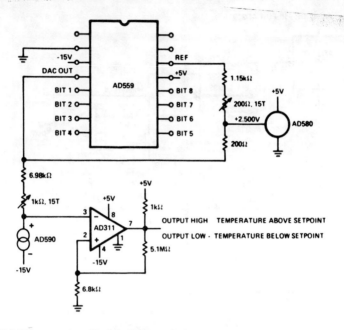

8-bit D/A converter with digitally controlled setpoint (courtesy Analog Devices, Inc.).

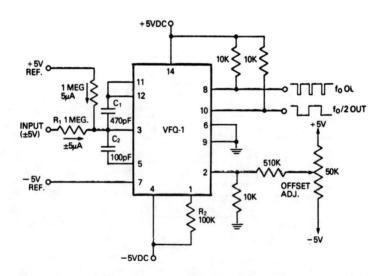

Voltage-to-frequency converter for bipolar operation. Full scale output is 20 kHz. Chip used is the Datel VFQ-1 14-pin DIP. Outputs at fo and fo/2 are possible (courtesy Datel Systems, Inc.).

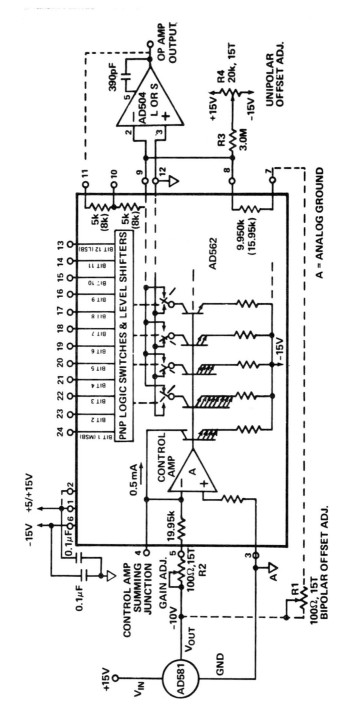

Precision 12-bit D/A converter (courtesy Analog Devices, Inc.).

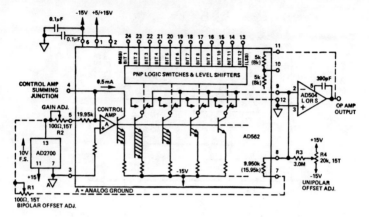

12-bit D/A converter with precision 10-volt reference (courtesy Analog Devices, Inc.).

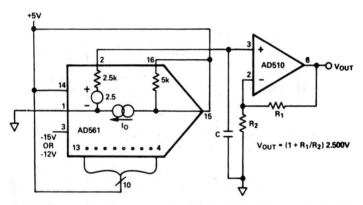

10-bit D/A converter with precision low-noise reference (courtesy Analog Devices, Inc.).

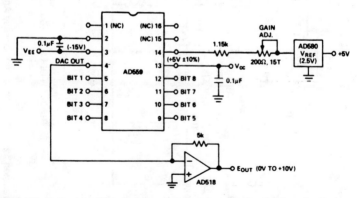

8-bit D/A converter with unipolar output (courtesy Analog Devices, Inc.).

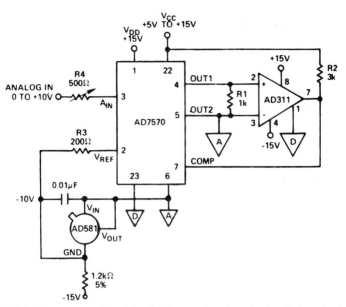

CMOS A/D converter with −10 volt reference (courtesy Analog Devices, Inc.).

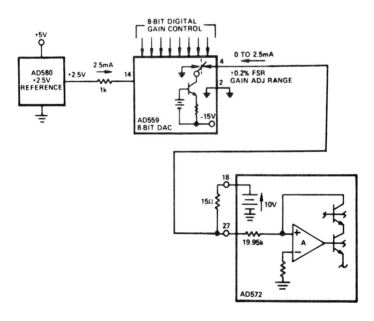

Digital gain control of a 12-bit successive approximation A/D converter (AD572) using an AD559 8-bit D/A converter (courtesy Analog Devices, Inc.).

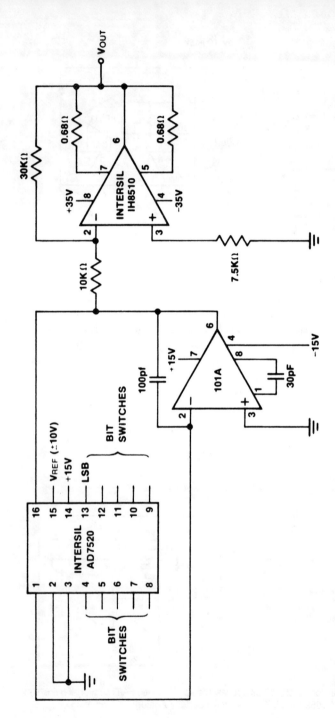

Power D/A converter using the AD7520. The Intersil AD7520 is an 18-pin multiplying D/A converter. This circuit is designed for 8-bit accuracy and 10-bit resolution. The Intersil IH8510 power amplifier (1A continuous output) is driven by the AD7520. A summing amplifier between the AD7520 and IH8510 is used to separate the gain block containing the AD7520 on-chip resistors from the power amplifier stage, whose gain is set only by the external resistors. This approach minimizes drift, otherwise the AD7520 can be directly connected to the IH8510 by using a 25-volt reference for the D/A converter (courtesy Intersil, Inc.).

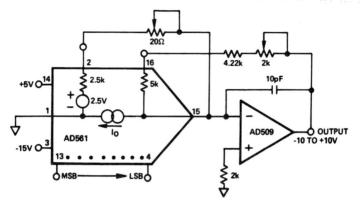

10-bit D/A converter with ±10-volt buffered output (courtesy Analog Devices, Inc.).

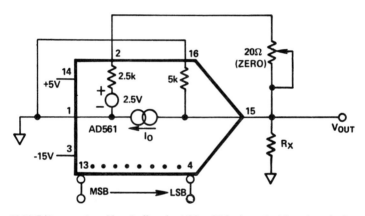

10-bit D/A converter with unbuffered ±1.66-volt bipolar output (courtesy Analog Devices, Inc.).

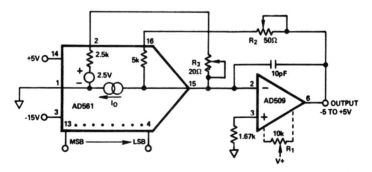

10-bit D/A converter with ±5-volt buffered bipolar output (courtesy Analog Devices, Inc.).

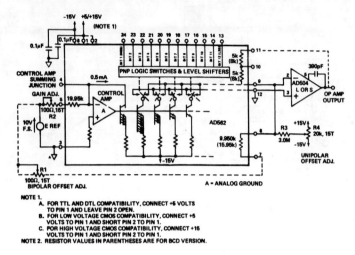

D/A converter in unipolar and bipolar hookup (courtesy Analog Devices, Inc.).

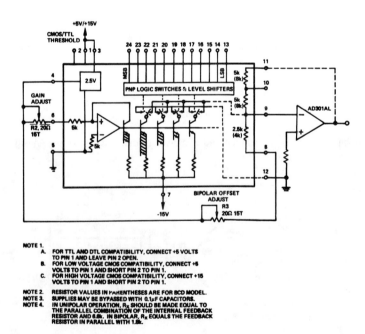

D/A converter in bipolar hookup using an AD563 (courtesy Analog Devices, Inc.).

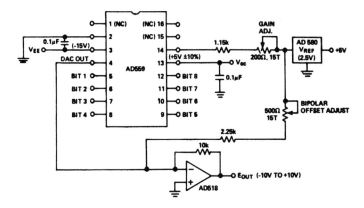

8-bit D/A converter with bipolar output (courtesy Analog Devices, Inc.).

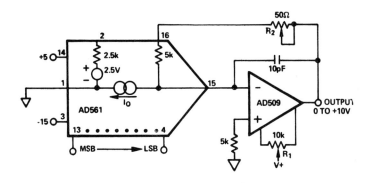

10-bit D/A converter with 0 to +10-volt unipolar output (courtesy Analog Devices, Inc.).

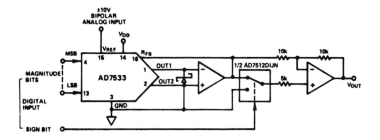

10-bit and sign multiplying D/A converter (courtesy Analog Devices, Inc.).

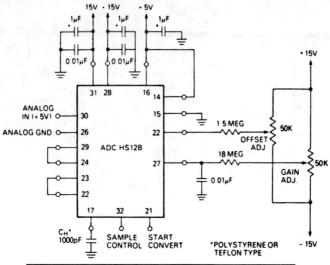

INPUT VOLTAGE RANGE			COMP. OFFSET BINARY	
±10V	±5V	±2.5V	MSB	LSB
+9.9951V	+4.9976V	+2.4988V	0000 0000	0000
+7.5000	+3.7500	+1.8750	0001 1111	1111
+5.0000	+2.5000	+1.2500	0011 1111	1111
0.0000	0.0000	0.0000	0111 1111	1111
−5.0000	−2.5000	−1.2500	1011 1111	1111
−7.5000	−3.7500	−1.8750	1101 1111	1111
−9.9951	−4.9976	−2.4988	1111 1111	1110
−10.0000	−5.0000	−2.5000	1111 1111	1111

A/D converter with sample-and-hold circuit for bipolar operation, ±5 volts. For ±2.5-volt operation jumper pins 22 and 25 with the connections shown. For ±10-volt operation change the jumper at pin 29 from pin 24 to pin 25 and leave all others as shown (courtesy Datel Systems, Inc.).

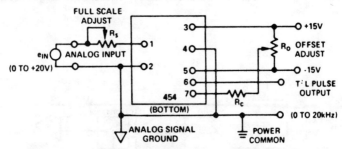

EXTERNAL CONNECTIONS FOR VOLTAGE TO FREQUENCY OPERATION, WITH OPTIONAL INPUT OFFSET VOLTAGE AND FULL SCALE OUTPUT FREQUENCY ADJUSTMENTS

R_s = FULL SCALE ADJUSTMENT; 500Ω (Use low T.C. Cermet, <50ppm/°C or equivalent)
R_o = INPUT OFFSET ADJUSTMENT; 50kΩ (Use low T.C. Cermet, <50ppm/°C or equivalent)
R_c = 20MΩ

CAUTION: DO NOT SHORT OUTPUT TO -15V

V/F converter with 20 kHz full scale output (courtesy Analog Devices, Inc.).

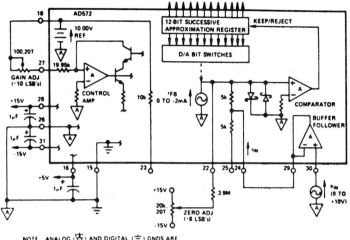

12-bit successive approximation A/D converter for unipolar 0 to +10-volt input range with buffer follower. A/D chip is the AD572 (courtesy Analog Devices, Inc.).

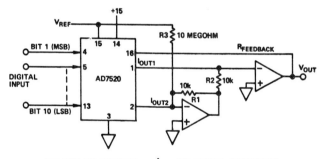

DIGITAL INPUT	ANALOG OUTPUT
1 1 1 1 1 1 1 1 1 1	$-V_{REF} (1 - 2^{-9})$
1 0 0 0 0 0 0 0 0 1	$-V_{REF} (2^{-9})$
1 0 0 0 0 0 0 0 0 0	0
0 1 1 1 1 1 1 1 1 1	$V_{REF} (2^{-9})$
0 0 0 0 0 0 0 0 0 1	$V_{REF} (1 - 2^{-9})$
0 0 0 0 0 0 0 0 0 0	V_{REF}

NOTE: 1 LSB = $2^{-9} V_{REF}$

10-bit D/A converter connected for four-quadrant multiplication (courtesy Analog Devices, Inc.).

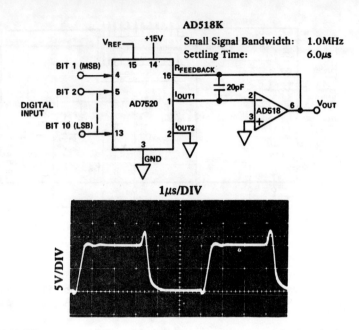

10-bit D/A converter with an AD518 op amp (courtesy Analog Devices, Inc.).

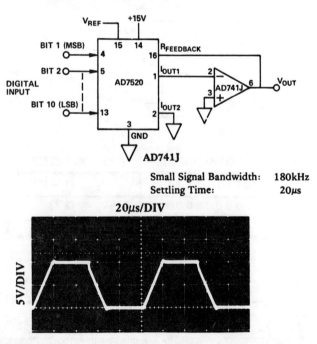

10-bit D/A converter with an AD741 op amp (courtesy Analog Devices, Inc.).

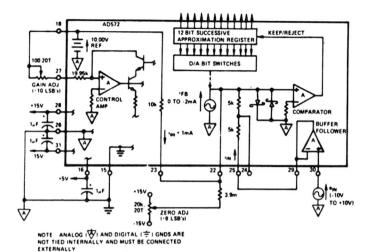

12-bit successive approximation A/D converter for bipolar −10 volt to +10 volt input range with buffer follower. The A/D chip is the AD572 (courtesy Analog Devices, Inc.).

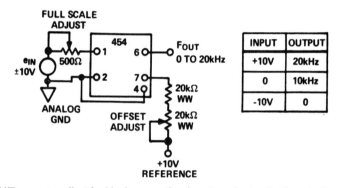

V/F converter offset for bipolar operation (courtesy Analog Devices, Inc.).

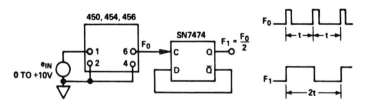

V/F converter using a flip-flop to obtain a square wave output. The full scale output for the 450 and 456 is 10 kHz and 20 kHz for the 454 (courtesy Analog Devices, Inc.).

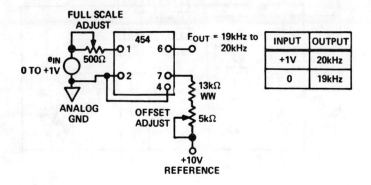

V/F converter with 20 kHz full scale output offset to improve dynamic response (courtesy Analog Devices, Inc.).

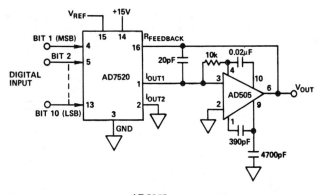

AD505J

Small Signal Bandwidth: 1.0MHz
Settling Time: 2.5µs

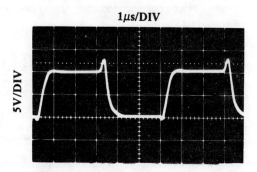

10-bit D/A converter with an AD505 op amp (courtesy Analog Devices, Inc.).

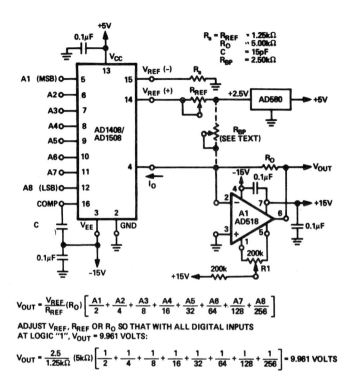

$$V_{OUT} = \frac{V_{REF}}{R_{REF}}(R_O)\left[\frac{A1}{2} + \frac{A2}{4} + \frac{A3}{8} + \frac{A4}{16} + \frac{A5}{32} + \frac{A6}{64} + \frac{A7}{128} + \frac{A8}{256}\right]$$

ADJUST V_{REF}, R_{REF} OR R_O SO THAT WITH ALL DIGITAL INPUTS AT LOGIC "1", V_{OUT} = 9.961 VOLTS:

$$V_{OUT} = \frac{2.5}{1.25k\Omega}(5k\Omega)\left[\frac{1}{2} + \frac{1}{4} + \frac{1}{8} + \frac{1}{16} + \frac{1}{32} + \frac{1}{64} + \frac{1}{128} + \frac{1}{256}\right] = 9.961\text{ VOLTS}$$

8-bit D/A converter with voltage output and using a fixed reference. For RREF at 1K, 2.5K or 5K the minimum value of C should be 15 pF, 37 pF or 75 pF respectively (courtesy Analog Devices, Inc.).

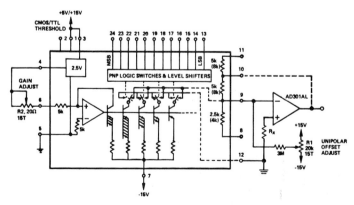

D/A converter in unipolar hookup using AD563 12-bit D/A converter chip (courtesy Analog Devices, Inc.).

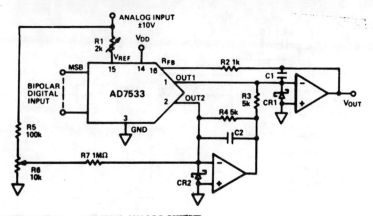

| DIGITAL INPUT | NOMINAL ANALOG OUTPUT |
MSB LSB	(V_{OUT} as shown in Figure 2)
1111111111	$-V_{REF}\left(\frac{511}{512}\right)$
1000000001	$-V_{REF}\left(\frac{1}{512}\right)$
1000000000	0
0111111111	$+V_{REF}\left(\frac{1}{512}\right)$
0000000001	$+V_{REF}\left(\frac{511}{512}\right)$
0000000000	$+V_{REF}\left(\frac{512}{512}\right)$

NOTES:

1. Nominal Full Scale Range for the circuit of Figure 6 is given by FSR = $V_{REF}\left(\frac{1023}{512}\right)$

2. Nominal LSB magnitude for the circuit of Figure 6 is given by LSB = $V_{REF}\left(\frac{1}{512}\right)$

10-bit multiplying D/A converter in bipolar operation. In this configuration the D/A converter is set up for four-quadrant multiplication (courtesy Analog Devices, Inc.).

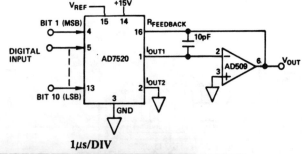

AD509K

Small Signal Bandwidth: 1.6MHz
Settling Time: 2.0μs

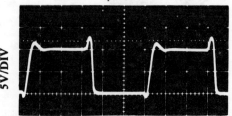

10-bit D/A converter with an AD509 op amp (courtesy Analog Devices, Inc.).

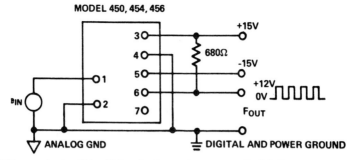

V/F converter for driving high-noise-immunity logic and CMOS. Full scale output is 10 kHz for the 450 and 456 and 20 kHz for the 454 (courtesy Analog Devices, Inc.).

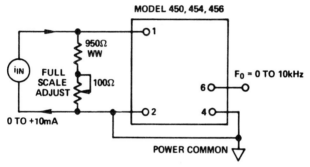

I/F converter using a V/F converter chip. Full scale output is 10 kHz for the 450 and 456 and 20 kHz for the 454 (courtesy Analog Devices, Inc.).

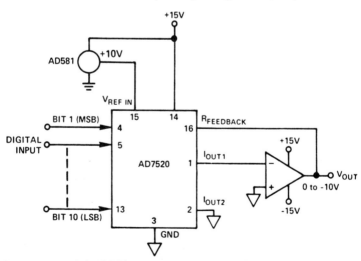

Low-power 10-bit CMOS D/A converter with 10-volt reference (courtesy Analog Devices, Inc.).

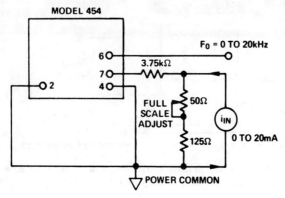

I/F converter using the 454 V/F converter chip with a 20 kHz full scale output (courtesy Analog Devices, Inc.).

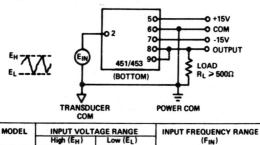

MODEL	INPUT VOLTAGE RANGE		INPUT FREQUENCY RANGE (F_{IN})
	High (E_H)	Low (E_L)	
451	+1.45V to +12V	-12V to +1.35V	0 to 11kHz
453	+1.5V to +12V	-12V to +1.3V	0 to 110kHz

F/V converter with +10-volt full scale output. Table shows input specs for both models (courtesy Analog Devices, Inc.).

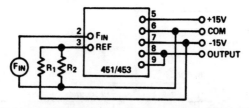

MODEL	R_1	R_2	THRESHOLD (V_T)	HYSTERESIS (V_H)	TRIGGER LEVELS ($V_T \pm V_H$)
451	∞	∞	+1.4V	±50mV	+1.45V; +1.35V
	100kΩ	∞	0V	±20mV	±20mV (40mV p-p)
	100kΩ	1kΩ	0V	±5mV	±5mV (10mV p-p)
453	∞	∞	+1.4V	±100mV	+1.5V; +1.3V
	100kΩ	∞	0V	±60mV	±60mV (120mV p-p)
	100kΩ	1kΩ	0V	±15mV	±15mV (30mV p-p)

F/V converter with +10-volt full scale output and decreased threshold for increased triggering sensitivity for low-level input signals. The input frequency range is 0 to 11 kHz for the 451 and 110 kHz for the 453 (courtesy Analog Devices, Inc.).

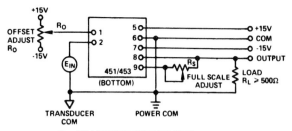

F/V converter with +10.000-volt full scale output and fine trim adjustment (courtesy Analog Devices, Inc.).

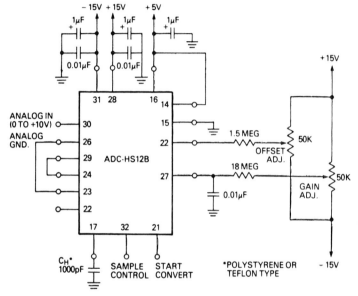

INPUT RANGE		COMP. BINARY CODING		
0 TO +10V	0 TO +5V	MSB		LSB
+9.9976V	+4.9988V	0000	0000	0000
+8.7500	+4.3750	0001	1111	1111
+7.5000	+3.7500	0011	1111	1111
+5.0000	+2.5000	0111	1111	1111
+2.5000	+1.2500	1011	1111	1111
+1.2500	+0.6250	1101	1111	1111
+0.0024	+0.0012	1111	1111	1110
0.0000	0.0000	1111	1111	1111

A/D converter with sample-and-hold circuit for unipolar operation, 0 to 10 volts. For operation of 0 to 5 volts jumper pins 22 and 25 with the connection shown (courtesy Datel Systems, Inc.).

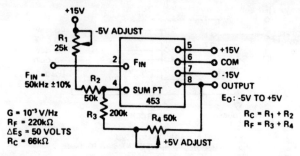

F/V converter used for FM demodulation. The example shown here has the carrier at 50 kHz with ±5 kHz modulating signal. The terms shown are defined as follows: G is gain, R_F is the external gain resistor, ΔE_S is the output offset shift and R_C is the offset current resistor (courtesy Analog Devices, Inc.).

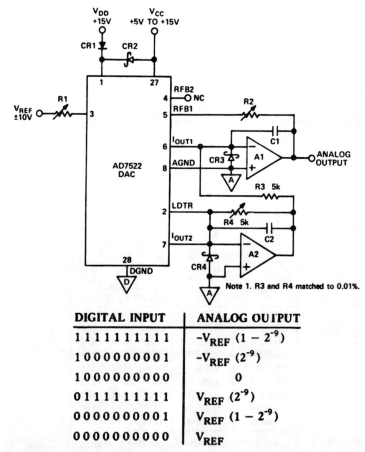

DIGITAL INPUT	ANALOG OUTPUT
1 1 1 1 1 1 1 1 1 1	$-V_{REF}(1 - 2^{-9})$
1 0 0 0 0 0 0 0 0 1	$-V_{REF}(2^{-9})$
1 0 0 0 0 0 0 0 0 0	0
0 1 1 1 1 1 1 1 1 1	$V_{REF}(2^{-9})$
0 0 0 0 0 0 0 0 0 1	$V_{REF}(1 - 2^{-9})$
0 0 0 0 0 0 0 0 0 0	V_{REF}

10-bit buffered multiplying D/A converter in bipolar operation (courtesy Analog Devices, Inc.).

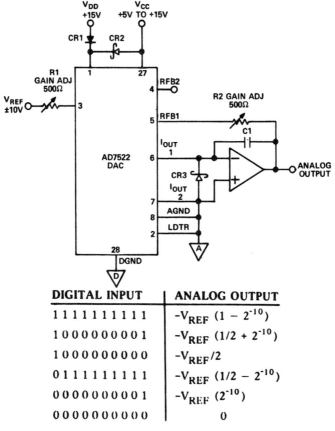

DIGITAL INPUT	ANALOG OUTPUT
1 1 1 1 1 1 1 1 1 1	$-V_{REF}(1 - 2^{-10})$
1 0 0 0 0 0 0 0 0 1	$-V_{REF}(1/2 + 2^{-10})$
1 0 0 0 0 0 0 0 0 0	$-V_{REF}/2$
0 1 1 1 1 1 1 1 1 1	$-V_{REF}(1/2 - 2^{-10})$
0 0 0 0 0 0 0 0 0 1	$-V_{REF}(2^{-10})$
0 0 0 0 0 0 0 0 0 0	0

10-bit buffered multiplying D/A converter for two-quadrant multiplication (courtesy Analog Devices, Inc.).

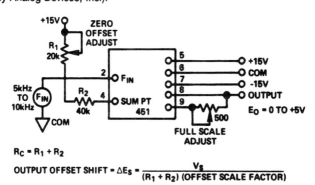

$R_C = R_1 + R_2$

OUTPUT OFFSET SHIFT = $\Delta E_S = \dfrac{V_S}{(R_1 + R_2)(\text{OFFSET SCALE FACTOR})}$

(OFFSET SCALE FACTOR: -56µA/V, MODEL 451; -45µA/V, MODEL 453)

F/V converter having 0 to +5-volt output with a 5 kHz to 10 kHz input (courtesy Analog Devices, Inc.).

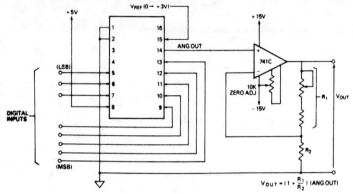

Multiplying D/A converter using a Datel DAC-MC8B 16-pin DIP. Pin functions of the DAC-MC8B as follows: pin 1, ground; pin 2, logic select; pin 3, reset; pin 4, strobe; pin 5, bit 8 (LSB); pin 6, bit 7; pin 7, bit 6; pin 8, +Vcc; pin 9, bit 5; pin 10, bit 4; pin 11, bit 3; pin 12, bit 2; pin 13, bit 1(MSB); pin 14, analog output; pin 15, V_{REF} input; pin 16, V_{REF} output (courtesy Datel Systems, Inc.).

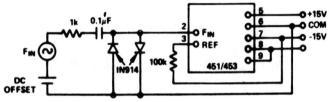

F/V converter with input diode protection against high-voltage transients. Model 451 has a 0 to 10 kHz full scale input range and model 453 has a 0 to 100 kHz full scale input range. Pins 8 and 9 are the output. Maximum input is ±10 volts (courtesy Analog Devices, Inc.).

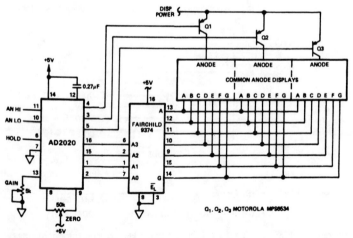

3-digit I^2L A/D converter with LED (courtesy Analog Devices, Inc.).

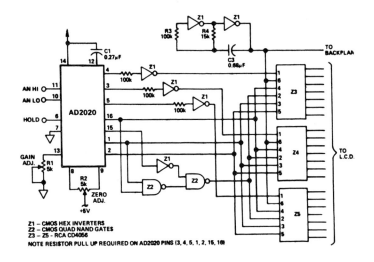

3-digit I²L A/D converter with LCD (courtesy Analog Devices, Inc.).

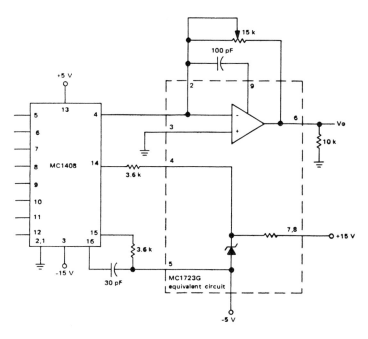

8-bit D/A converter with voltage regulator (courtesy Motorola Semiconductor Products Inc.).

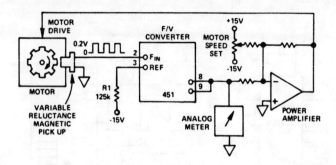

F/V converter controlling and monitoring motor speed in a closed loop system. The F/V converter is a 451/453 10/100 kHz F/V converter with a +10-volt full scale output (courtesy Analog Devices, Inc.).

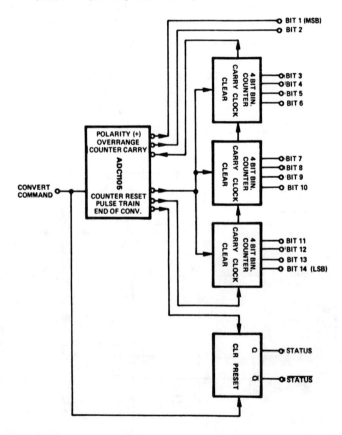

14-bit sign-magnitude binary converter using an ADC1105 (courtesy Analog Devices, Inc.).

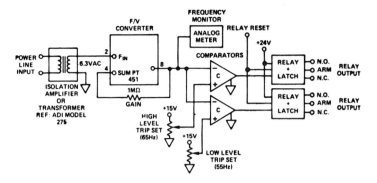

F/V converter monitoring the 60-hertz line frequency. The gain of the F/V converter has been set to 0.1 volt per hertz, which results in a 100-hertz full scale frequency range (courtesy Analog Devices, Inc.).

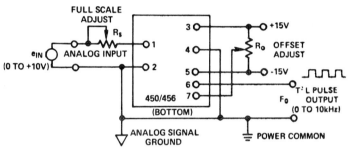

EXTERNAL CONNECTIONS FOR VOLTAGE TO FREQUENCY OPERATION, WITH OPTIONAL INPUT OFFSET VOLTAGE AND FULL SCALE OUTPUT FREQUENCY ADJUSTMENTS

R_s = FULL SCALE ADJUSTMENT; 500Ω (Use low T.C. Cermet, <50ppm/°C or equivalent)
R_o = INPUT OFFSET ADJUSTMENT; 50kΩ (Use low T.C. Cermet, <50ppm/°C or equivalent)
R_c = 20MΩ

CAUTION: DO NOT SHORT OUTPUT TO -15V

V/F converter with 10 kHz full scale output. The 450 is a high-performance version of the economy 456 (courtesy Analog Devices, Inc.).

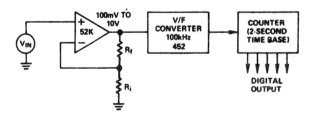

V/F converter used as a nearly 18-bit A/D converter. Resolution is 1 pulse in 200,000, or 0.05% of the smallest input signal (courtesy Analog Devices, Inc.).

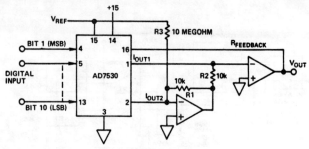

DIGITAL INPUT	ANALOG OUTPUT
1 1 1 1 1 1 1 1 1 1	$-V_{REF}(1 - 2^{-9})$
1 0 0 0 0 0 0 0 0 1	$-V_{REF}(2^{-9})$
1 0 0 0 0 0 0 0 0 0	0
0 1 1 1 1 1 1 1 1 1	$V_{REF}(2^{-9})$
0 0 0 0 0 0 0 0 0 1	$V_{REF}(1 - 2^{-9})$
0 0 0 0 0 0 0 0 0 0	V_{REF}

NOTE: 1 LSB = $2^{-9} V_{REF}$

10-bit D/A converter in bipolar operation (courtesy Analog Devices, Inc.).

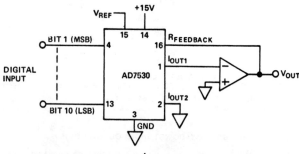

DIGITAL INPUT	ANALOG OUTPUT
1 1 1 1 1 1 1 1 1 1	$-V_{REF}(1 - 2^{-10})$
1 0 0 0 0 0 0 0 0 1	$-V_{REF}(1/2 + 2^{-10})$
1 0 0 0 0 0 0 0 0 0	$\dfrac{-V_{REF}}{2}$
0 1 1 1 1 1 1 1 1 1	$-V_{REF}(1/2 - 2^{-10})$
0 0 0 0 0 0 0 0 0 1	$-V_{REF}(2^{-10})$
0 0 0 0 0 0 0 0 0 0	0

NOTE: 1 LSB = $2^{-10} V_{REF}$

10-bit D/A converter in unipolar binary operation (courtesy Analog Devices, Inc.).

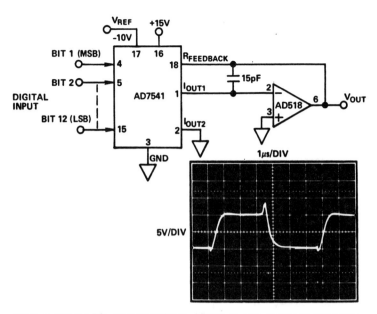

12-bit multiplying D/A converter with an AD518 op amp (courtesy Analog Devices, Inc.).

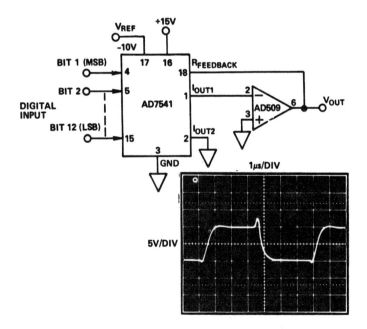

12-bit D/A multiplying converter with an AD509 op amp (courtesy Analog Devices, Inc.).

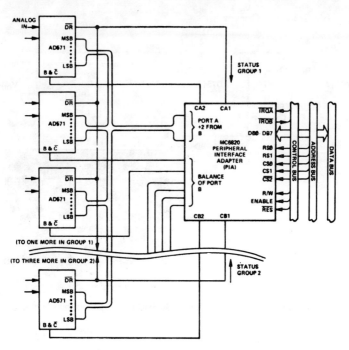

Eight 10-bit A/D converters multiplexed with the 6800 microprocessor through a single PIA (courtesy Analog Devices, Inc.).

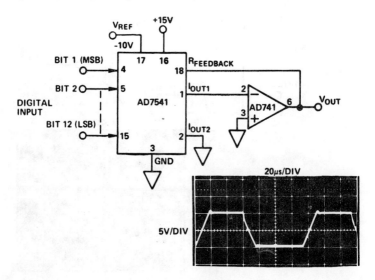

12-bit multiplying D/A converter with an AD741 op amp (courtesy Analog Devices, Inc.).

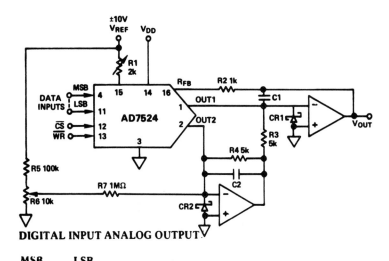

DIGITAL INPUT ANALOG OUTPUT

MSB LSB	
11111111	$-V_{REF}\left(\dfrac{127}{128}\right)$
10000001	$-V_{REF}\left(\dfrac{1}{128}\right)$
10000000	0
01111111	$+V_{REF}\left(\dfrac{1}{128}\right)$
00000001	$+V_{REF}\left(\dfrac{127}{128}\right)$
00000000	$+V_{REF}\left(\dfrac{128}{128}\right)$

NOTES:
1. R3/R4 MATCH 0.1% OR BETTER.
2. R1, R2 USED ONLY IF GAIN ADJUSTMENT IS REQUIRED.
3. R5–R7 USED TO ADJUST V_{OUT} = 0V AT INPUT CODE 10000000.
4. CR1 & CR2 PROTECT AD7524 AGAINST NEGATIVE TRANSIENTS. SEE "CAUTION" NOTE 3.
5. C1 AND C2 PHASE COMPENSATION (10–15pF) IS REQUIRED WHEN USING FAST AMPLIFIERS TO PREVENT RINGING OR OSCILLATION.

Note: $1LSB = (2^{-7})(V_{REF}) = \left(\dfrac{1}{128}\right)(V_{REF})$

8-bit buffered multiplying CMOS D/A converter in bipolar operation (courtesy Analog Devices, Inc.).

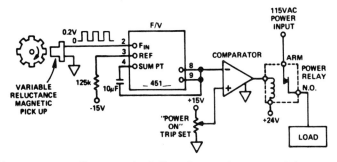

F/V converter controlling power load. The relay remains open until the preset power on trip level is reached (courtesy Analog Devices, Inc.).

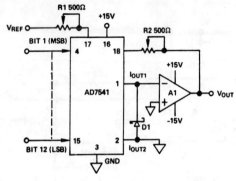

DIGITAL INPUT	NOMINAL ANALOG OUTPUT
111111111111	$-0.99975\ V_{REF}$
100000000000	$-0.50000\ V_{REF}$
010000000000	$-0.49975\ V_{REF}$
000000000000	0

12-bit multiplying D/A converter in unipolar operation. In this configuration the converter is set up for two-quadrant multiplication (courtesy Analog Devices, Inc.).

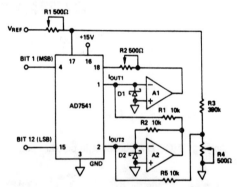

DIGITAL INPUT	NOMINAL ANALOG OUTPUT
111111111111	$-0.99951\ V_{REF}$
100000000001	$-0.00049\ V_{REF}$
100000000000	0
010000000000	$+0.50000\ V_{REF}$
000000000000	$+1.00000\ V_{REF}$

12-bit multiplying D/A converter in bipolar operation. In this configuration the converter is set up for four-quadrant multiplication (courtesy Analog Devices Inc.).

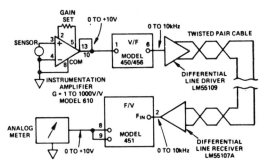

F/V and V/F converters used in a two-wire data transmission system Model 610 instrumentation amplifier amplifies the low-level differential transducer signal to the 10-volt full scale of models 450 and 456 10 kHz V/F converters. A differential line driver is used to drive the twisted pair through a noisy environment. A differential line receiver is used to drive model 451 10 kHz F/V converter (courtesy Analog Devices, Inc.).

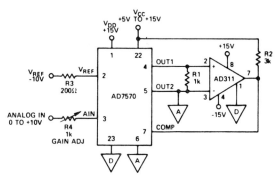

NOTE:
IF POSITIVE V_{REF} IS USED, THE ANALOG INPUT RANGE IS 0 TO $-V_{REF}$, AND THE COMPARATOR'S (−) INPUT SHOULD BE CONNECTED TO OUT1 (PIN 4) OF THE AD7570.

Analog Input (AIN) Notes 1, 2, 3	Digital Output Code MSB LSB
FS − 1LSB	1 1 1 1 1 1 1 1 1 1
FS − 2LSB	1 1 1 1 1 1 1 1 1 0
3/4 FS	1 1 0 0 0 0 0 0 0 0
1/2 FS + 1LSB	1 0 0 0 0 0 0 0 0 1
1/2 FS	1 0 0 0 0 0 0 0 0 0
1/2 FS − 1LSB	0 1 1 1 1 1 1 1 1 1
1/4 FS	0 1 0 0 0 0 0 0 0 0
1LSB	0 0 0 0 0 0 0 0 0 1
0	0 0 0 0 0 0 0 0 0 0

NOTES:
1. Analog inputs shown are nominal center values of code.
2. "FS" is full scale, i.e., $(-V_{REF})$.
3. For 8-bit operation, 1LSB equals $(-V_{REF})(2^{-8})$; for 10-bit operation, 1LSB equals $(-V_{REF})(2^{-10})$.

10-bit CMOS A/D converter in unipolar binary operation (courtesy Analog Devices, Inc.).

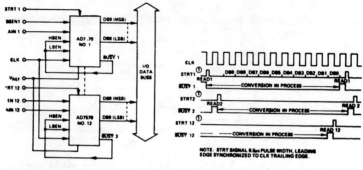

Twelve 10-bit CMOS A/D converters onto a data bus (courtesy Analog Devices, Inc.).

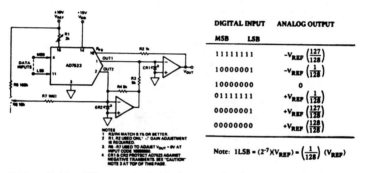

DIGITAL INPUT	ANALOG OUTPUT
MSB LSB	
11111111	$-V_{REF}\left(\frac{127}{128}\right)$
10000001	$-V_{REF}\left(\frac{1}{128}\right)$
10000000	0
01111111	$+V_{REF}\left(\frac{1}{128}\right)$
00000001	$+V_{REF}\left(\frac{127}{128}\right)$
00000000	$+V_{REF}\left(\frac{128}{128}\right)$

Note: $1 LSB = (2^{-7})(V_{REF}) = \left(\frac{1}{128}\right)(V_{REF})$

8-bit multiplying D/A converter in bipolar operation (courtesy Analog Devices, Inc.).

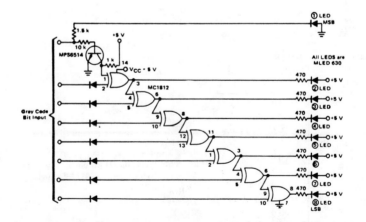

Parallel gray-to-binary converter (courtesy Motorola Semiconductor Products Inc.).

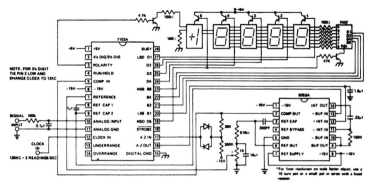

4½-digit (2.000-volt) A/D converter with LED readout using the 8052A/7103A digital pair (courtesy Intersil, Inc.).

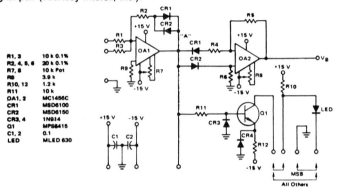

Cyclic converter (courtesy Motorola Semiconductor Products Inc.).

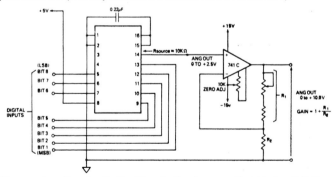

D/A converter with output buffer. V_{OUT} is directly proportional to the digital input. R2 should be less than or equal to 650K to assure good temperature compensation. Metal film 1% resistors and 100 PPM/°C trimmers are recommended. For fast settling time use a Datel AM-452 buffer. To calibrate apply continuous start commands to the start input (pin 3). Set all bits to logic zero and vary the zero adjust pot until V_{OUT} is zero. Set all bits to logic one and vary the gain adjust pot until V_{OUT} is equal to the nominal FS-1 LSB, where LSB is equal to FSR/256 (courtesy Datel Systems, Inc.).

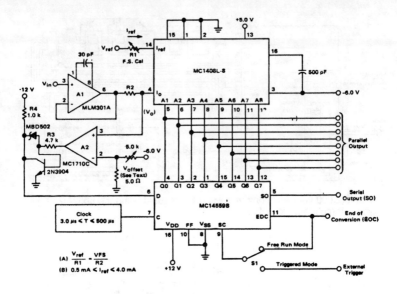

High-speed 8-bit successive-approximation A/D converter (courtesy Motorola Semiconductor Products Inc.).

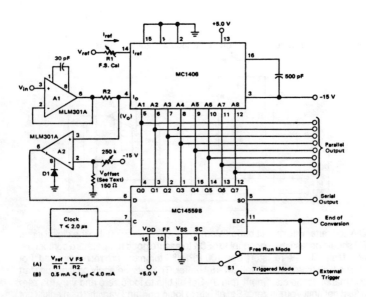

8-bit successive-approximation A/D converter (courtesy Motorola Semiconductor Products Inc.).

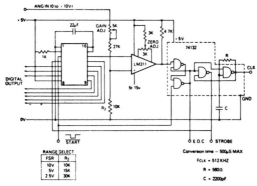

8-bit A/D converter using an ADC-MC8B 16-pin DIP. Pin functions of the Datel ADC-MC8B chip are as follows: pin 1, ground; pin 2, logic select; pin 3, reset; pin 4, strobe; pin 5, bit 8 (LSB); pin 6, bit 7; pin 7, bit 6; pin 8, +Vcc; pin 9, bit 5; pin 10, bit 4; pin 11, bit 3; pin 12, bit 2; pin 13, bit 1(MSB); pin 14, analog out; pin 15, V_{REF} input; pin 16, V_{REF} output. To calibrate, apply continous start commands to the start input. For the zero adjustment ground the analog input and vary the zero adjust pot until the LSB flickers between one and zero with all other inputs at logic zero. For the gain adjustment apply FS-½ LSB to the analog input and vary the gain adjust pot until LSB flickers between one and zero with all other bits at logic one (courtesy Datel Systems, Inc.).

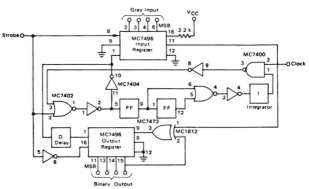

Serial gray-to-binary converter above (courtesy Motorola Semiconductor Products Inc.).

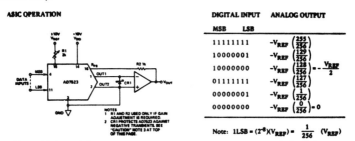

8-bit multiplying D/A converter in unipolar binary operation below (courtesy Analog Devices, Inc.).

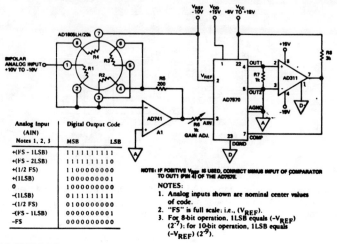

10-bit CMOS A/D converter in bipolar (offset binary) operation. This is a modified 2s complement operation (courtesy Analog Devices, Inc.).

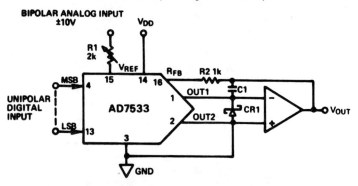

NOTES:
1. R1 AND R2 USED ONLY IF GAIN ADJUSTMENT IS REQUIRED.
2. SCHOTTKY DIODE CR1 (HP5082-2811 OR EQUIV) PROTECTS OUT1 TERMINAL AGAINST NEGATIVE TRANSIENTS. SEE "CAUTION" NOTE 3.
3. C1 PHASE COMPENSATION (5 – 15pF) MAY BE REQUIRED WHEN USING HIGH SPEED AMPLIFIER.

DIGITAL INPUT		NOMINAL ANALOG OUTPUT (V_{OUT} as shown in Figure 1)
MSB	LSB	
1111111111		$-V_{REF}\left(\frac{1023}{1024}\right)$
1000000001		$-V_{REF}\left(\frac{513}{1024}\right)$
1000000000		$-V_{REF}\left(\frac{512}{1024}\right) = -\frac{V_{REF}}{2}$
0111111111		$-V_{REF}\left(\frac{511}{1024}\right)$
0000000001		$-V_{REF}\left(\frac{1}{1024}\right)$
0000000000		$-V_{REF}\left(\frac{0}{1024}\right) = 0$

NOTES:
1. Nominal Full Scale for the circuit of Figure 5 is given by FS = $-V_{REF}\left(\frac{1023}{1024}\right)$
2. Nominal LSB magnitude for the circuit of Figure 5 is given by LSB = $V_{REF}\left(\frac{1}{1024}\right)$

10-bit multiplying D/A converter in unipolar operation (courtesy Analog Devices, Inc.).

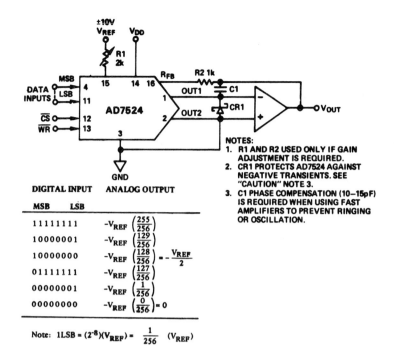

8-bit buffered multiplying CMOS D/A converter in unipolar binary operation (courtesy Analog Devices, Inc.).

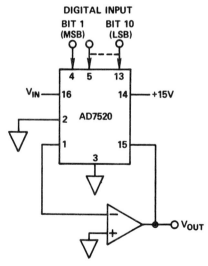

A/D divider using an AD7520 10-bit multiplying D/A converter (courtesy Analog Devices, Inc.).

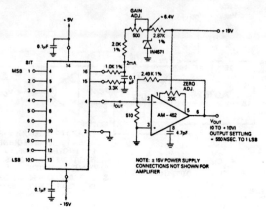

D/A converter with fast unipolar output using the Datel DAC-IC10BC and AM-452 (courtesy Datel Systems, Inc.).

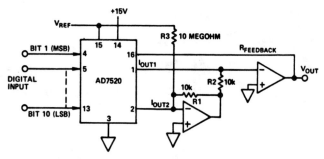

D/A converter in bipolar operation (courtesy Analog Devices, Inc.).

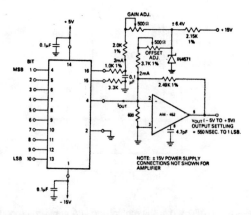

D/A converter with fast bipolar output using the Datel DAC-IC10BC and AM-452. See coding table (courtesy Datel Systems, Inc.).

Power Supplies, Regulators, Etc.

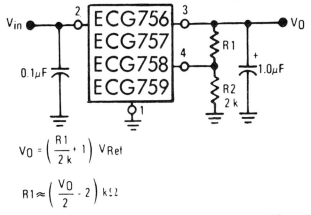

5-volt regulator. Input voltage should not exceed 22 volts for the ECG758 and ECG759 or 38 volts for the ECG756 and ECG757. All of these units are 4-lead packages (courtesy GTE Sylvania Incorporated).

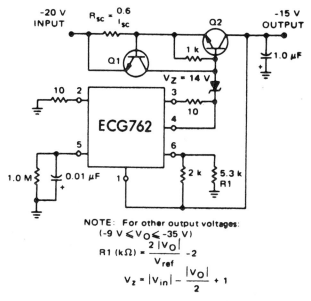

15-volt regulator with current limit (courtesy GTE Sylvania Incorporated).

5-volt 5-ampere regulator with remote sensing PNP current boost (courtesy GTE Sylvania Incorporated).

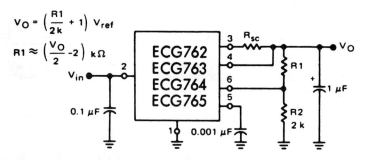

$$V_O = \left(\frac{R1}{2k} + 1\right) V_{ref}$$

$$R1 \approx \left(\frac{V_O}{2} - 2\right) k\Omega$$

15-volt regulator. Input voltage for ECG762 and ECG763 should not exceed 38 volts and for the ECG764 and ECG765 should not exceed 22 volts (courtesy GTE Sylvania Incorporated).

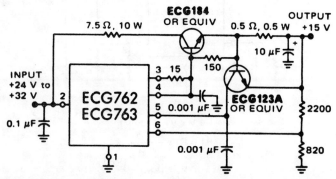

15-volt 1-ampere regulator with short-circuit protection (courtesy GTE Sylvania Incorporated).

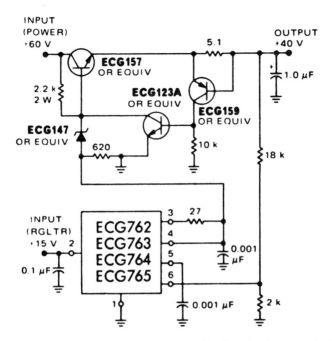

Voltage-boosted 40-volt 100 mA regulator with short-circuit current limiting (courtesy GTE Sylvania Incorporated).

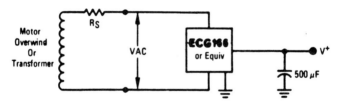

12-volt power supply using an overwinding from a phonograph motor. With 16 volts AC the output is approximately 12 volts DC. Rs is the series resistance of the winding (courtesy GTE Sylvania Incorporated).

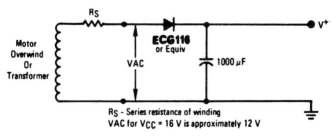

12-volt power supply using an overwinding of a phonograph motor or a transformer. With 16 volts AC, the output will be approximately 12 volts DC. Rs is the series resistance of the winding (courtesy GTE Sylvania Incorporated).

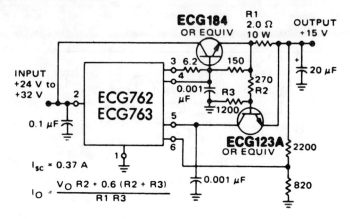

15-volt 2-ampere regulator with current foldback (courtesy GTE Sylvania Incorporated).

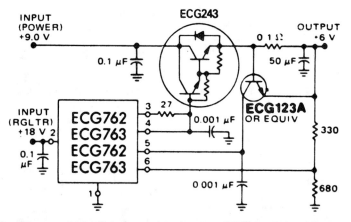

6-volt 5-ampere high-efficiency regulator (courtesy GTE Sylvania Incorporated).

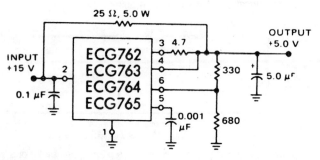

Current bypass with load current range from 400 to 500 mA (courtesy GTE Sylvania Incorporated).

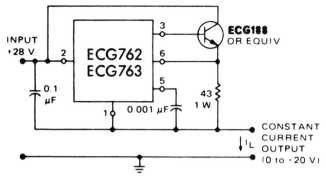

100 mA constant-current source (courtesy GTE Sylvania Incorporated).

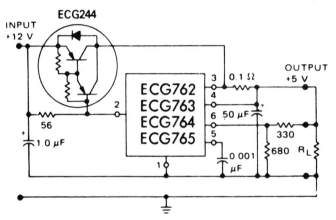

5-volt 5-ampere regulator with remote sensing and PNP current boost (courtesy GTE Sylvania Incorporated).

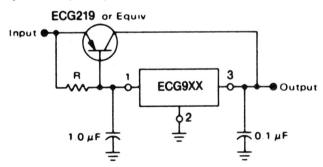

Current boost regulator for 5 amperes using the ECG9XX series of regulators. The ratings of the ECG9XX series are as follows: ECG960, 5 volts; ECG962, 6 volts; ECG966, 12 volts; ECG968, 15 volts; and ECG972, 24 volts. Resistor R in conjunction with the V_{BE} of the transistor determines when the transistor begins to conduct. This circuit is not short-circuit proof. Input-output differential voltage minimum is increased by the V_{BE} of the transistor (courtesy GTE Sylvania Incorporated).

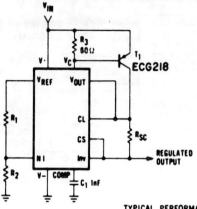

TYPICAL PERFORMANCE

Regulated Output Voltage	+5 V
Line Regulation ($\Delta V_{in} = 3$ V)	0.5 mV
Load Regulation ($\Delta I_L = 1$ A)	5 mV

Positive precision voltage regulator (5 volts) using an ECG915 or ECG915D IC. For a ±5% fixed output R1 is 2.15 ohms, R2 is 4.99 ohms and Rsc is 10 ohms (courtesy GTE Sylvania Incorporated).

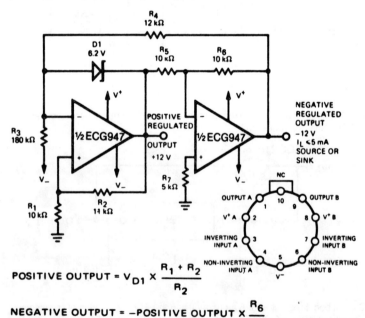

$$\text{POSITIVE OUTPUT} = V_{D1} \times \frac{R_1 + R_2}{R_2}$$

$$\text{NEGATIVE OUTPUT} = -\text{POSITIVE OUTPUT} \times \frac{R_6}{R_5}$$

Tracking positive and negative voltage reference using an ECG947 dual operational amplifier. The ECG947 is short-circuit protected and requires no external components for frequency compensation (courtesy GTE Sylvania Incorporated).

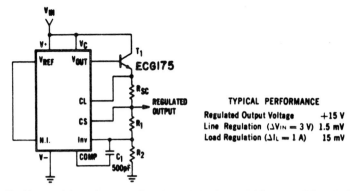

Positive precision voltage regulator (15 volts) using an ECG915 or ECG915D IC. R1 is 7.87 ohms, R2 is 7.15 ohms and Rsc is 10 ohms for a ±5% fixed output (courtesy GTE Sylvania Incorporated).

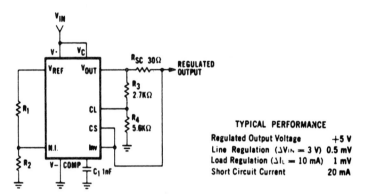

Positive voltage regulator (5 volts) with foldback current limiting using an ECG915 or ECG915D. For a ±5% fixed output R1 is 2.15 ohms and R2 is 4.99 ohms (courtesy GTE Sylvania Incorporated).

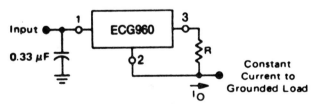

Current regulator using the ECG9XX series of fixed voltage regulators. The series and ratings are as follows: ECG960, 5.0 volts; ECG962, 6.0 volts; ECG966, 12 volts; ECG968, 15 volts; and ECG972, 24 volts. It is recommended that the regulator input be bypassed if the regulator is at a distance from the power supply filters. A 0.33 μF or larger tantalum or Mylar should be used. If an aluminum type is used it should be 1.0 μF or larger. Using the 5-volt ECG960 as an example, resistor R determines the current as follows: Io = (5 volts/R) + Iq, where Iq is 1.5 mA over line and load changes (courtesy GTE Sylvania Incorporated).

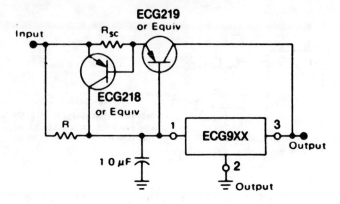

Short-circuit protected current boost regulator using the ECG9XX series of regulators. The ratings of the ECG9XX series are as follows: ECG960, 5 volts; ECG962, 6 volts; ECG966, 12 volts; ECG968, 15 volts; and ECG972, 24 volts. The current-sensing ECG218 transistor must be able to handle the current of the three-terminal regulator; therefore, a 4-ampere transistor is specified. Resistor R in conjunction with the V_{BE} of the pass transistor determines when the transistor begins to conduct. Input-output differential voltage minimum is increased by the V_{BE} of the pass transistor (courtesy GTE Sylvania Incorporated).

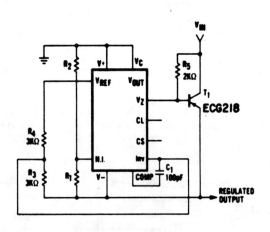

TYPICAL PERFORMANCE

Regulated Output Voltage	−15 V
Line Regulation ($\Delta V_{IN} = 3$ V)	1 mV
Load Regulation ($\Delta I_L = 100$ mA)	2 mV

Negative voltage regulator (15 volts) using an ECG923 or ECG923D precision regulator IC. For metal can applications, where V_Z is required, an external 6.2-volt zener should be connected in series with the regulated output. R1 is 3.65 ohms and R2 is 11.5 ohms for a ±5% fixed output (courtesy GTE Sylvania Incorporated).

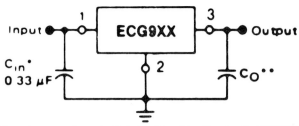

Positive voltage regulators for 5, 6, 12, 15, or 24 volts using the ECG9XX series. Rating for the three-terminal ECG9XX positive voltage regulators are as follows: ECG960, 5 volts; ECG962, 6 volts; ECG966, 12 volts; ECG968, 15 volts; and ECG972, 24 volts. A common ground is required between the input and output voltages. The input voltage must remain typically 2 volts above the output voltage even during the low point on the ripple (courtesy GTE Sylvania Incorporated).

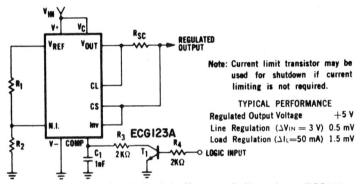

Remote shutdown regulator (+5 volts) with current limiting using an ECG915 or ECG915D. For a ±5% fixed output resistor R1 is 2.15 ohms and R2 is 4.99 ohms. Resistor Rsc is 10 ohms (courtesy GTE Sylvania Incorporated).

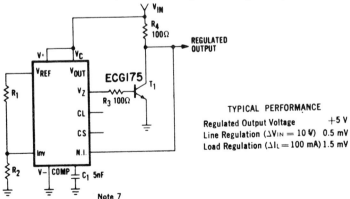

Shunt regulator (+5 volts) using an ECG915 or ECG915D. For a ±5% fixed output resistor R1 is 2.15 ohms and R2 is 4.99 ohms. For metal can applications where Vz is required, a 6.2-volt zener should be connected in series with the regulated output (courtesy GTE Sylvania Incorporated).

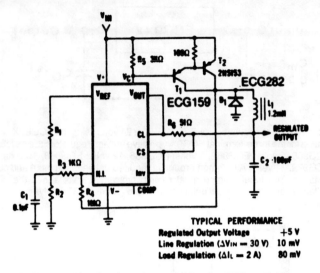

Positive switching regulator (5 volts) using an ECG915 or ECG915D. For a ±5% fixed output R1 is 1.15 ohms and R2 is 4.99 ohms. For metal can applications where Vz is required connect a 6.2-volt zener in series with the regulated output (courtesy GTE Sylvania Incorporated).

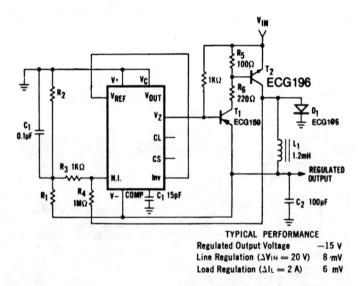

Negative switching regulator (15 volts) using an ECG915 or ECG915D IC. For a ±5% fixed output R1 is 3.65 ohms and R2 is 11.5 ohms. In metal can applications where Vz is required, connect a 6.2-volt zener in series with the regulated output. L1 is forty turns of AWG #20 enameled copper wire wound on Ferroxcube P36/22-387 pot core or equivalent with 0.009-inch air gap (courtesy GTE Sylvania Incorporated).

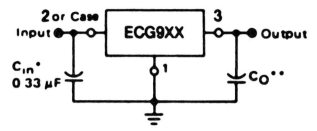

Negative voltage regulators for 5, 6, 12, or 15 volts using the ECG9XX series. Ratings for the three-terminal ECG9XX negative voltage regulators are as follows: ECG961, 5 volts; ECG963, 6 volts; ECG967, 12 volts; ECG969, 15 volts. A common is required between the input and output voltages. The input must remain 2 volts more negative than the output even during the high point on the input ripple voltage. These devices will handle up to 1 ampere without any heat sink. The input capacitor should be 0.33 μF if a tantalum or Mylar type is used. If an aluminum capacitor is used it should be 1.0 μF or larger (courtesy GTE Sylvania Incorporated).

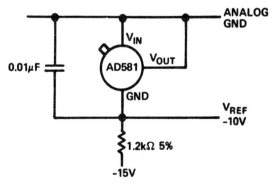

Two-component precision current limiter (courtesy Analog Devices, Inc.).

Two-terminal −10 volt reference (courtesy Analog Devices, Inc.).

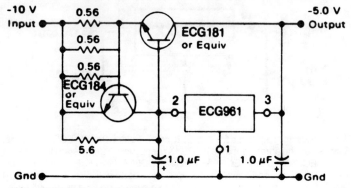

*Mounted on common heat sink

Current boost regulator for −5 volts at 4 amperes with 5-ampere current limiting. When a boost transistor is used short-circuit current is equal to the sum of the series pass and regulator limits, which are measured at 3.2 amperes and 1.8 amperes respectively. Series pass limiting is approximately equal to 0.6 volts/Rsc (courtesy GTE Sylvania Incorporated).

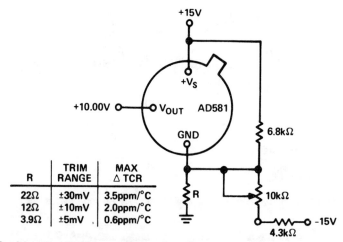

R	TRIM RANGE	MAX Δ TCR
22Ω	±30mV	3.5ppm/°C
12Ω	±10mV	2.0ppm/°C
3.9Ω	±5mV	0.6ppm/°C

Precision 10-volt reference with fine trim adjustment (courtesy Analog Devices, Inc.).

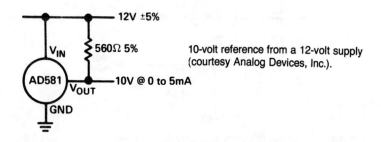

10-volt reference from a 12-volt supply (courtesy Analog Devices, Inc.).

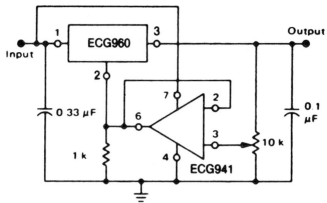

Adjustable output regulators for 5, 6, 12, 15, or 24 volts. The input capacitor is valued at 0.33 µF if Mylar or tantalum; if aluminum, it should be 1.0 µF or larger. Although the circuit shown is for a 7-volt output, voltages of 8, 14, 17, or 26 volts can be obtained by substituting ECG962, ECG966, ECG968 or ECG972, respectively, for the ECG960. These devices are three-terminal 1-ampere devices. The minimum voltage obtainable is 2 volts greater than the regulator voltage (courtesy GTE Sylvania Incorporated).

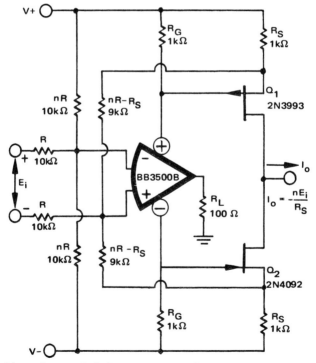

Precision voltage-controlled current source using an op amp and two FETs (courtesy Burr-Brown Research Corporation).

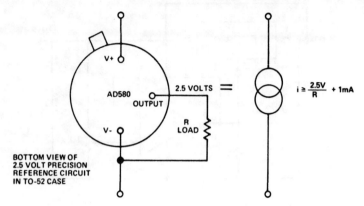

Two-component precision current limiter (courtesy Analog Devices, Inc.).

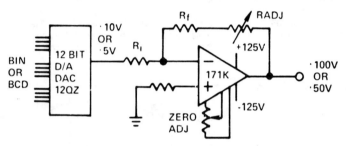

Programmable power supply with ±100-volt and ±50-volt output (courtesy Analog Devices, Inc.).

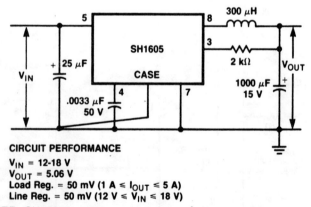

CIRCUIT PERFORMANCE
V_{IN} = 12-18 V
V_{OUT} = 5.06 V
Load Reg. = 50 mV (1 A ≤ I_{OUT} ≤ 5 A)
Line Reg. = 50 mV (12 V ≤ V_{IN} ≤ 18 V)

NOTE: SH1605 must be mounted on a heat sink with a maximum thermal resistance of ϕ_{CA} ≤ 4° C/W.

Step-down switching regulator using an SH1605. Output voltage is +5 volts. Available current is between 1 and 5 amperes. Input voltage is 12 to 18 volts. Line regulation is 2% with load regulation at 2%. Maximum ripple is 0.1 volt peak to peak (courtesy Fairchild Semiconductor).

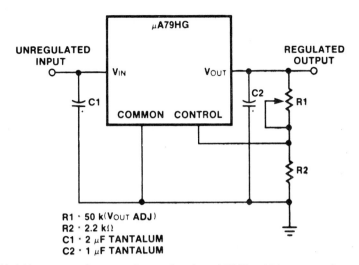

R1 · 50 k(V_OUT ADJ)
R2 · 2.2 kΩ
C1 · 2 μF TANTALUM
C2 · 1 μF TANTALUM

Variable output voltage regulator using the μA79HG, which can supply a minimum of 5 amperes at voltages from −2.3 volts to −24 volts (courtesy Fairchild Semiconductor).

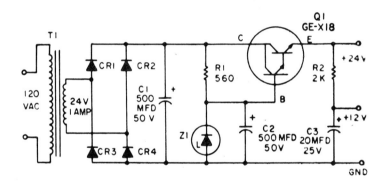

Parts List

C1, C2 — 500-mfd, 50-volt electrolytic capacitor
C3 — 20-mfd, 25-volt electrolytic capacitor
CR1 through CR4 — GE-504A rectifier diode
Q1 — GE-X18 power transistor
R1 — 560-ohm, 1/2-watt resistor
R2 — 2000-ohm, 1/2-watt resistor
T1 — Power transformer, 120-volt primary; 24-volt, 1 amp secondary; Stancor No. P6469, or equivalent
Z1 — Z4X12 and Z4X14 zener diodes in series, or single 27-volt zener (GE ZD-27)

24-volt DC regulated power supply with 500 mA rating. Output ripple is less than 4.5 mV. R2 can be changed to obtain other secondary output voltages as desired (courtesy General Electric Company).

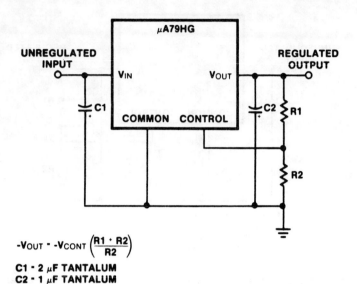

$$-V_{OUT} = -V_{CONT}\left(\frac{R1 + R2}{R2}\right)$$

C1 - 2 μF TANTALUM
C2 - 1 μF TANTALUM

Fixed negative voltage regulator using the μA79HG, which can supply a minimum of 5 amperes at voltages from −2.3 volts to −24 volts (courtesy Fairchild Semiconductor).

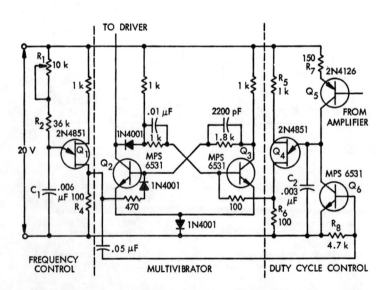

Duty cycle and frequency control circuitry for a switching mode regulator. Q1 sets the basic operating frequency of the regulator at 5 kHz, while Q4 controls the duty cycle (courtesy Motorola Semiconductor Products Inc.).

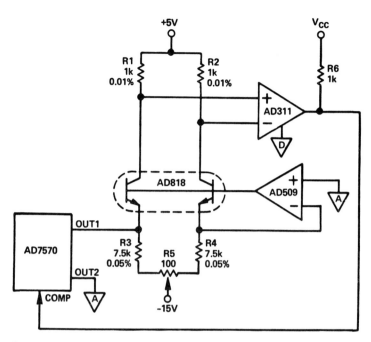

Current comparator with low input impedance using an AD7570 10-bit A/D converter (courtesy Analog Devices, Inc.).

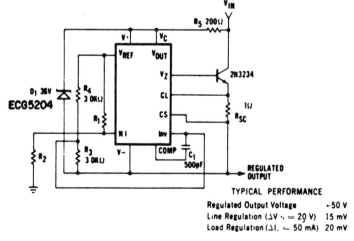

Positive floating regulator (50 volts) using an ECG915 or ECG915D IC. For a ±5% fixed output R1 is 3.57 ohms and R2 is 48.7 ohms. For metal can applications where Vz is required, an external 6.2-volt zener should be connected in series with the regulated output (courtesy GTE Sylvania Incorporated).

223

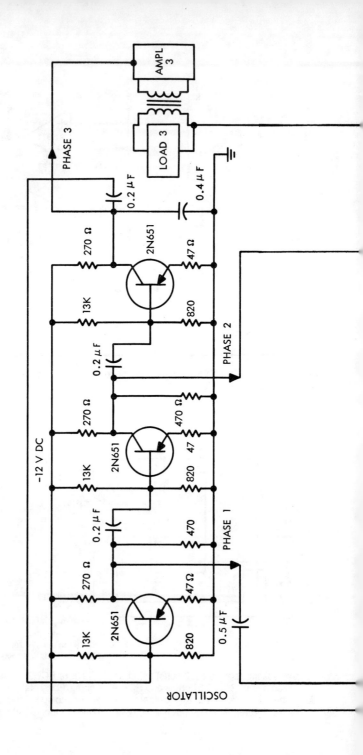

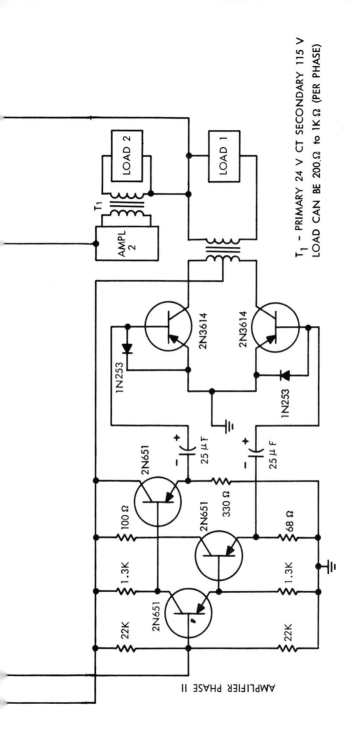

20-watt 3-phase inverter with 12-volt DC input and 115-volt 400-hertz AC output (courtesy Motorola Semiconductor Products Inc.).

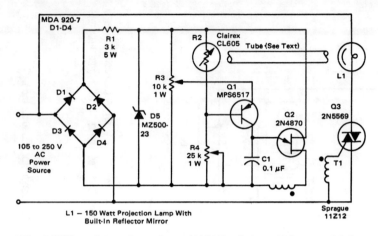

100-volt RMS regulator using a triac and UJT. To eliminate 60-hertz modulation of the photocell it is mounted at one end of a black tube with the other end directed at the back side of the lamp reflector. The reflector glows red and since it has a relatively large mass it cannot respond to the supply frequency. This provides a form of integration (courtesy Motorola Semiconductor Products Inc.).

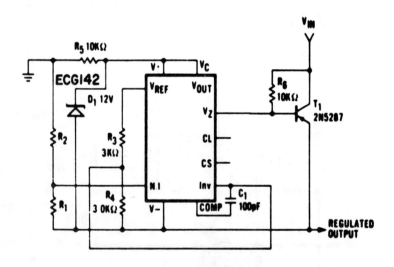

TYPICAL PERFORMANCE

Regulated Output Voltage   −100 V
Line Regulation (ΔV_{IN} = 20 V)   30 mV
Load Regulation (ΔI_L = 100 mA)   20 mV

Negative floating regulator (100 volts) using an ECG915 or ECG915D IC. For a ±5% fixed output, R1 should be 3.57 ohms and R2 should be 97.6 ohms (courtesy GTE Sylvania Incorporated).

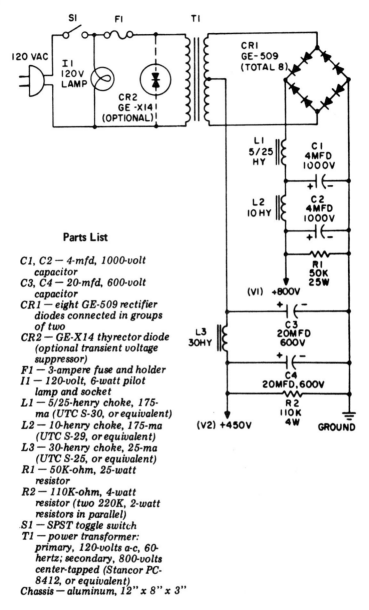

Parts List

C1, C2 — 4-mfd, 1000-volt capacitor
C3, C4 — 20-mfd, 600-volt capacitor
CR1 — eight GE-509 rectifier diodes connected in groups of two
CR2 — GE-X14 thyrector diode (optional transient voltage suppressor)
F1 — 3-ampere fuse and holder
I1 — 120-volt, 6-watt pilot lamp and socket
L1 — 5/25-henry choke, 175-ma (UTC S-30, or equivalent)
L2 — 10-henry choke, 175-ma (UTC S-29, or equivalent)
L3 — 30-henry choke, 25-ma (UTC S-25, or equivalent)
R1 — 50K-ohm, 25-watt resistor
R2 — 110K-ohm, 4-watt resistor (two 220K, 2-watt resistors in parallel)
S1 — SPST toggle switch
T1 — power transformer: primary, 120-volts a-c, 60-hertz; secondary, 800-volts center-tapped (Stancor PC-8412, or equivalent)
Chassis — aluminum, 12" x 8" x 3"

Dual-voltage transmitter power supply with 800-volt and 450-volt taps. This circuit will handle a 100-watt transmitter. The 800-volt tap is rated at 175 mA intermittent duty with 1% ripple and 16% load rejection. The 450-volt tap is rated at 25 mA with 0.02% ripple. The voltage at V2 can be lowered to 375 volts by removing C3. The current rating of V2 can be increased by selecting L3 with a higher current rating (courtesy General Electric Company).

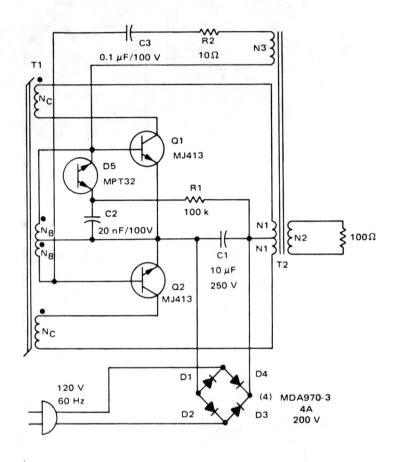

TRANSFORMERS

T1 — CORE — MAGNETICS INC. #80623-1/2D-080
   N_B 15 TURNS OF AWG #26 WIRE
   N_C 3 TURNS OF AWG #22 WIRE
T2 — CORE — ARNOLD GT-5800-D1
   N1 100 TURNS OF AWG #22 WIRE BYFILAR
   N2 104 TURNS OF AWG #19 WIRE BYFILAR
   N3 7 TURNS OF AWG #26 WIRE

Line operated 15 kHz inverter with 120-volt output. This circuit can be used in ultrasonic applications and for high-frequency fluorescent lights. The output waveform is a 120-volt amplitude-modulated 15 kHz square wave (courtesy Motorola Semiconductor Products Inc.).

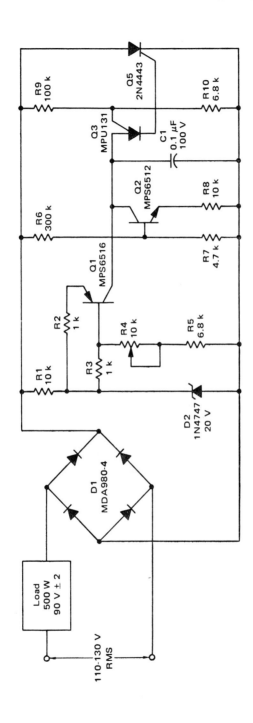

RMS voltage regulator using an SCR and a PUT. This circuit provides 90 ±2 volts at 500 watts for an input of 110 to 130 volts RMS (courtesy Motorola Semiconductor Products Inc.).

229

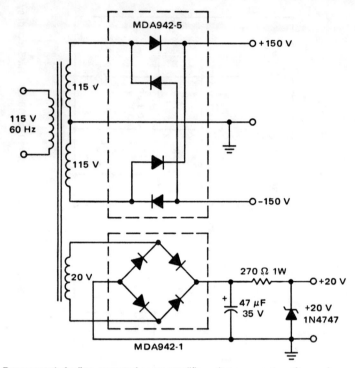

Power supply for line-operated servo amplifier using a power transformer (courtesy Motorola Semiconductor Products Inc.).

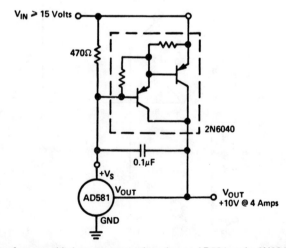

10-volt reference with 4-ampere capacity using an AD581 and a 2N6040 (courtesy Analog Devices, Inc.).

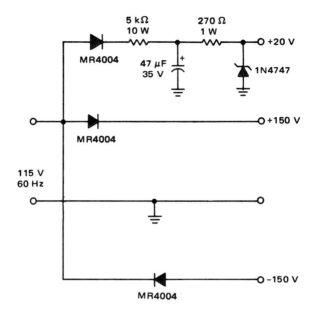

Direct-line-operated power supply for line-operated servo amplifier (courtesy Motorola Semiconductor Products Inc.).

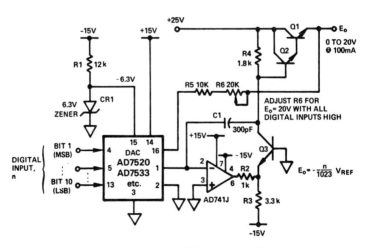

Digitally programmed power supply. A D/A converter with an op amp supplies voltage in accordance with a digital code (courtesy Analog Devices, Inc.).

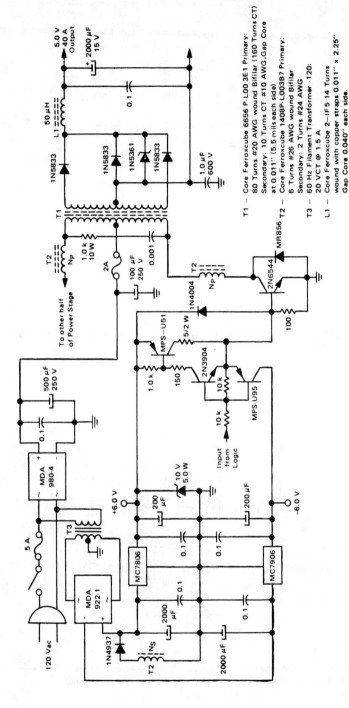

5-volt 40-ampere power inverter supply (courtesy Motorola Semiconductor Products Inc.).

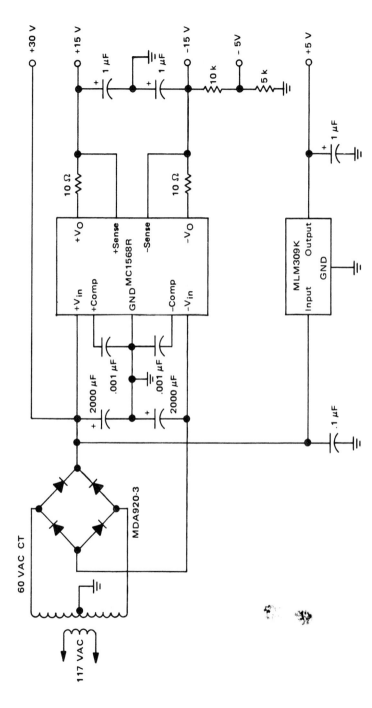

Multioutput low-voltage power supply for TTL and CMOS (courtesy Motorola Semiconductor Products Inc.).

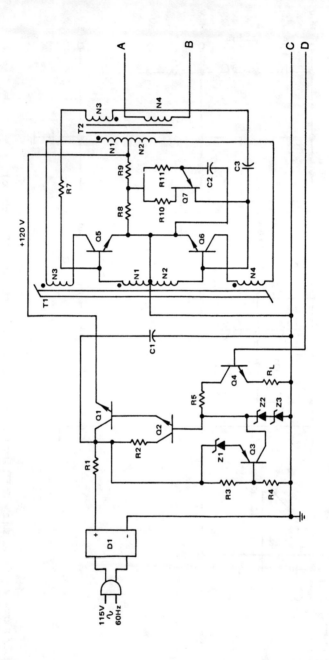

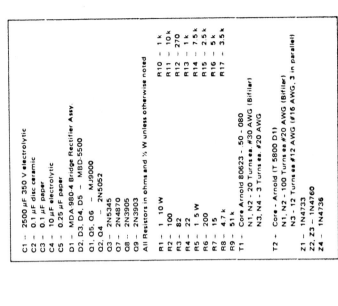
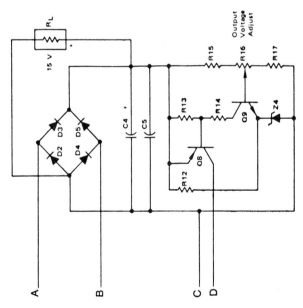

Parts List

C1 — 2500 μF 350 V electrolytic
C2 — 0.1 μF disc ceramic
C3 — 0.1 μF paper
C4 — 10 μF electrolytic
C5 — 0.25 μF paper
D1 — MDA-980-4 Bridge Rectifier Assy.
D2, D3, D4, D5 — MBD-5500
Q1, Q5, Q6 — MJ9000
Q2, Q4 — 2N5052
Q3 — 2N5345
Q7 — 2N4870
Q8 — 2N3905
Q9 — 2N3903

All Resistors in ohms and ½ W unless otherwise noted

R1 — 1 10 W
R2 — 100
R3 — 82
R4 — 22
R5 — 1 5 W
R6 — 200
R7 — 15
R8 — 4.7 k
R9 — 51 k
R10 — 1 k
R11 — 10 k
R12 — 270
R13 — 1 k
R14 — 7.5 k
R15 — 2.5 k
R16 — 5 k
R17 — 3.5 k

T1 — Core Arnold 80623 - 50 - 080
N1, N2 - 20 Turns ea. #30 AWG (Bifilar)
N3, N4 - 3 Turns ea. #20 AWG

T2 — Core - Arnold (T 5800 D1)
N1, N2 - 100 Turns ea #20 AWG (Bifilar)
N3 - 12 Turns ea #12 AWG (#16 AWG, 3 in parallel)

Z1 — 1N4733
Z2, Z3 — 1N4760
Z4 — 1N4736

Line-operated inverter with 15-volt DC 225-watt output (courtesy Motorola Semiconductor Products Inc.).

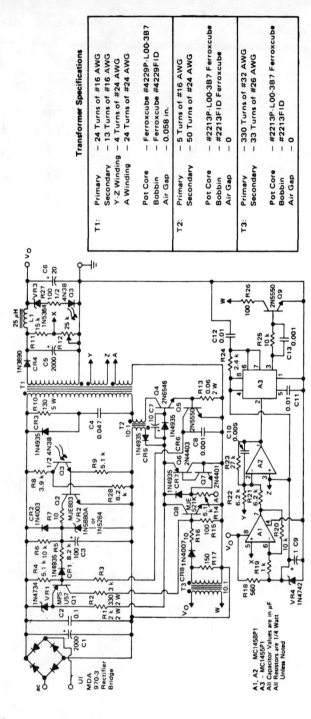

80-watt switching regulator supply for CATV applications with 24-volt 3-ampere output. This circuit operates above 18 kilohertz from a 40- to 60-volt 60-hertz square wave or from a DC standby source with input/output isolation (courtesy Motorola Semiconductor Products Inc.).

(-5.0 V @ 4.0 A, with 5.0 A current limiting)

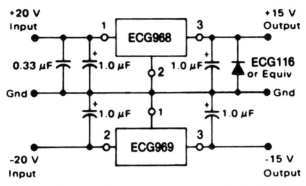

Operational amplifier ±15-volt 1-ampere power supply. The ECG968 and ECG969 positive and negative regulators can be connected as shown to obtain a dual power supply. A clamp diode should be used at the output of the ECG968 to prevent latch-up problems. The regulators have thermal-overload and short-circuit protection. The input capacitors should be 0.33 μF if tantalum or Mylar, or 1.0 μF if aluminum. Bypassing the output is also recommended. The same type of dual power supply can be constructed with other ratings of the ECG9XX series of regulator. Ratings are as follows: ECG960, +5 volts; ECG961, −5 volts; ECG962, +6 volts; ECG963, −6 volts; ECG966, +12 volts; ECG967, −12 volts; ECG968; +15 volts, ECG969, −15 volts; and ECG972, +24 volts (courtesy GTE Sylvania Incorporated).

Readouts

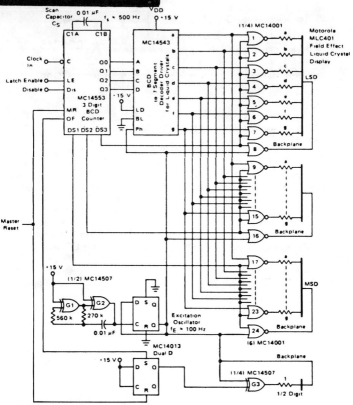

3½-digit multiplexed MLC401 field-effect LCD (courtesy Motorola Semiconductor Products Inc.).

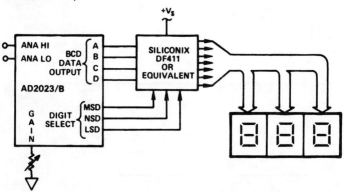

3-digit I²L DPM with LCD interface (courtesy Analog Devices, Inc.).

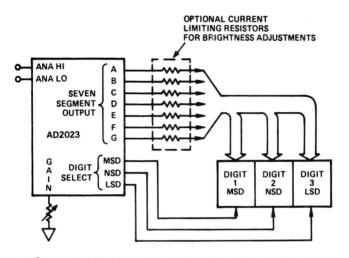

3-digit I²L DPM with LED interface (courtesy Analog Devices, Inc.).

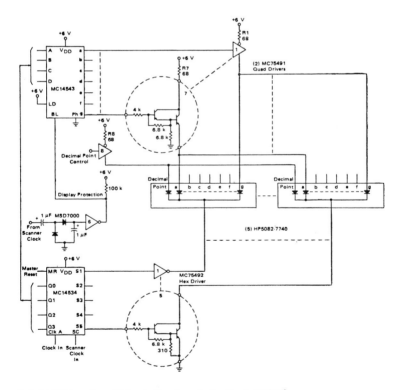

5-digit LED (courtesy Motorola Semiconductor Products Inc.).

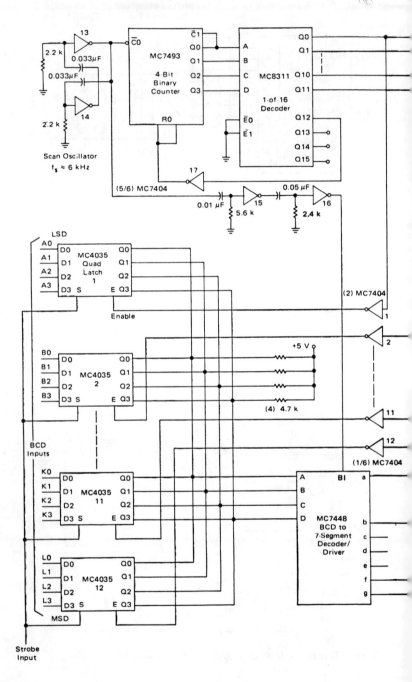

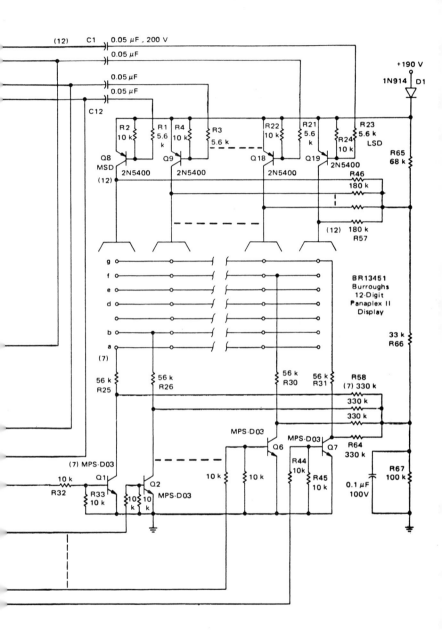

12-digit TTL multiplexed planar gas discharge display (courtesy Motorola Semiconductor Products Inc.).

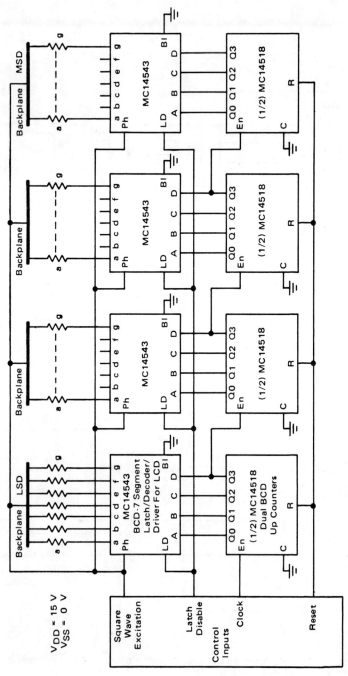

4-digit direct drive LCD (courtesy Motorola Semiconductor Products Inc.).

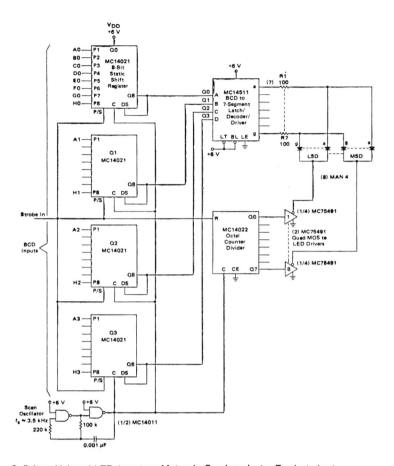

8-digit multiplexed LED (courtesy Motorola Semiconductor Products Inc.).

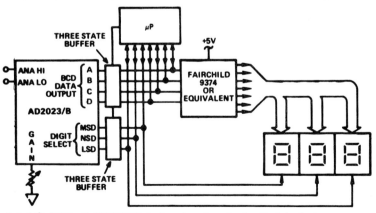

3-digit I²L DPM with LED/CPU interface (courtesy Analog Devices, Inc.).

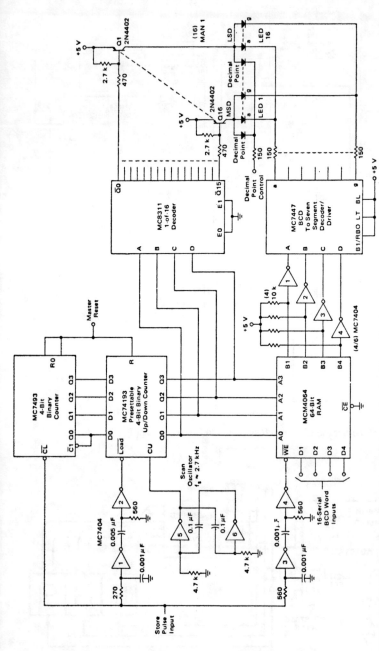

16-digit multiplex circuitry for LED (courtesy Motorola Semiconductor Products Inc.).

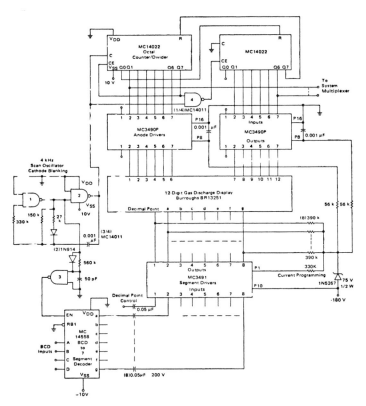

12-digit CMOS gas discharge display (courtesy Motorola Semiconductor Products Inc.).

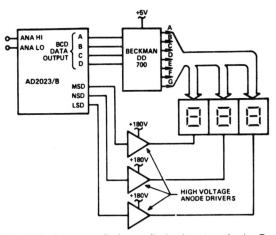

3-digit I²L DPM with Beckman gas discharge display (courtesy Analog Devices, Inc.).

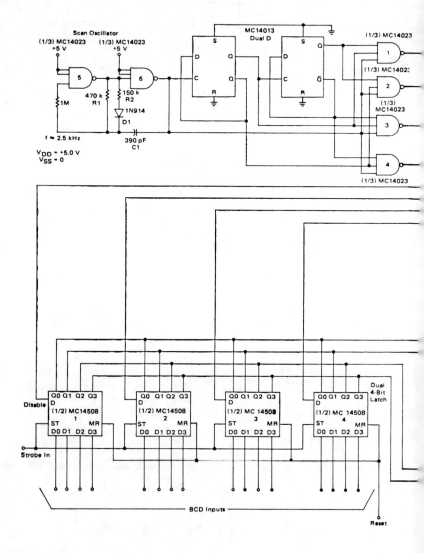

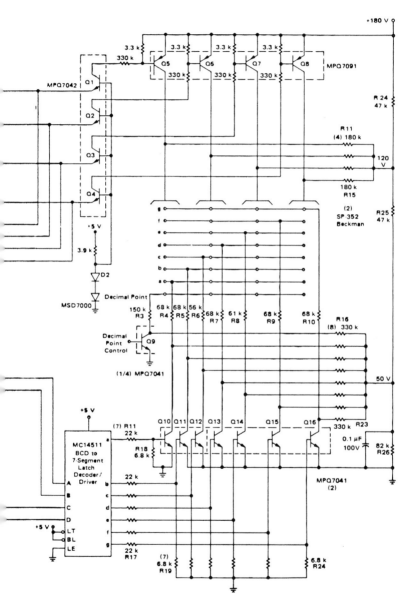

4-digit CMOS gas discharge display (courtesy Motorola Semiconductor Products Inc.).

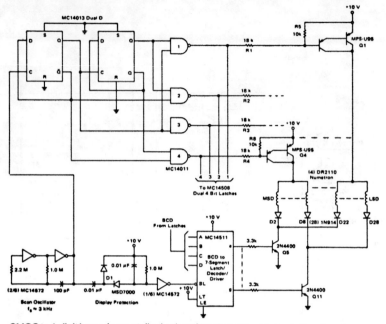

CMOS to 4-digit incandescent display interface (courtesy Motorola Semiconductor Products Inc.).

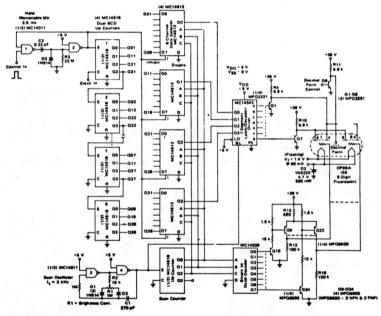

8-digit fluorescent triode display (courtesy Motorola Semiconductor Products Inc.).

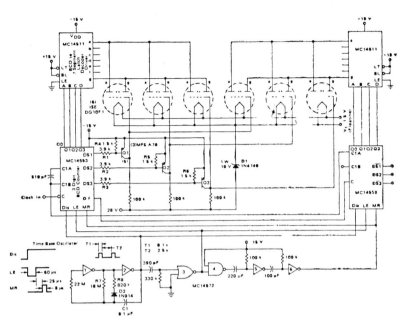

6-digit fluorescent triode display (courtesy Motorola Semiconductor Products Inc.).

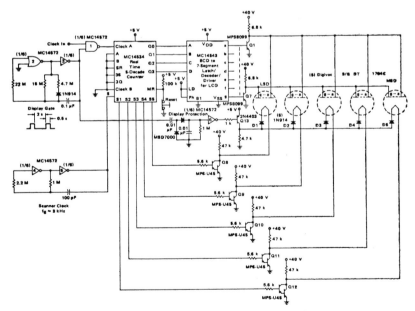

Real-time 5-digit fluorescent diode display (courtesy Motorola Semiconductor Products Inc.).

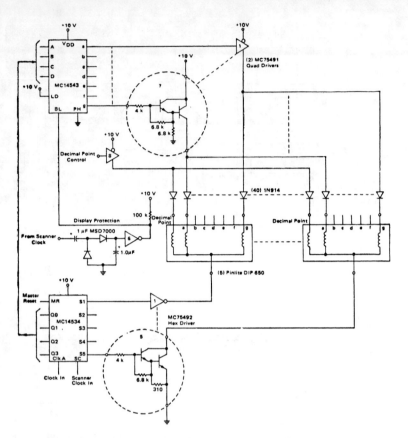

5-digit incandescent display (courtesy Motorola Semiconductor Products Inc.).

Interface Circuits

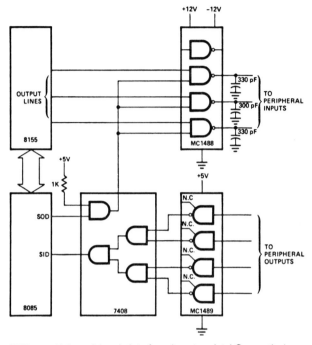

8085 CPU to multiple peripherals interface (courtesy Intel Corporation).

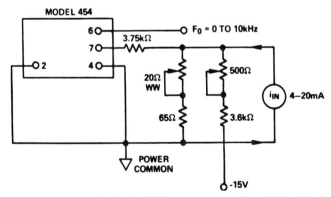

4-to-20 mA interface using the 454 V/F converter chip with a 10 kHz full scale output (courtesy Analog Devices, Inc.).

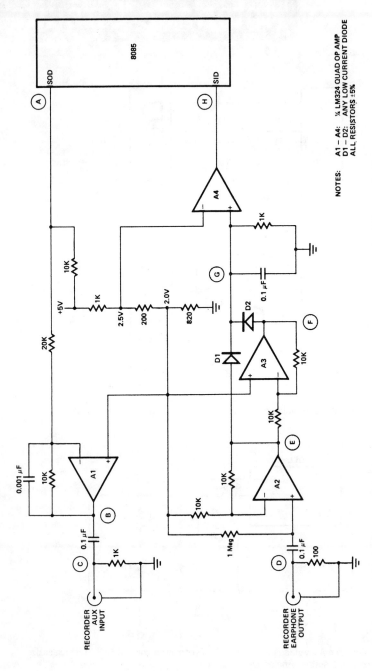

8085 CPU to cassette tape recorder interface using one LM324 quad op amp (courtesy Intel Corporation).

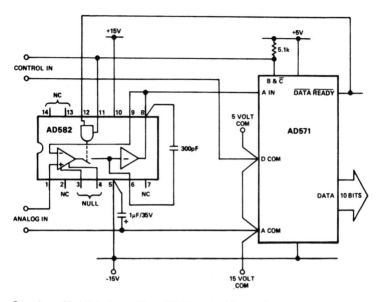

Sample-and-hold interface with an AD571 10-bit A/D converter (courtesy Analog Devices, Inc.).

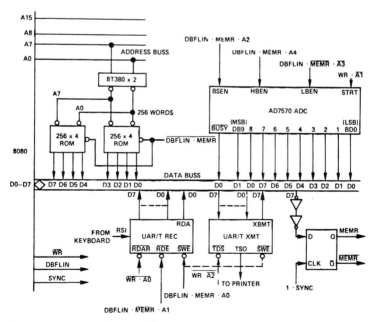

Microprocessor-controlled TTY-to-A/D converter interface. The AD7570 is a 10-bit CMOS A/D converter (courtesy Analog Devices, Inc.).

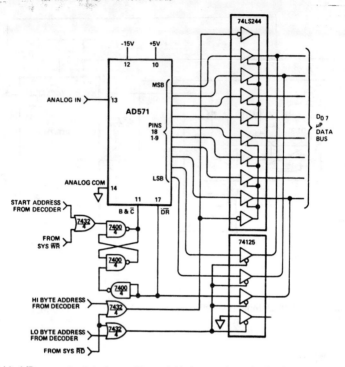

10-bit A/D converter interface with an 8-bit bus such as in the 8080 control structure (courtesy Analog Devices, Inc.).

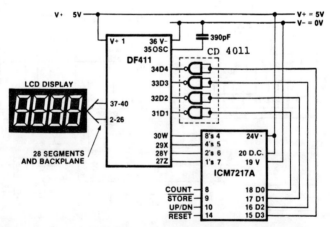

LCD interface using an ICM7217A, a DF411 and a CD4011. Total system power consumption is less than 5 mW. Common-cathode devices should be used since the digit drivers are CMOS, while in common-anode devices the digit drivers are NPN devices and will not provide full logic swing (courtesy Intersil, Inc.).

Power down to SBC 80/20 interface. The SBC 80/20 is a single board computer containing an 8259 programmable interrupt controller (courtesy Intel Corporation).

255

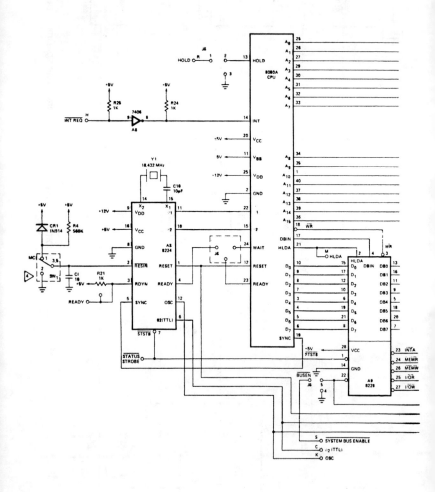

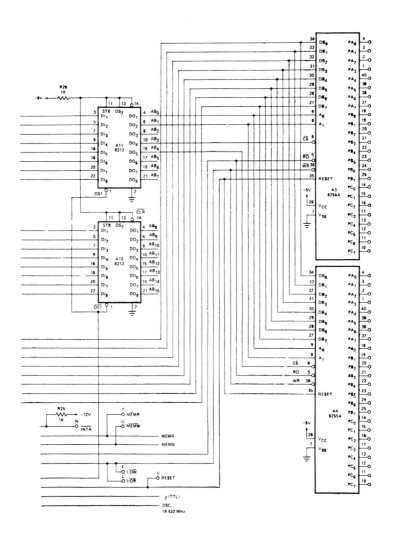

8080A CPU module to 8255A interface. This circuit comes complete as the Intel SDK 80 kit board. All of the 8255A interface lines are directly driven by the CPU (courtesy Intel Corporation).

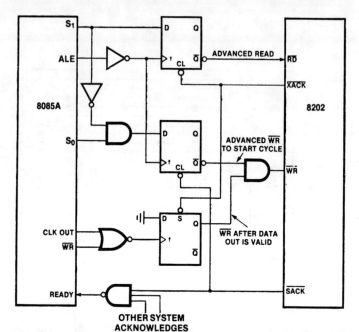

NOTE: IF YOU CAN BE *GUARANTEED* THAT EVERY REFRESH WILL BE AN EXTERNAL REFRESH, YOU CAN IGNORE THE SACK LINE IN THE ABOVE CONFIGURATION AND USE WR DIRECTLY FROM 8085A.

8085A CPU to 8202 interface with no read or write wait states (courtesy Intel Corporation).

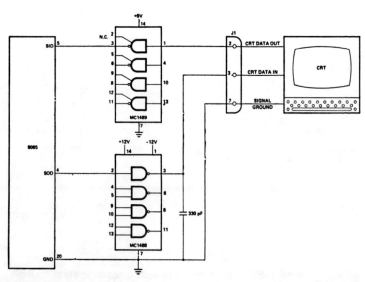

8085 CPU to RS-232C interface using an MC1488 and an MC1489 (courtesy Intel Corporation).

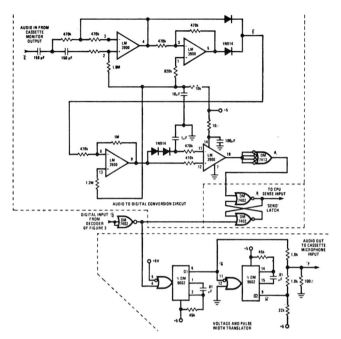

SC/MP to cassette tape recorder interface (courtesy National Semiconductor Corporation).

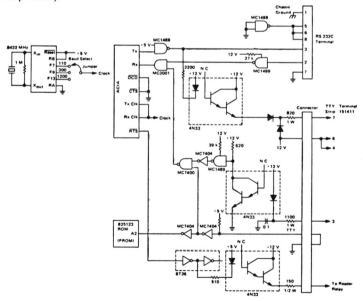

Teletypewriter/RS-232C terminal to MC6850 ACIA interface (courtesy Motorola Semiconductor Products Inc.).

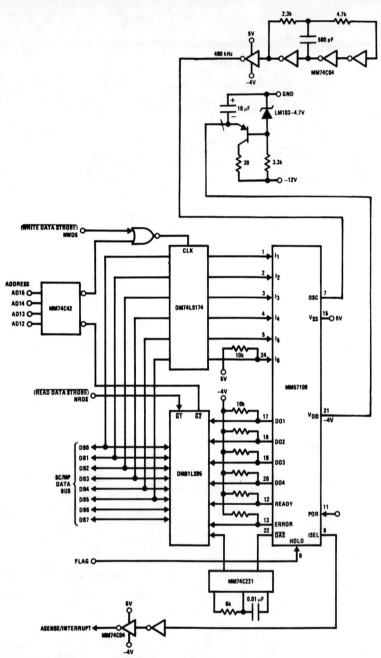

Number cruncher interface with the SC/MP microprocessor. The number cruncher unit (NCU) shown is an MM57109 (courtesy National Semiconductor Corporation).

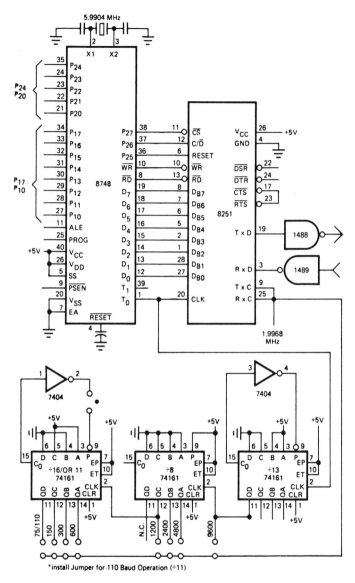

8748 to 8251 interface. The 8748 is an MCS-48 processor and the 8251 is a USART (courtesy Intel Corporation).

Chopper Circuits

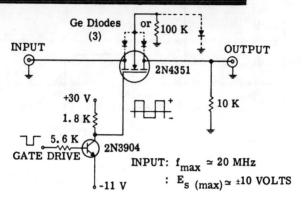

MOSFET analog switching circuit (chopper) for large input voltages (courtesy Motorola Semiconductor Products Inc.).

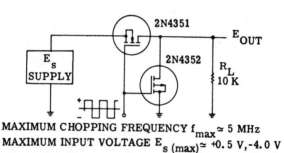

Series-shunt chopper for high-frequency applications using complementary enhancement mode MOSFETs (courtesy Motorola Semiconductor Products Inc.).

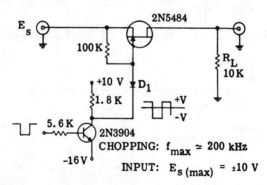

Series chopper for large input voltages using an N-channel JFET (courtesy Motorola Semiconductor Products Inc.).

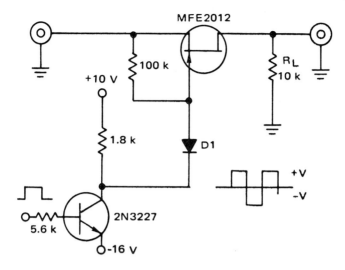

JFET chopper with extended range of ±10 volts (courtesy Motorola Semiconductor Products Inc.).

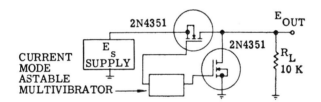

MAXIMUM CHOPPING FREQUENCY $f_{(max)} \simeq 5$ MH$_z$

MINIMUM INPUT VOLTAGE $E_{s\,(min)} \simeq \pm 10 \mu V$

Series-shunt chopper for low input voltages (courtesy Motorola Semiconductor Products Inc.).

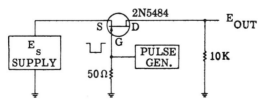

MAXIMUM CHOPPING FREQUENCY $f_{(max)} \simeq 200$ kHz

MAXIMUM INPUT VOLTAGE $E_{s\,(max)} \simeq +2$ V, -0.4 V

Series chopper using an N-channel JFET (courtesy Motorola Semiconductor Products Inc.).

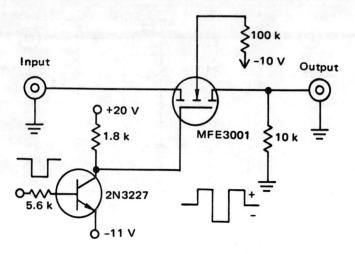

MOSFET chopper with extended range of ±3 volts (courtesy Motorola Semiconductor Products Inc.).

Indicator Circuits

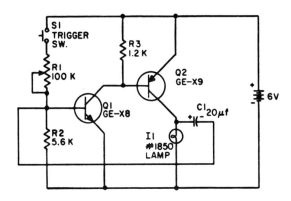

Parts List

- C1 — 20-mfd, 6-volt electrolytic capacitor
- I1 — 6-volt, GE No. 1850 lamp and socket
- Q1 — GE-X8 transistor
- Q2 — GE-X9 transistor
- R1 — 100K-ohm, 2-watt potentiometer
- R2 — 5.6K-ohm, 1/2-watt resistor
- R3 — 1.2K-ohm, 1/2-watt resistor
- S1 — Single-pole push-button trigger switch
- Battery — 6-volt dry pack

Trigger switch flasher. The flasher circuit is activated whenever pushbutton switch S1 is pressed. The circuit is a two-stage direct-coupled amplifier working as a multivibrator (courtesy General Electric Company).

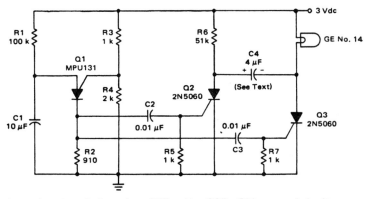

Low-voltage lamp flasher using a PUT and two SCRs. C4 is a nonpolarized type capacitor (courtesy Motorola Semiconductor Products Inc.).

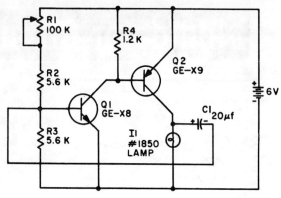

Parts List

- C1 — 20-mfd, 6-volt electrolytic capacitor
- I1 — 6-volt, GE No. 1850 lamp and socket
- Q1 — GE-X8 transistor
- Q2 — GE-X9 transistor
- R1 — 100K-ohm, 2-watt potentiometer
- R2, R3 — 5.6K-ohm, 1/2-watt resistor
- R4 — 1.2K-ohm, 1/2-watt resistor
- Battery — 6-volt dry pack

Warning flasher light. This device is handy to have around boats, cars, and camp sites. The circuit is a two-stage direct-coupled transistor amplifier connected as a free-running multivibrator. Both flash duration and flash interval can be adjusted by R1 (courtesy General Electric Company).

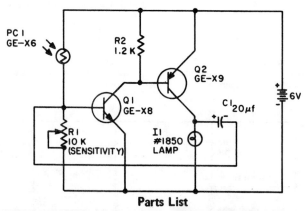

Parts List

- C1 — 20-mfd, 6-volt electrolytic capacitor
- I1 — 6-volt, GE No. 1850 lamp and socket
- PC1 — GE-X6 photoconductive cell
- Q1 — GE-X8 transistor
- Q2 — GE-X9 transistor
- R1 — 10K-ohm, 2-watt potentiometer
- R2 — 1.2K-ohm, 1/2-watt resistor
- Battery — 6-volt dry pack

Light target flasher. Photoresistor PC1 starts the flasher circuit whenever light hits it. Sensitivity control R1 is adjusted so that the flasher stops whenever the light source is removed (courtesy General Electric Company).

Parts List

B1 — 12-volt battery
C1 — 15- to 20-mfd, 350-volt capacitor
C2 — 0.22-mfd, 150-volt capacitor
C3 — 10-mfd, 25-volt capacitor
CR1 thru CR3 — GE-504A rectifier diode
I1 — GE No. 1125 lamp and socket
PC1 — GE-X6 photoconductive cell
Q1 — GE-X10 unijunction transistor
R1 — 1-megohm, 2-watt potentiometer
R2 — 500K-ohms, 2-watt potentiometer
R3, R5, R6 — 47K-ohm, 1/2-watt resistor
R4 — 47-ohm, 1/2-watt resistor
R7 — 100-ohm, 1/2-watt resistor
R8 — 1K-ohm, 1/2-watt resistor
R9 — 680-ohm, 1/2-watt resistor
R10 — 220-ohm, 1/2-watt resistor
SCR1, SCR2 — GE-X1 silicon controlled rectifier

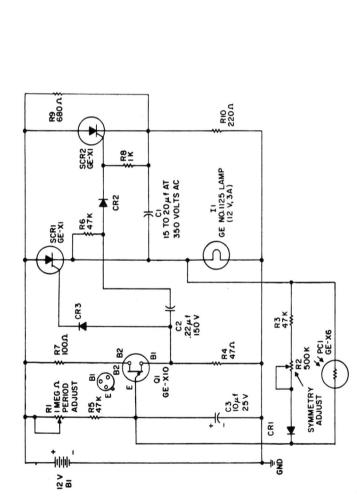

High-power battery-operated flasher with 40-watt output. SCR1 and SCR2 form a basic DC flip-flop (courtesy General Electric Company).

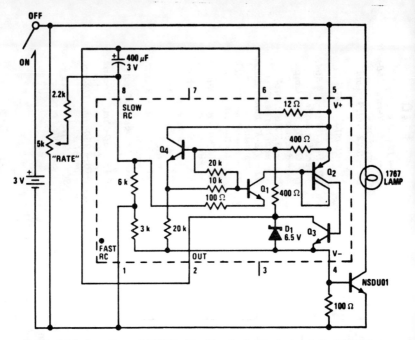

3-volt ministrobe using an LM3909 chip. Circuitry inside dashes is the LM3909. Flash rate can be varied from no flash to continuously on (courtesy National Semiconductor Corporation).

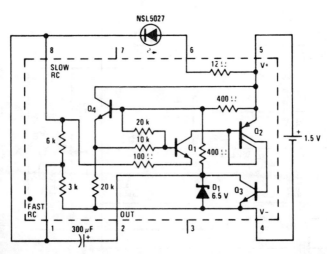

1.5-volt LED flasher using an LM3909. Flashing rate is about 1 hertz. Circuitry inside dashed lines is the LM3909 (courtesy National Semiconductor Corporation).

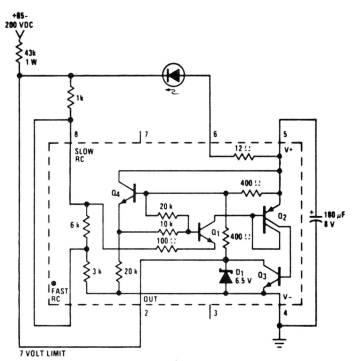

Safe high-voltage flasher using an LM3909. Circuitry inside dashes is the LM3909. If the 43K dropping resistor shown is employed the IC and LED will be about 7 volts above ground (courtesy National Semiconductor Corporation).

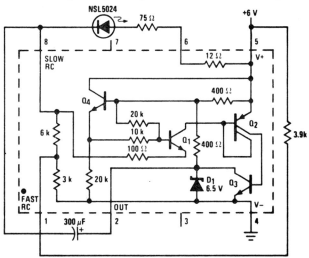

6-volt flasher using an LM3909 chip. Flashing rate is about 1 hertz. Circuitry inside dashed lines is the LM3909 (courtesy National Semiconductor Corporation).

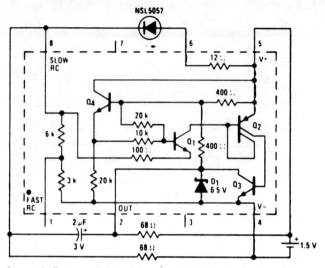

Continuous indicator using an LM3909 chip. Circuitry inside dashes is the LM3909. The circuit actually flashes at a 2 kHz rate, which cannot be detected by the human eye. This indicator is not intended as a long life system since current drain is 12 mA (courtesy National Semiconductor Corporation).

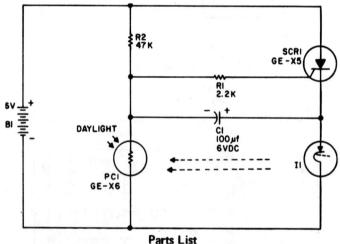

Parts List

C1 — 100-mfd, 6-volt electrolytic capacitor
B1 — 6-volt dry-pack lantern battery
I1 — GE No. 407 flasher lamp
R1 — 2.2K-ohm, 1/2-watt resistor
R2 — 47K-ohm, 1/2-watt resistor
SCR1 — GE-X5 silicon controlled rectifier

Automatic flasher light for use around marine buoys, piers, or towers. It starts working at dark and stops at dawn. When constructing this circuit be sure to place the photoresistor so that it receives as much light as possible without interfering with the flasher light (courtesy General Electric Company).

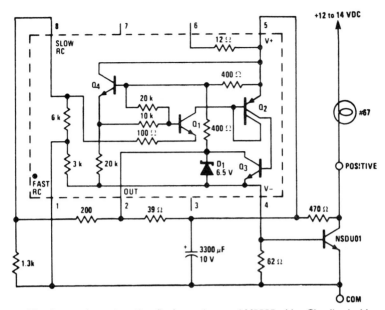

12-volt two-wire automotive flasher using an LM3909 chip. Circuitry inside dashes is the LM3909. Rate is about 1 hertz. Since it is a two-wire circuit it can be adapted for either negative or positive ground systems. In positive ground systems place the lamp between the common and the negative battery terminal (courtesy National Semiconductor Corporation).

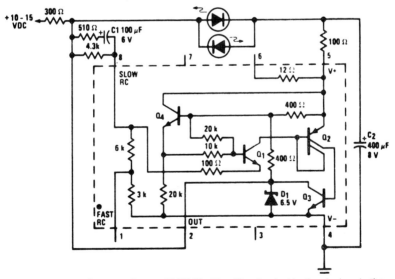

Alternating flasher using an LM3909 chip. Circuitry inside the dashes is the LM3909. This circuit is a relaxation oscillator flashing two LEDs sequentially. With a 12-volt DC supply the repetition rate is about 2.5 hertz (courtesy National Semiconductor Corporation).

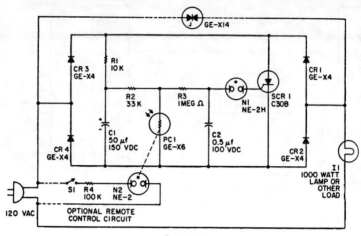

Parts List

C1 — 50-mfd, 150-volt electrolytic capacitor
C2 — 0.5-mfd, 100-volt capacitor
CR1, CR2, CR3, CR4 — GE-X4 rectifier diode
CR5 — GE-X14 thyrector diode (optional transient voltage suppressor)
I1 — 1000-watt lamp or other load
N1 — G-E Type NE-2H neon lamp
N2 — G-E Type Ne-2 neon lamp (optional)
PC1 — GE-X6 cadmium sulfide photoconductor
R1 — 10000-ohm, 1/2-watt resistor
R2 — 33000-ohm, 1/2-watt resistor
R3 — 1-megohm, 1/2-watt resistor (optional)
R4 — 100K-ohms, ½-watt
S1 — SPST switch
SCR1 — G-E Type C30B silicon controlled rectifier
Note: For 300-watt maximum load, use GE-X1 SCR.

1-kilowatt flasher with photoelectric control. This circuit can be used to control warning lights on towers, piers or construction hazards. Remote control can be attained by adding neon lamp N2 and masking PC1 so that it sees only N2 (courtesy General Electric Company).

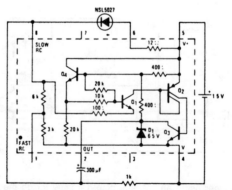

1.5-volt fast blinker using an LM3909. Flashing rate is about 3 hertz. Circuitry inside dashed lines is the LM3909 chip (courtesy National Semiconductor Corporation).

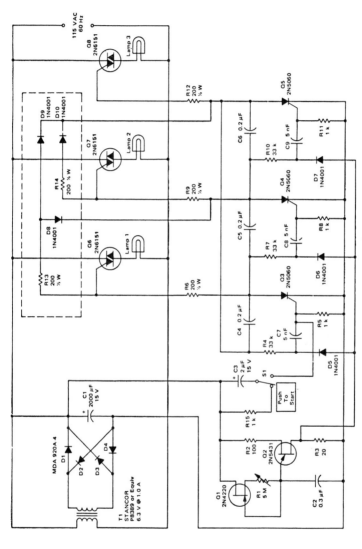

Sequential AC flasher using SCRs, triacs and a UJT. Only three stages are shown but many additional stages can be added. By adding the circuitry inside the dashed lines the circuitry can be modified so that the previous lamps remain on when a new lamp sequences on (courtesy Motorola Semiconductor Products Inc.).

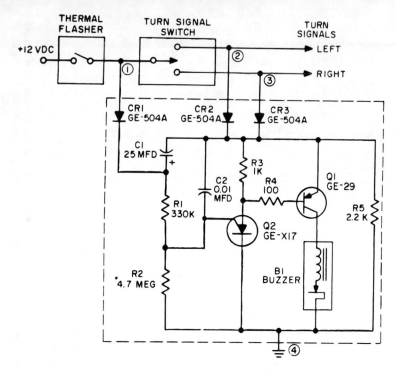

Parts List

B1 — 12-volt d-c buzzer
C1 — 25-mfd, 15-volt electrolytic capacitor
C2 — 0.01-mfd, 50-volt capacitor
CR1, CR2, CR3 — GE-504A rectifier diode
Q1 — GE-29 transistor
Q2 — GE-X17 programmable unijunction transistor

R1 — 330K-ohm, 1/2-watt resistor
R2 — 4.7-megohm, 1/2-watt resistor
R3 — 1K-ohm, 1/2-watt resistor
R4 — 100-ohm, 1/2-watt resistor
R5 — 2.2K-ohm, 1/2-watt resistor

Car turn signal reminder. This circuit is designed for 12-volt negative ground systems only. For cars with a 6-volt negative ground system use a 6-volt buzzer. The circuit cannot readily be adapted for positive ground systems (courtesy General Electric Company).

Audio Amplifiers

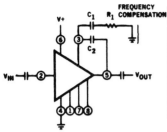

Voltage Gain	C_1	C_2	R_1	Decouple Pins:
10	68 pF	39 pF	75 Ω	1
20	50 pF	27 pF	75 Ω	8
100	None	3 pF	None	1, 7
200	None	3 pF	None	7, 8

General-purpose fixed-gain low-distortion amplifier, using an ECG716, that can be employed for such things as a telephone or headset AF amplifier. Fixed voltage gain of 10, 20, 100, and 200 can be obtained by referring to the table. Supply voltage is 27 volts, input voltage is ±5 volts, and power dissipation is 600 mW. Voltage gain falls off sharply above 1 MHz (courtesy GTE Sylvania Incorporated).

R_1	R_2	R_3	Voltage Gain	V_{in} for P_e = 1W
10 k	1.0 M	100 k	47	0.1
47 k	1.0 M	100 k	16	0.28
100 k	1.0 M	100 k	7.8	0.6
330 k	1.0 M	100 k	2.4	2.0

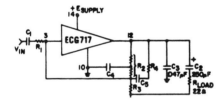

R_1	R_2	R_3	R_4	C_1	C_4	C_5
120 k	two 390 k	82 k	3.3 M	.01 μF	.05 μF	150 pF

Gain performance will be as follows:

Voltage Gain	12
V_{in} for P_o = 1 W	.4 V
Bass Boost	3 dB at 150 Hz

1-watt power amplifier. Refer to the table to select the desired voltage gain. In the circuit the AC feedback path is different from the DC path. Resistor R2 is bypassed to ground, taking it out of the AC feedback loop. The voltage gain is determined by the ratio of R4/R1. Another feature of this circuit is that you can introduce bass boost by the selection of capacitor values for C1, C4 and C5. With the values shown in the table at the bottom of the figure, gain performance is as follows: voltage gain is 12, V_{IN} for Po at 1W is 0.4 volt, and bass boost is 3 dB at 150 Hz (courtesy GTE Sylvania Incorporated).

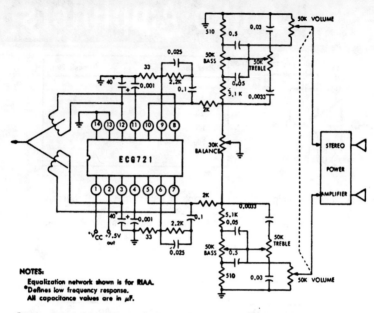

Stereo record player AF preamplifier, with bass and treble controls, for magnetic pickup. Current drain is typically 16 mA at 12 volts. Voltage gain is 68 dB. Channel separation is typically 60 dB (courtesy GTE Sylvania Incorporated).

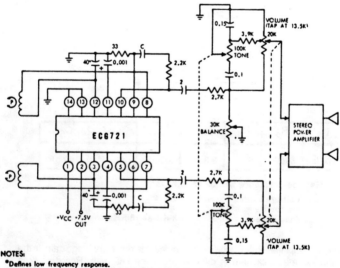

NOTES:
*Defines low frequency response.
$C = 0.02\mu F$ for 7½ ips (19CM/S); $0.033\mu F$ for 3¾ ips (9.5 CM/S).
All capacitance values are in microfarads.

Stereo tape system AF preamplifier with simple tone control. Current drain at 12 volts is typically 16 mA. Voltage gain is 68 dB. Channel separation is typically 60 dB (courtesy GTE Sylvania Incorporated).

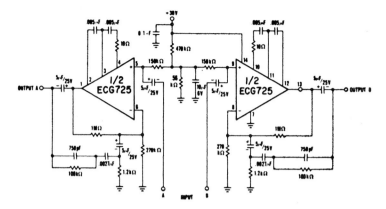

TYPICAL PERFORMANCE

Gain 40 dB at 1 kHz, RIAA equalized
Input overload point, 80 mV rms
Noise level, 2μV referred to input
Signal to noise ratio, 74 dB below 10 mV
Channel separation @ 1 kHz, 80 dB

Stereo phono AF preamplifier with RIAA equalization. Gain is 40 dB at 1 kHz with the input overload point set at 80 mV RMS. Noise level is 2 μV referred to the input. Signal to noise ratio is 74 dB below 10 mV. Channel separation at 1 kHz is 80 dB (courtesy GTE Sylvania Incorporated).

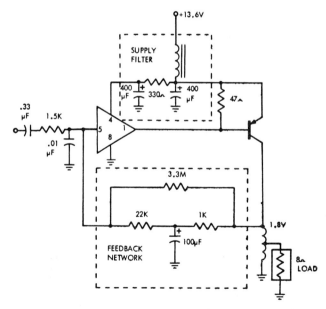

4-watt AF amplifier using an ECG735 class A driver and an ECG153 audio output. This is a typical automotive tape player amplifier (courtesy GTE Sylvania Incorporated).

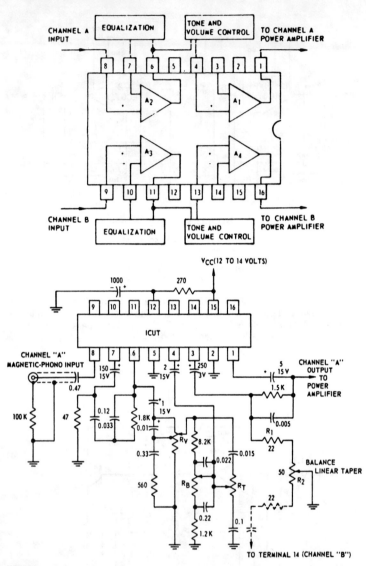

NOTES:

1) Resistor values are in ohms, capacitance values are in microfarads, unless otherwise specified.
2) R_1 and R_2 resistor values are selected for a sensitivity of 3 mV input at 1 kHz.
3) R_V, volume control potentiometer, 15000 ohms tap at 6000 ohms.
4) R_B, bass control potentiometer, 25000 ohms.
5) R_T, treble control potentiometer, 25000 ohms.

Stereo phono AF preamplifier for magnetic pickups using the ECG727. The schematic shows only one channel. Duplicate components must be used for channel B. Refer to the block diagram for pin identification. Voltage gain at 1 kHz is 47 dB. The ECG727 has four internal AC amplifiers (courtesy GTE Sylvania Incorporated).

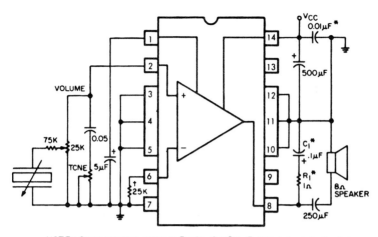

NOTE: Compensation network $R_l = 1\Omega$, $C_l = 0.1\mu F$. However, actual values are dependent upon circuit layout.

* For stability with high current loads.
† For optimum output centering.

2-watt AF phono amplifier using an ECG740A 14-pin DIP. Supply voltage is typically 18 volts. Voltage gain is 34 dB. Minimum audio power output is 2 watt, with 2.5 watts typical. When driving a 16-ohm load the supply voltage at pin 14 should be 24 volts (courtesy GTE Sylvania Incorporated).

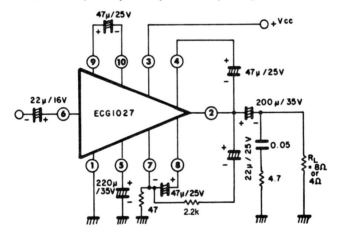

Supply Voltage..................32 V	Distortion (f = 1 kHz, P_o = 0.1 W) .... 0.1% Typ.
Load...................... 8 Ohms	Voltage Gain (f = 1 kHz, P_o = 0.1 W) .... 33 dB Typ.
Output Power (KF = 1%)...... 13 W Typ.	Power Bandwidth (KF = 1%, ± 3 dB) .... 20 Hz - 30 kHz

13-watt AF power amplifier designed for an 8-ohm or 4-ohm load. The circuit needs only one power supply rated at 32 volts, 50 mA (courtesy GTE Sylvania Incorporated).

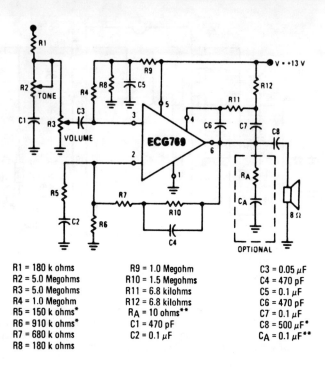

R1 = 180 k ohms
R2 = 5.0 Megohms
R3 = 5.0 Megohms
R4 = 1.0 Megohm
R5 = 150 k ohms*
R6 = 910 k ohms*
R7 = 680 k ohms
R8 = 180 k ohms

R9 = 1.0 Megohm
R10 = 1.5 Megohms
R11 = 6.8 kilohms
R12 = 6.8 kilohms
R_A = 10 ohms**
C1 = 470 pF
C2 = 0.1 μF

C3 = 0.05 μF
C4 = 470 pF
C5 = 0.1 μF
C6 = 470 pF
C7 = 0.1 μF
C8 = 500 μF*
C_A = 0.1 μF**

*For 16-ohm load change R5 to 100 k ohms, R6 to 820 k ohms and C8 to 250 μF.
**Optional

1-watt phonograph amplifier with an 8-ohm speaker. Intended for use with a ceramic cartridge. Sensitivity approximately 450 mV. The gain of the amplifier is determined by (R7 + R10)/R5. To change the gain, alter the value of R5 and hold (R7 + R10) between 1M and 2.2M. Bass boost can be eliminated by removing C4 and replacing (R7 + R10) with a fixed resistor valued at 2.2M (courtesy GTE Sylvania Incorporated).

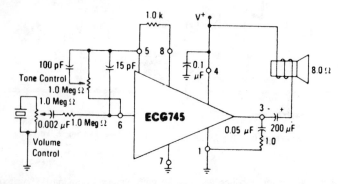

0.5-watt AF amplifier for a phonograph with a ceramic cartridge. The ECG745 is an 8-pin DIP (courtesy GTE Sylvania Incorporated).

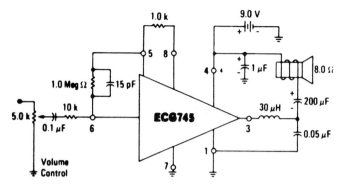

0.5-watt audio amplifier for an AM/FM radio. The ECG745 is an 8-pin DIP (courtesy GTE Sylvania Incorporated).

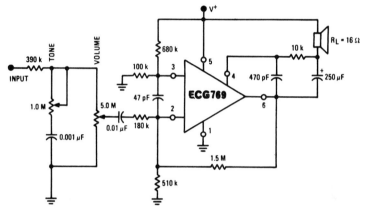

1-watt AF amplifier for phonographs with ceramic cartridges, using a 16-ohm speaker. Sensitivity is 100 mV for 1-watt output. Supply voltage should be 16 volts (courtesy GTE Sylvania Incorporated).

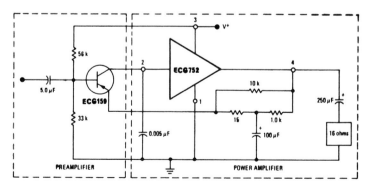

0.25-watt AF amplifier using a 16-ohm speaker. Supply voltage should be approximately 12 volts (courtesy GTE Sylvania Incorporated).

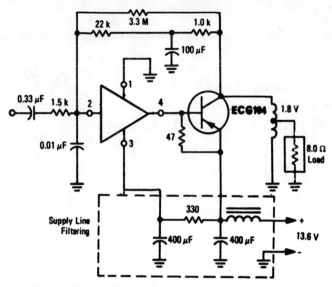

4-watt class A audio amplifier using an ECG755 audio driver IC. This audio amplifier is ideal for automotive applications. The ECG755 is a 4-lead package and is designed to drive a PNP power output transistor such as the ECG104 (courtesy GTE Sylvania Incorporated).

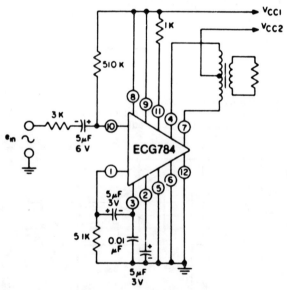

1-watt class B audio amplifier with preamplifier. Supply voltage V_{CC1} and V_{CC2} are 9.0 volts and 12.0 volts, respectively. Sensitivity is 45 mV. Power gain is 75 dB. Input resistance is 55K. Signal-to-noise ratio is 66 dB. Use a standard class B audio output transformer with the appropriate speaker load (courtesy GTE Sylvania Incorporated).

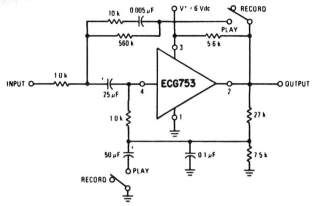

Record/playback amplifier for cassette and portable tape recorders (courtesy GTE Sylvania Incorporated).

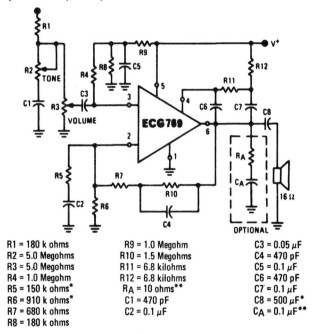

R1 = 180 k ohms	R9 = 1.0 Megohm	C3 = 0.05 µF
R2 = 5.0 Megohms	R10 = 1.5 Megohms	C4 = 470 pF
R3 = 5.0 Megohms	R11 = 6.8 kilohms	C5 = 0.1 µF
R4 = 1.0 Megohm	R12 = 6.8 kilohms	C6 = 470 pF
R5 = 150 k ohms*	R_A = 10 ohms**	C7 = 0.1 µF
R6 = 910 k ohms*	C1 = 470 pF	C8 = 500 µF*
R7 = 680 k ohms	C2 = 0.1 µF	C_A = 0.1 µF**
R8 = 180 k ohms		

*For 16-ohm load change R5 to 100 k ohms, R6 to 820 k ohms and C8 to 250 µF.
**Optional

1-watt phonograph amplifier for ceramic cartridges with 16-ohm speaker. See table for component values. Sensitivity is approximately 450 mV. The gain of this amplifier is determined by (R7 + R10)/R5. To alter the gain, change R5 and hold (R7 + R10) between 1M and 2.2M. Bass boost can be eliminated by removing C4 and replacing (R7 + R10) with a fixed resistor value at 2.2M (courtesy GTE Sylvania Incorporated).

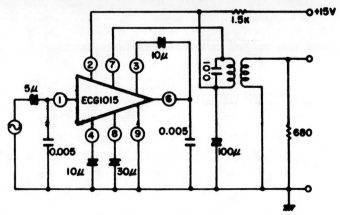

Low-noise AF preamplifier. The preamplifier provides 0.9 volts output at 1 kHz with a 2 mV input. ECG1015 is a hybrid module (courtesy GTE Sylvania Incorporated).

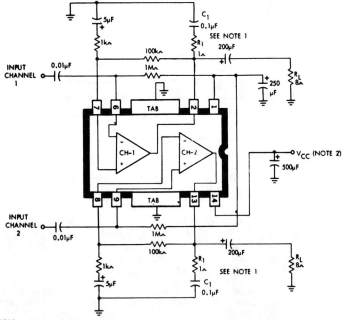

NOTES:
1. Compensation network: R_1, C_1 values are dependent upon circuit layout.
2. When an unregulated supply voltage is used, the actual voltage present at pin 14 during full signal conditions should not drop below the nominal supply voltage level if full power output is to be maintained.
3. Closed loop gain should be limited to 30dB min. to 60dB max. to maintain stable circuit operation.

2-watts-per-channel dual audio amplifier using an ECG804 IC. For 8-ohm speaker loads use a 20-volt supply, for 16-ohm speaker loads use a 24-volt supply and for 4-ohm speaker loads use a 12-volt supply (courtesy GTE Sylvania Incorporated).

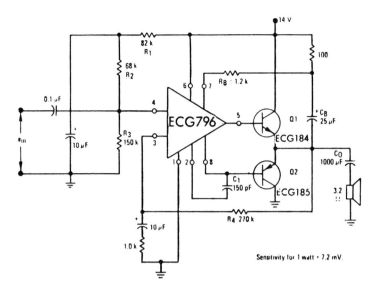

4-watt audio power amplifier with a 7 mV sensitivity for automobile applications. By changing the output transistors the output can be increased up to 20 watts (courtesy GTE Sylvania Incorporated).

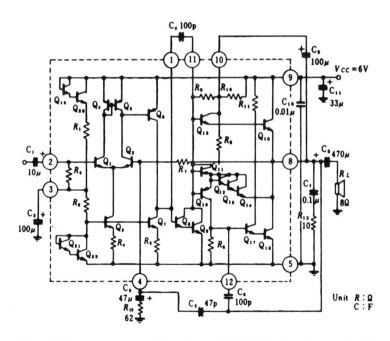

0.5-watt OTL audio power amplifier using an ECG1031. Typical voltage gain at 1 kHz is 50 dB. Input impedance is 20K (courtesy GTE Sylvania).

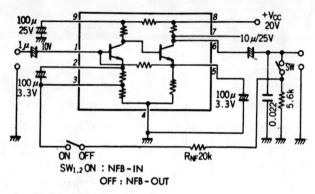

Low-noise flat-frequency-response preamplifier using an ECG1020 thin-film hybrid module. At 1 kHz with the switch off typical voltage gain is 66 dB; with the switch on typical voltage gain is 42.5 dB (courtesy GTE Sylvania Incorporated).

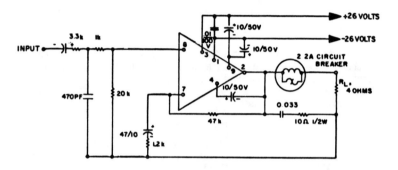

15-watt AF power amplifier using an ECG1028 module. An ECG1090 module can be substituted for the ECG1028 by omitting the 20K resistor at pin 8 and the 47K resistor between pins 7 and 2 (courtesy GTE Sylvania Incorporated).

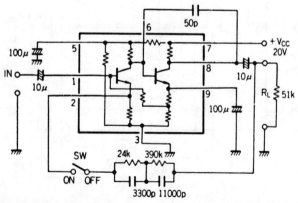

Low-noise equalizing amplifier using an ECG1019 thin-film hybrid module (courtesy GTE Sylvania Incorporated).

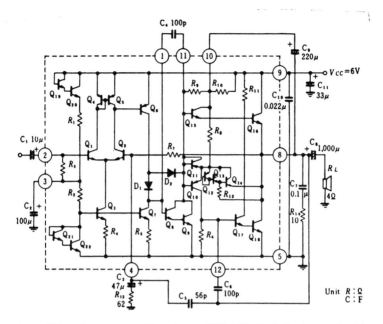

1-watt OTL audio power amplifier using an ECG1032. Typical voltage gain at 1 kHz is 50 dB (courtesy GTE Sylvania Incorporated).

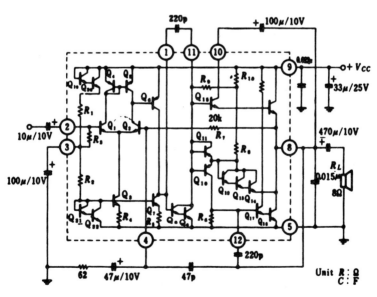

2-watt OTL audio power amplifier using an ECG1034. Recommended supply voltage is 12 volts. Typical voltage gain at 1 kHz is 50 dB. Input impedance is 20K. Frequency response is from 60 hertz to 50 kilohertz (courtesy GTE Sylvania Incorporated).

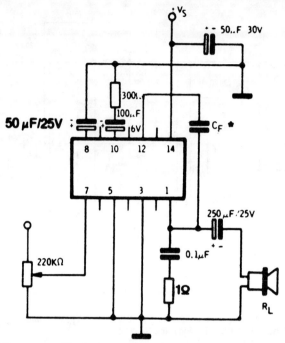

4-watt AF power amplifier for low-cost record players using an ECG1111 14-pin DIP. The amplifier is designed to drive a 16-ohm load. Recommended supply voltage is 24 volts. Typical voltage gain is 74 dB. Input impedance is 110K. Capacitor C_F can be either 510 pF or 820 pF, depending on the frequency response desired. For 510 pF the frequency response extends beyond 10 kHz. With C_F at 820 pF the response falls off at about 8 kHz (courtesy GTE Sylvania Incorporated).

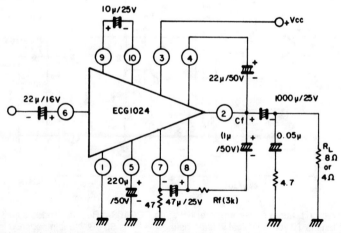

8-watt AF power amplifier for 8-ohm or 4-ohm speaker load. Typical supply voltage is 25 volts (courtesy GTE Sylvania Incorporated).

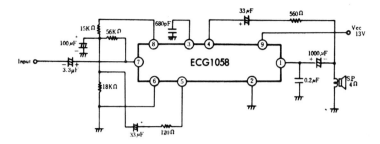

4.4-watt audio power amplifier for a 4-ohm load. The ECG1058 consists of a differential amplifier, a driver amplifier, a ripple filter, an automatic operating point stabilizer and a quasi-complementary SEPP OTL power amplifier circuit. This circuit is particularly good for automotive stereo applications since it is powered by a 13-volt supply. Typical voltage gain at 1 kHz is 45 dB (courtesy GTE Sylvania Incorporated).

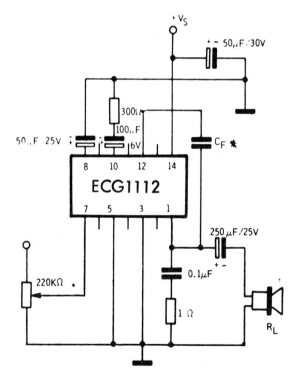

2.2-watt AF power amplifier for low-cost phonographs using an ECG1112 14-pin DIP. The 2.2-watt rating is based on a 16-ohm load. Recommended supply voltage is 18 volts. Input impedance is 150K. Typical voltage gain is 72 dB. Current drain at maximum output is 175 mA. Capacitor C_F can be either 510 pF or 820 pF. With C_F at 820 pF the frequency response falls off at about 8 kHz; at 510 pF it extends up to 10 kHz (courtesy GTE Sylvania Incorporated).

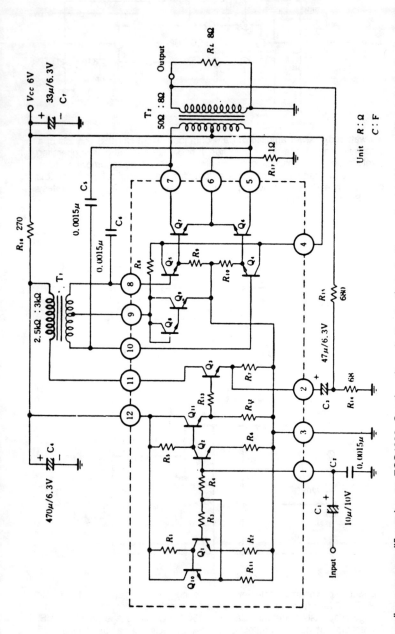

1-watt audio power amplifier using an ECG1033. Supply voltage is 6 volts. Typical voltage gain at 1 kHz is 47 dB. Packaging for the ECG1033 is a 12-pin plastic pack with metal tabs (courtesy G1E Sylvania Incorporated).

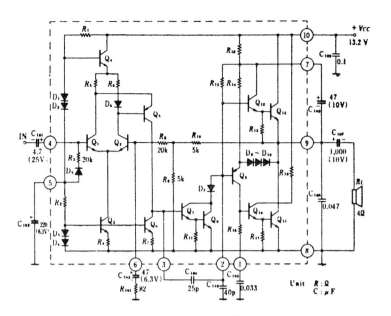

5.5-watt OTL audio power amplifier using an ECG1037. Typical voltage gain is 55 dB. Frequency response is from 30 hertz and 50 kilohertz. Input impedance is 36K (courtesy GTE Sylvania Incorporated).

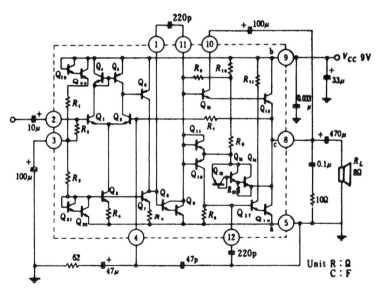

1-watt OTL audio power amplifier using an ECG1035. Packaging on the ECG1035 is a 12-pin plastic flat pack with metal tabs. Recommended supply voltage is 9 volts. Voltage gain is 50 dB. Frequency response is from 50 hertz to 100 kilohertz (courtesy GTE Sylvania Incorporated).

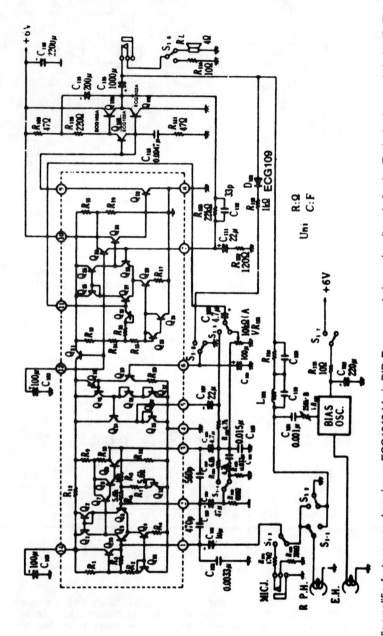

Audio amplifier for tape recorder using an ECG1043 14-pin DIP. Recommended supply voltage is 6 volts. Typical voltage gain is 85 dB. This circuit has an input impedance of 22K (courtesy GTE Sylvania Incorporated).

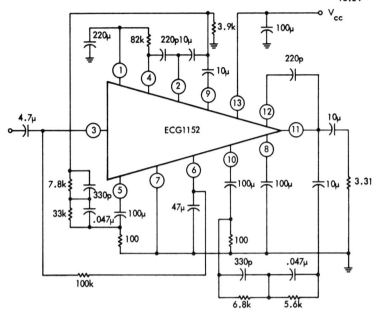

Low-noise preamplifier using an ECG1152 13-pin module. The ECG1152 is a dual two-stage direct-coupled amplifier. Typical voltage gain is 75 dB. Input resistance is 50K (courtesy GTE Sylvania Incorporated).

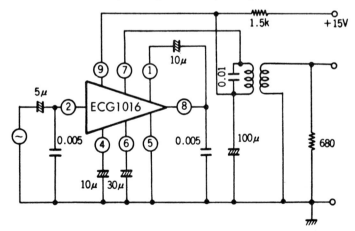

AF preamplifier with a thin-film hybrid module. Output is between 0.42 and 0.90 volts at 1 kHz with a 2 mV input. Distortion is less than 3%. The AF transformer is standard and can be purchased at Radio Shack (courtesy GTE Sylvania Incorporated).

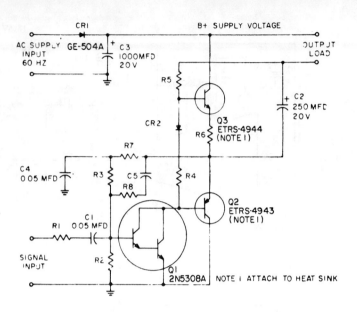

Parts List

Components	One Watt Output 8-ohms Load	One Watt Output 16-ohms Load	Two Watts Output 16-Ohms Load
B + supply	12 Vdc	19 Vdc	22 Vdc
C1	0.05 Mfd	0.05 Mfd	0.05 Mfd
C2	250 Mfd, 20V	250 Mfd, 20V	250 Mfd, 20V
C3	1000 Mfd, 20V	100 Mfd, 20V	1000 Mfd, 20V
C4	0.05 Mfd	0.05 Mfd	0.05 Mfd
C5	270 Picofarads	150 Picofarads	150 Picofarads
CR1	GE-504A	GE-504A	GE-504A
CR2	ETRS-4946*	ETRS-4946*	ETRS-4946*
Q1	2N5308A	2N5308A	2N5308A
Q2	ETRS-4943*	ETRS-4943*	ETRS-4943*
Q3	ETRS-4944*	ETRS-4944*	ETRS-4944*
R1 (sensitivity 0.5V)	270K ohms	330K ohms	220K ohms
R1 (sensitivity 1.0V)	510K ohms	680K ohms	470K ohms
R1 (sensitivity 1.5V)	750K ohms	1 megohm	680K ohms
R2	330K ohms	560K ohms	330K ohms
R3	820K ohms	1.8 megohms	1.8 megohms
R4	27 ohms	82 ohms	39 ohms
R5	560 ohms	1.8K ohms	1000 ohms
R6	0.68 ohm	2.7 ohms	2.7 ohms
R7	470K ohms	1.8 megohms	1 megohm
R8	1.5 megohms	2.7 megohms	2.7 megohms

* *Available from General Electric Co., Dept B, 3800 North Milwaukee Avenue, Chicago, Ill. 60641*

NOTE: All resistors 1/2 watt

2-watt AF power amplifier with bass boost for a 16-ohm load (courtesy General Electric Company).

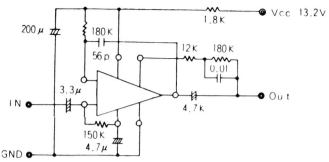

High-gain AF preamplifier using an EGC1135 7-pin module. This circuit is ideal for automotive stereo applications. No-signal supply current is 1.5 mA. Input impedance is 120K, while output impedance is 5 ohms (courtesy GTE Sylvania Incorporated).

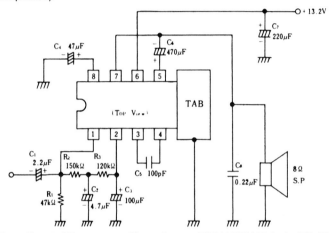

1.5-watt/2-watt AF power amplifier using an ECG1141/1140 8-pin DIP. This amplifier is ideal for automotive applications since the recommended supply voltage is 13.2 volts. Typical voltage gain is between 51 and 56 dB. No signal supply current is 12 mA (courtesy GTE Sylvania Incorporated).

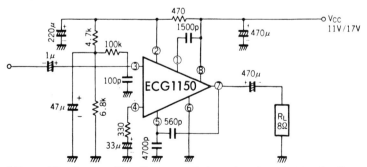

2.5-watt AF power amplifier. Recommended supply voltage is 17 volts. No-signal current is 24 mA. Input impedance is 85K (courtesy GTE Sylvania Incorporated).

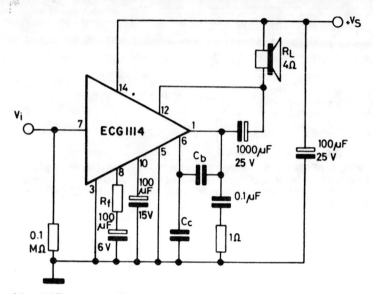

4.5-watt AF power amplifier for radio and TV. Recommended supply voltage is 14 volts. Current drain at maximum power output is 485 mA. Typical voltage gain is 46 dB. Input impedance is 3M (courtesy GTE Sylvania Incorporated).

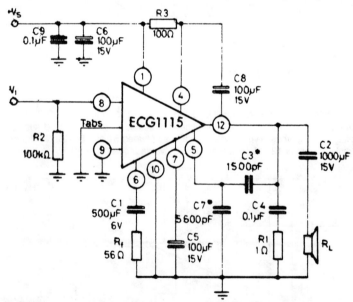

7-watt AF power amplifier featuring thermal shutdown with load connected to ground. For the rated output of 7 watts a 16-volt supply voltage is required. At a supply voltage of 9 volts the power output is 2.5 watts. Input voltage is 220 mV RMS. Input resistance is 5M. The ECG1115 is a 12-pin QIP with two metal tabs (courtesy GTE Sylvania Incorporated).

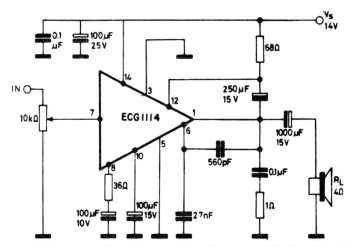

4.5-watt AF power amplifier with grounded load for radio and TV. The ECG1114 is a 14-pin QIP. Current drain at maximum output is 485 mA. Typical voltage gain is 46 dB. Input impedance is 3M (courtesy GTE Sylvania Incorporated).

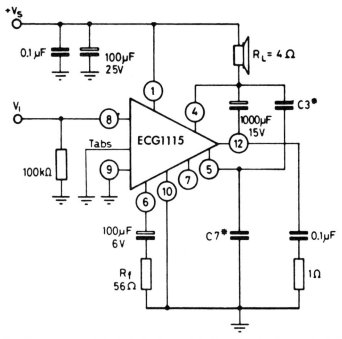

7-watt AF power amplifier featuring thermal shutdown with load connected to the supply voltage. Capacitor C3 is 1500 pF typically with C7 being 5600 pF. Generally C7 should be five times greater than C3. Typical supply voltage is 16 volts for the rated output. Input resistance is 5M. Input voltage is 220 mV RMS. With a supply voltage of 9 volts the output power will be 2.5 watts (courtesy GTE Sylvania Incorporated).

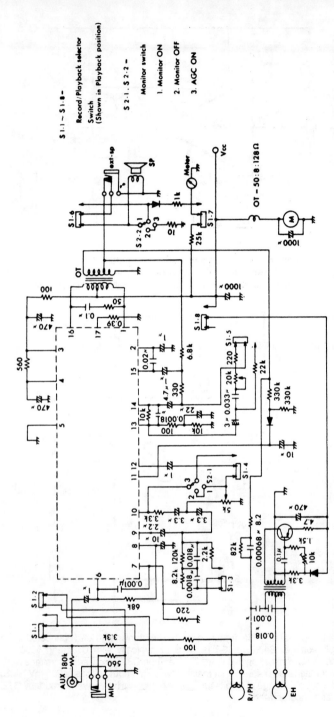

Complete cassette record/playback circuit using an ECG1110 16-pin DIP that provides 2 watts of audio power. This IC includes a preamplifier, tone amplifier and power amplifier. The 2-watt rating is for an 8-ohm load. Recommended supply voltage is 9 volts. Typical voltage gains for each section are as follows: preamplifier, 60 dB; tone amplifier, 40 dB; and power amplifier, 40 dB (courtesy GTE Sylvania Incorporated).

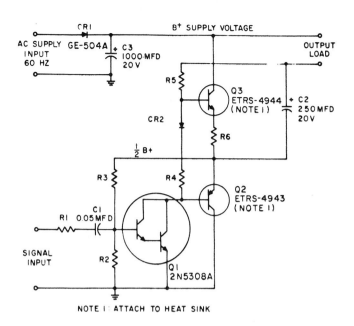

Parts List

Components	One Watt Output 8-ohms Load	One Watt Output 16-ohms Load	Two Watts Output 16-ohms Load
Battery (B+) Voltage and current OR	12 Vdc @ 160 MA	19 Vdc @ 110 MA	22 Vdc @ 160 MA
AC Supply Input (approximate)	9 — 12.6 Vac	14 — 18 Vac	16 — 20 Vac
C1	0.05 Mfd	0.05 Mfd	0.05 Mfd
C2	250 Mfd, 20V	250 Mfd, 20V	250 Mfd, 20V
C3	1000 Mfd, 20V	1000 Mfd, 20V	1000 Mfd, 20V
CR1	GE-504A	GE-504A	GE-504A
CR2	ETRS-4946*	ETRS-4946*	ETRS-4946*
Q1	2N5308A	2N5308A	2N5308A
Q2	ETRS-4943*	ETRS-4943*	ETRS-4943*
Q3	ETRS-4944*	ETRS-4944*	ETRS-4944*
R1 (sensitivity 0.5V)	220K ohms	470K ohms	270K ohms
R1 (sensitivity 1.0V)	470K ohms	820K ohms	510K ohms
R1 (sensitivity 1.5V)	680K ohms	1.2 megohms	750K ohms
R2	390K ohms	560K ohms	330K ohms
R3	1.5 megohms	3.6 megohms	2.7 megohms
R4	22 ohms	82 ohms	22 ohms
R5	560 ohms	1.8K ohms	1000 ohms
R6	0.68 ohm	2.7 ohms	2.7 ohms

* Available from General Electric Co., Dept B, 3800 North Milwaukee Avenue, Chicago, Ill. 60641

2-watt AF power amplifier for a 16-ohm load (courtesy General Electric Company).

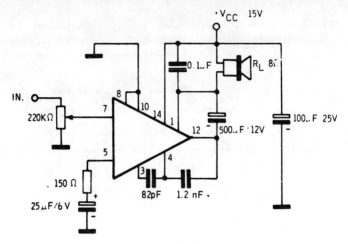

ELECTRICAL CHARACTERISTICS

Supply Voltage		15V
Load Resistance		8 Ω
Voltage Gain		34 dB
Sensitivity	Po = 50mW f = 1KHz	12.6 mV
	Po = 3.3W f = 1KHz	105 mV
Frequency Response	−3 dB	50 Hz to 15 KH
Total Current	Po = 0	4 mA
	Po = 3.3W	300 mA
Max Output Power	THD = 10% f = 1KHz	3.3 W
Distortion	Po = 50mW to 2W f = 1KHz	0.5%
Efficiency	Po = 3.3W	75%
Input Noise Voltage	R_S = 220K BW = 15 KHz	13 μV
DC Output Voltage	R_S = 0 to 20 KΩ	7.9 V
Supply Voltage Rejection (referred to the input)		40 dB

3.3-watt AF power amplifier using an ECG1118. This circuit is intended for phonograph applications (courtesy GTE Sylvania Incorporated).

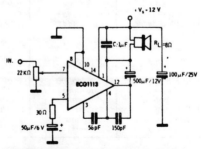

2.1-watt AF power amplifier for radio. The 2.1-watt rating is for an 8-ohm load. The ECG1113 is a 14-pin QIP. Recommended supply voltage is 12 volts, which makes it ideal for automotive applications. Current drain at maximum output is 235 mA. Typical voltage gain is 70 dB (courtesy GTE Sylvania Incorporated).

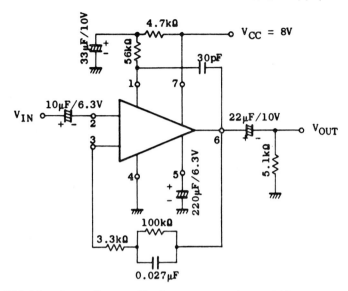

NAB 9.8 cm/s equalizer amplifier for car stereo using an ECG1087 module. Typical voltage gain is 35 dB (courtesy GTE Sylvania Incorporated).

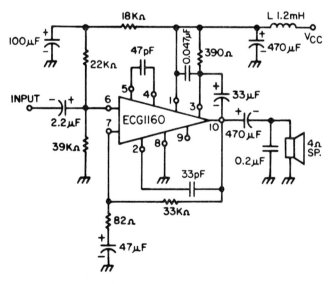

5.2-watt AF power amplifier for automotive applications. Recommended supply voltage is 13.2 volts. No-signal current drain is typically 28 mA. Voltage gain is 51.5 dB (courtesy GTE Sylvania Incorporated).

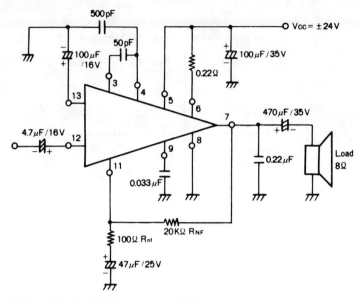

6.5-watt AF power amplifier using an ECG1078 chip powered by a single 24-volt supply. Frequency response is from 100 hertz to 20 kilohertz (courtesy GTE Sylvania Incorporated).

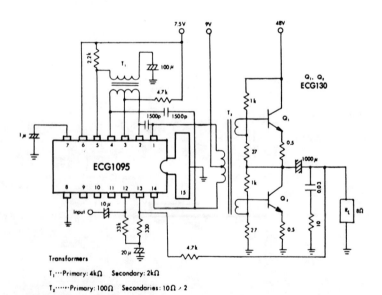

20-watt single-ended AF power amplifier. The ECG1095 is a 14-pin DIP with a tab and built-in AGC. The two output transistors are ECG130 bipolars (courtesy GTE Sylvania Incorporated).

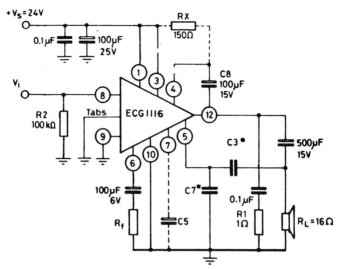

5-watt AF power amplifier with 16-ohm load connected to ground and using bootstrap. For low voltage operation (e.g., 9 to 14 volts) 150 ohms is connected between pins 1 and 4. To improve power supply ripple rejection capacitor C5 (10 to 100 μF at 25 volts) is connected between pin 7 and ground. Capacitor C3 is typically 330 pF and C7 is 1500 pF. Generally C7 is five times greater than C3. Frequency response is from 40 hertz to 20 kilohertz (courtesy GTE Sylvania Incorporated).

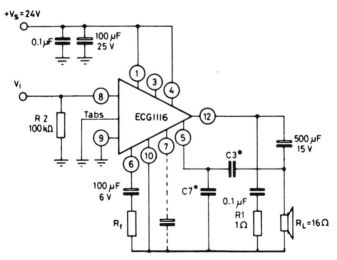

5-watt AF power amplifier with 16-ohm load connected to ground using no bootstrap. The ECG1116 is a 12-pin QIP. To obtain the rated output heatsinking is required. Capacitor C3 is typically 330 pF, while C7 is 1500 pF. Generally C7 is five times greater than C3. Supply voltage ripple rejection is improved by connecting a capacitor (10 to 100 μF at 25 volts) between pin 7 and ground. Frequency response is from 40 hertz to 20 kilohertz (courtesy GTE Sylvania Incorporated).

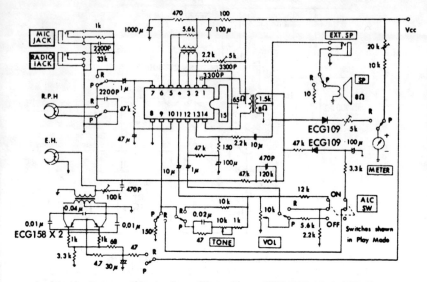

1-watt cassette record/playback amplifier using an ECG1095 14-pin DIP. The only external transistor outside of the ECG1095 are two ECG158s used for the AC bias circuit (courtesy GTE Sylvania Incorporated).

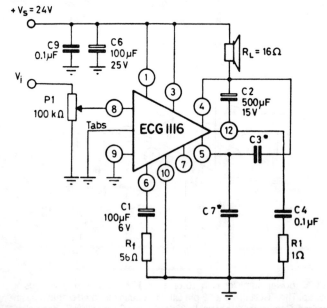

5-watt AF power amplifier with 16-ohm load connected to supply. Capacitor C3 is typically 330 pF and C7 is 1500 pF. Generally, C7 is five times greater than C3. The ECG1116 is a 12-pin QIP. To obtain the rated output heatsinking is necessary via the two tabs. Frequency response is from 40 hertz to 20 kilohertz. Input resistance is 5M (courtesy GTE Sylvania Incorporated).

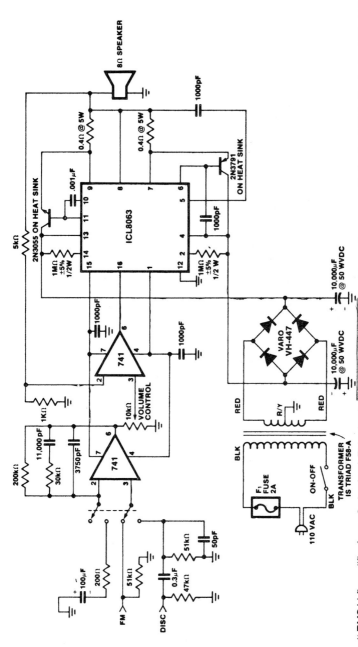

50-watt RMS hi-fi amplifier for an 8-ohm load. This circuit delivers about 56 volts peak to peak across an 8-ohm load. Distortion is about 1% at 20 kHz. The ganged switch at the input is for selecting either disc or FM radio. The input 741 stage is a preamplifier with RIAA equalization for records (courtesy Intersil, Inc.).

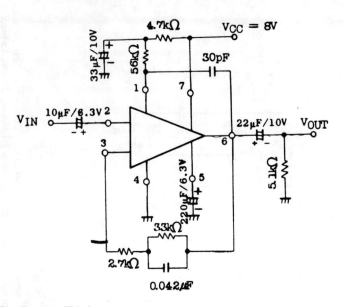

Equalizer amplifier for cassette tape recorders using an ECG1087 module. Typical voltage gain is 35 dB (courtesy GTE Sylvania Incorporated).

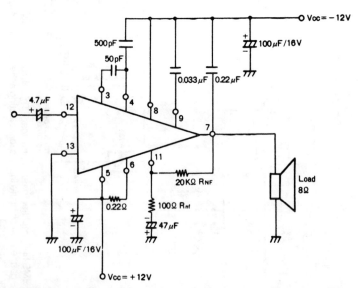

6.5-watt AF power amplifier using an ECG1078 chip powered by ±12-volt supplies. Frequency response is from 100 hertz to 20 kilohertz (courtesy GTE Sylvania Incorporated).

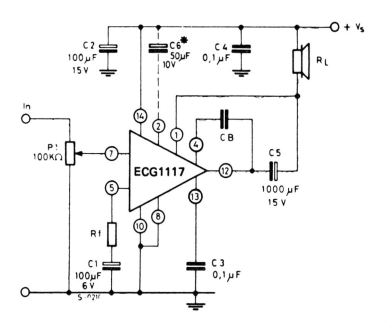

2-watt AF power amplifier with 8-ohm load connected to supply. Rated output of 2 watts is obtained with a supply voltage of 12 volts with an 8-ohm load. At 9 volts with the same 8-ohm load a rated output of 1.2 watts is obtained. At 9 volts with a 4-ohm load power output is rated at 1.6 watts. Capacitor C6 must be used when high ripple rejection is required. The ECG1117 is a 14-pin QIP (courtesy GTE Sylvania Incorporated).

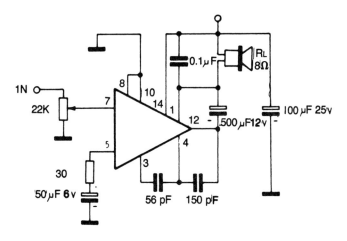

3.3-watt AF power amplifier using an ECG1118 14-pin QIP. This circuit is designed for radio application (courtesy GTE Sylvania Incorporated).

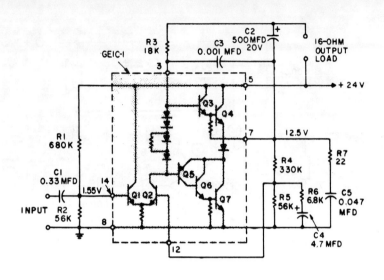

Parts List

IC audio amplifier — General Electric GEIC-1
C1 — 0.33-mfd, 50-volt capacitor
C2 — 500-mfd, 20-volt electrolytic capacitor
C3 — 0.001-mfd, 50-volt capacitor
C4 — 4.7-mfd, 25-volt electrolytic capacitor
C5 — 0.047-mfd, 50-volt capacitor

R1 — 680K-ohm, 1/2-watt resistor
R2 — 56K-ohm, 1/2-watt resistor
R3 — 18K-ohm, 1/2-watt resistor
R4 — 330K-ohm, 1/2-watt resistor
R5 — 56K-ohm, 1/2-watt resistor
R6 — 6.8K-ohm, 1/2-watt resistor
R7 — 22-ohm, 1/2-watt resistor

Sensitivity and Resistance for Input Impedance Matching

Resistor added in series with Capacitor C1	Input Resistance (ohms)	Two-watt Sensitivity
None	40K	120 Millivolts
68K	108K	300 Millivolts
120K	160K	450 Millivolts
330K	370K	1.0 Volts
470K	510K	1.4 Volts
680K	720K	2.0 Volts

2-watt AF power amplifier for a 16-ohm load using a GEIC-1. See table for sensitivity (courtesy General Electric Company).

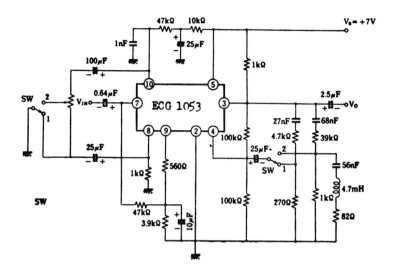

High-gain AF preamplifier using an ECG1053 chip. Frequency response is from 30 hertz to 15 kilohertz. Minimum voltage gain is 93 dB (courtesy GTE Sylvania Incorporated).

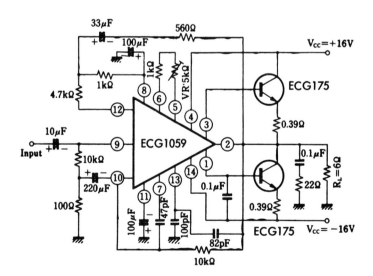

15-watt AF power amplifier using an ECG1059 chip and two ECG175 bipolar transistors. Minimum voltage gain is 63 dB. Maximum input voltage is 1 mV for the given output (courtesy GTE Sylvania Incorporated).

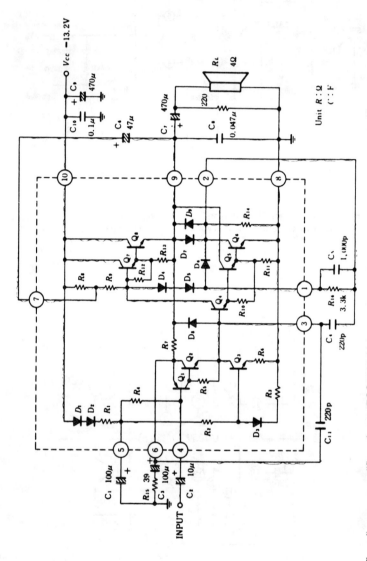

3.5-watt OTL audio power amplifier using an ECG1029. Voltage gain at 1 kHz is 44 dB (courtesy GTE Sylvania Incorporated).

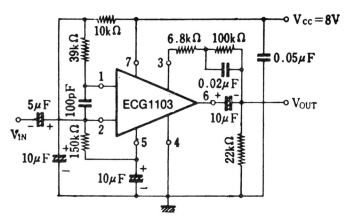

VOLTAGE GAIN 35.5 dB

INPUT RESISTANCE 120 k Ohms

General-purpose voltage preamplifier. Typical voltage gain is 35.5 dB. Input resistance is 120K. The ECG1103 is a 7-pin module (courtesy GTE Sylvania).

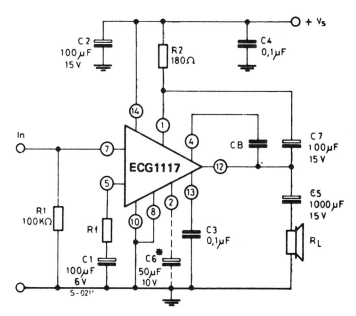

2-watt AF power amplifier with 8-ohm load connected to ground. Rated output of 2 watts is obtained with a power supply voltage of 12 volts. At 9 volts with the same load a rated output of 1.2 watts is obtained. With a 4-ohm load at 9 volts 1.6 watts is obtained. Capacitor C6 must be used when high ripple rejection is required. The ECG1117 is a 14-pin QIP (courtesy GTE Sylvania Incorporated).

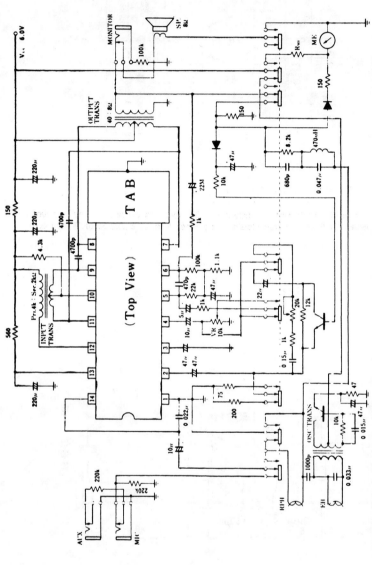

Complete cassette record/playback circuitry for low-cost devices using an ECG1093 14-pin DIP with tab. Typical power output is 1 watt. No signal current drain is 12 mA (courtesy GTE Sylvania Incorporated).

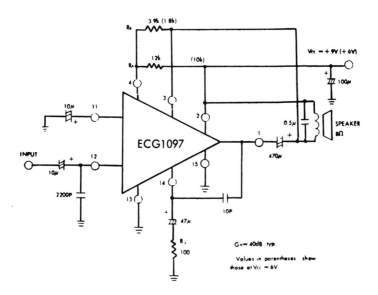

1-watt OTL audio power amplifier. This amplifier can be operated on 6 volts at a reduced audio output level. Values in parentheses are for 6-volt operation (courtesy GTE Sylvania Incorporated).

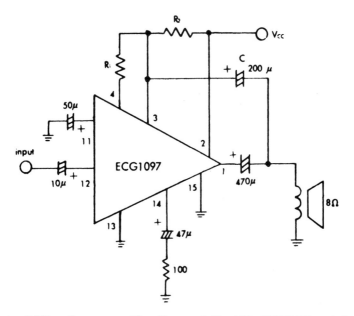

1-watt OTL audio power amplifier with grounded load. The ECG1097 is a 14-pin DIP with tab (courtesy GTE Sylvania Incorporated).

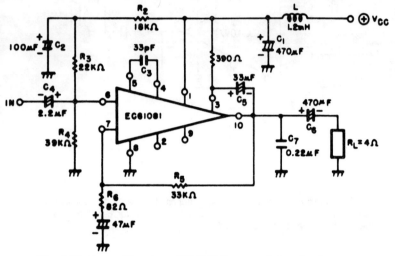

4.5-watt AF power amplifier using an ECG1081. Recommended supply voltage is 13.2 volts, which makes it ideal for automotive applications. Voltage gain at 400 hertz is 50 dB (courtesy GTE Sylvania Incorporated).

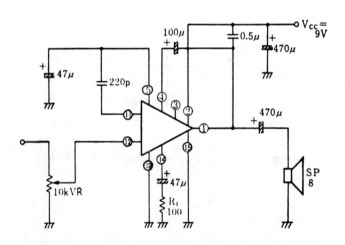

1-watt OTL AF power amplifier using an ECG1126 14-pin DIP with tab. Typical voltage gain is 40 dB. Maximum current drain is 35 mA (courtesy GTE Sylvania Incorporated.

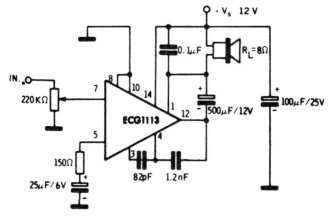

2.1-watt AF power amplifier for phonographs/tape recorders. It is recommended that a 12-volt supply be employed, which makes it ideal for automotive applications. THe ECG1113 is a 14-pin QIP. Current drain at maximum output is 235 mA (courtesy GTE Sylvania Incorporated).

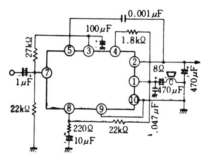

1.3-watt AF power amplifier using an ECG1137 10-pin TO-99. Recommended supply voltage is 10 volts and is to be applied across the 470 µF capacitor. Suggested load for the 1.3-watt output is 8 ohms (courtesy GTE Sylvania Incorporated).

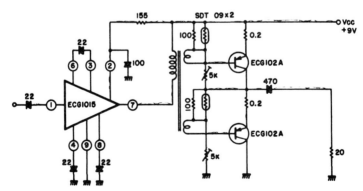

600 mW AF power amplifier for a 20-ohm load. The driver transformer can be purchased locally at Radio Shack (courtesy GTE Sylvania Incorporated).

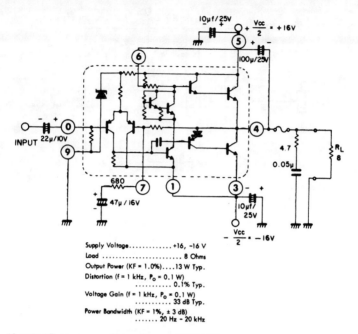

13-watt AF power amplifier using an ECG1139 module and powered by two 16-volt supplies. Use a 2-ampere fuse in the output. Voltage gain is 33 dB (courtesy GTE Sylvania Incorporated).

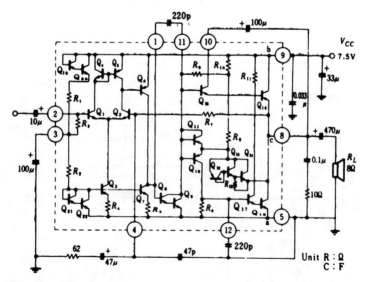

0.7-watt audio power amplifier using an ECG1036. Typical voltage gain at 1 kHz is 50 dB. Input impedance is 20K. Frequency response is from 50 hertz to 100 kilohertz (courtesy GTE Sylvania).

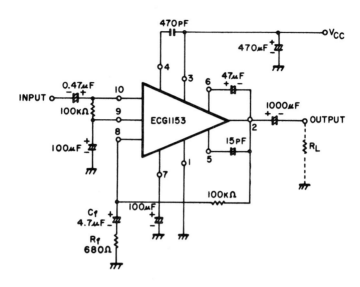

4.2-watt AF power amplifier using an ECG1153 10-pin SIP. Recommended supply voltage is 13.2 volts. The 4.2-watt rating is with a 4-ohm load. Typical voltage gain is 42 dB. Input resistance is 70K (courtesy GTE Sylvania Incorporated).

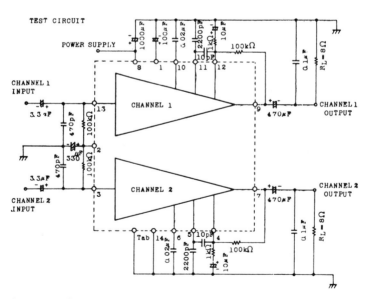

2-watts-per-channel dual AF power amplifier using an ECG1154 14-pin DIP. Recommended supply voltage is 14 volts. Input resistance is 80K. This circuit is intended for low-cost stereos. Frequency response is from 60 hertz to 30 kilohertz (courtesy GTE Sylvania Incorporated).

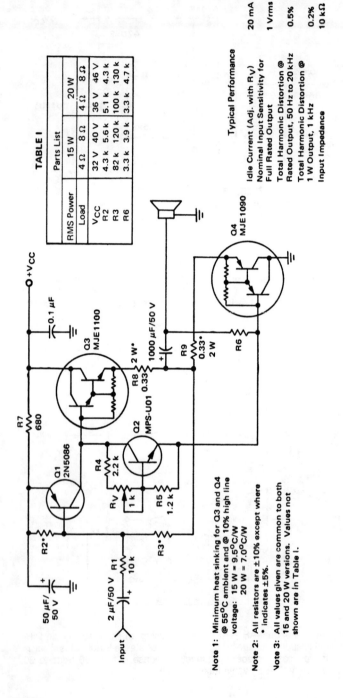

15-watt/20-watt AF amplifier using complementary Darlington output transistors. To ensure maximum signal swing the junction of R8 and R9 should be half-Vcc (courtesy Motorola Semiconductor Products Inc.).

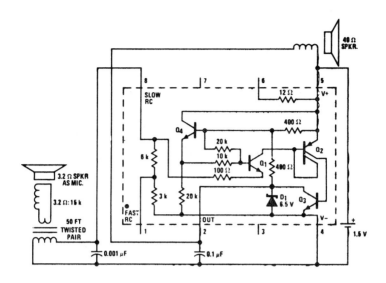

250 mW babysitter intercom using an LM3909 chip. Circuitry inside dashed lines is the LM3909. Operating current is 12 to 15 mA (courtesy National Semiconductor Corporation).

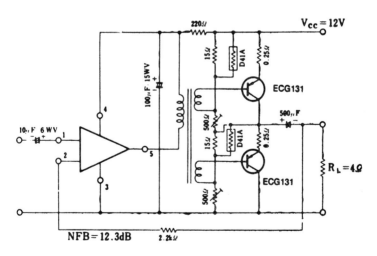

4-watt audio power amplifier using an ECG1102 as an AF driver. The ECG1102 is a 5-pin module. The two bias pots in the output stage should be adjusted to provide an idle current of 15 mA. Total current at maximum output is 450 mA. Voltage gain is 28 dB. Bandwidth is from 20 hertz to 30 kilohertz (courtesy GTE Sylvania Incorporated).

Recommended Transistors

TRANSISTOR		TYPE	POLARITY	$V_{CEO(sus)}$ Min V	h_{FE} Min	@ I_C A	f_T typ MHz	REMARKS
Front End	Q1, Q2	2N5961	npn	60	135	1 m	150	low noise
	Q3, Q4	PN4250A-18	pnp	60	250	100 μ	70	high gain down to μA, gain linearity
Multiplier	Q6	2N5961	npn	See Above				
Pre driver	Q5	2N5400	pnp	120	40	10 m	200	high voltage, rugged
	Q7	2N5830	npn	100	80	10 m	200	
Driver	Q8	FT317	npn	100	35	1	35	high voltage, fast 40 W, TO-220
	Q9	FT417	pnp	100	35	1	25	
Output	Q10	FT324	npn	140	20	5	6	200 W TO-3, rugged SOA; 50 V, 3.0 A
	Q11	FT424	pnp	140	20	5	6	medium speed
Protection Circuit	Q12	2N5831	npn	140	80	10 m	200	high voltage rugged
	Q13	2N5401	pnp	150	60	10 m	150	
	Q14	PN4250A-18	pnp	See Above				
	Q15	2N5961	npn	60	150	10 m	100	high gain

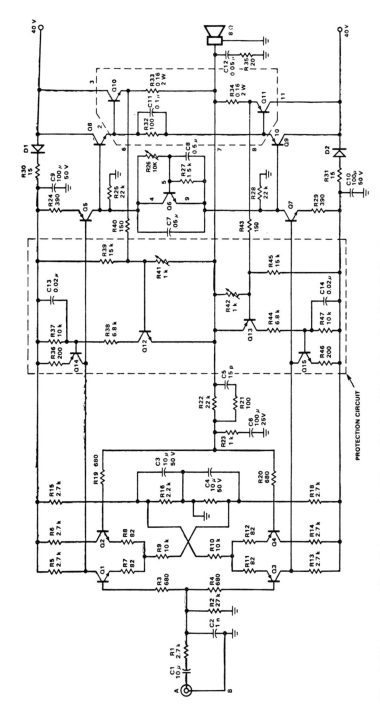

75-watt AF power amplifier with low-transient-intermodulation distortion (courtesy Fairchild Semiconductor).

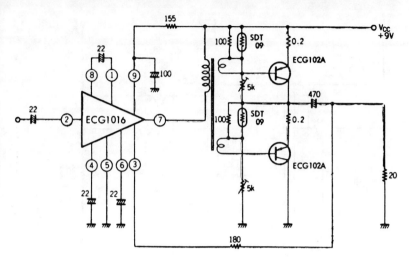

500 mW AF power amplifier for a 20-ohm speaker. Frequency response is from 70 hertz to 20 kilohertz. The ECG1016 is a thin-film hybrid module. The AF driver transistor is standard and can be purchased at Radio Shack (courtesy GTE Sylvania Incorporated).

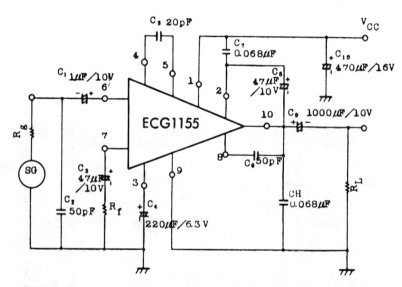

5.8-watt AF power amplifier for automotive applications. This circuit is ideal for car stereo or car radio outputs. In addition it can be employed for a CB modulator circuit. Recommended supply voltage is 13.2 volts. Input resistance is 40K. Voltage gain ranges between 52 and 58 dB (courtesy GTE Sylvania Incorporated).

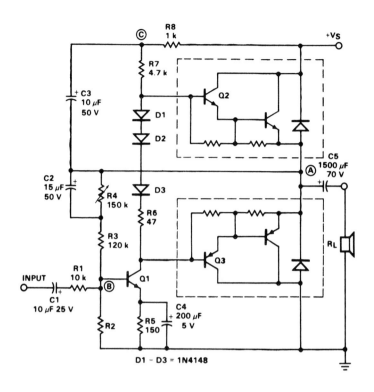

P_{OUT} Nominal	20	40	40	50	60	W
P_{OUT} (0.5% Distortion)	21.4	43	42	52	60	W
Supply Voltage, V_S	40	55	42	60	50	V
Load Resistance, R_L	8	8	4	8	4	Ω
Input Transistor, Q1	2N5961	2N5830	2N5830	2N5830	2N5830	
Output Transistor, Q2	SE9301	SE9304	SE9304	SE9304	SE9304	
Q3	SE9401	SE9404	SE9404	SE9404	SE9404	
Special SOA Selection of Q2 and Q3				45 V, 2.4 A	30 V, 5A	
Q2, Q3 Package	TO-220	TO-3	TO-3	TO-3	TO-3	
Max Permissible Case Temperature of Q2, Q3, $T_{C(max)}$	103	107	103	118	112	°C
Heat Sink for each Q2, Q3	6.5	3.5	2.3	4.0	2.2	max °C/W
Heat Sink for $T_{C(max)}$ 100°C	6.0	3.2	2.1	2.6	1.4	°C/W
Resistor R2	15	12	15	12	12	kΩ
Input Voltage for Maximum Output	1.2	1.6	1.25	1.75	1.5	V_{RMS}

20-/40-/50-/60-watt complementary Darlington AF power amplifier. See table for amplifier component specs (courtesy Fairchild Semiconductor).

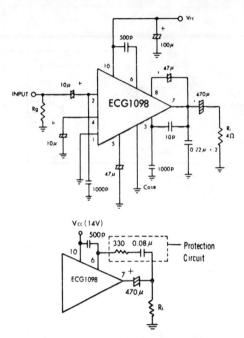

4-watt OTL power amplifier using 10-pin TO3. Typical supply voltage is 14 volts. The circuit is designed for a 4-ohm load and its efficiency is 60%. Typical voltage gain is 39 dB. Input impedance is 6.5K. Bandwidth is 70 hertz to 20 kilohertz. Although the ECG1098 can handle short-circuit conditions for short durations, it is recommended that the protection circuit shown be used if long periods with the speaker load shorted occur (courtesy GTE Sylvania Incorporated).

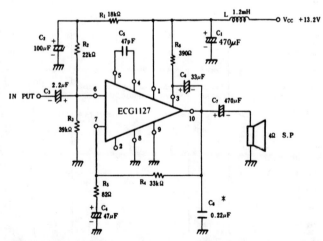

4.5-watt AF power amplifier for a 4-ohm load. Recommended supply voltage is 13.2 volts, which makes this circuit ideal for automotive applications. Typical voltage gain is 50 dB (courtesy GTE Sylvania Incorporated).

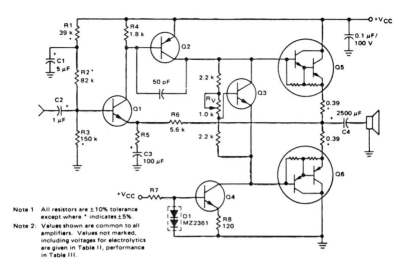

Note 1: All resistors are ±10% tolerance except where * indicates ±5%.

Note 2: Values shown are common to all amplifiers. Values not marked, including voltages for electrolytics are given in Table II, performance in Table III.

Idle Current (Adjusted with R_V)	20 mA
Input Impedance	50 kΩ
Nominal Input Sensitivity for Rated Power Output	1.0 Vrms
Total Harmonic Distortion at 1.0 kHz and any Power up to Full Rated Output (See Figure 3)	0.2%
Intermodulation Distortion 60 Hz with 2 kHz and 7 kHz Mixed 4:1 at 1/2 Maximum Rated Output Power	0.2%
Frequency Response (–1 dB Points)	20 Hz and 50 kHz
Maximum Safe Operating Frequency at Full Rated Power — 20 Watt Amplifier: 60 Watt Amplifier:	50 kHz 30 kHz

Power Watts (RMS)	15		20		25		35		50		60	
Load Impedance	4	8	4	8	4	8	4	8	4	8	4	8
V_{CC}	32 V	38 V	36 V	46 V	38 V	48 V	44 V	56 V	50 V	65 V	56 V	72 V
R5 (ohms)	620	510	560	470	560	390	470	330	390	270	330	220
R7 (ohms)	33 k	39 k	39 k	47 k	39 k	47 k	47 k	56 k	47 k	68 k	56 k	68 k
Q1	MPS A05	MPS A05	MPS A05	MPS A05	MPS A05	MPS A05	MPS A05	MPS A06	MPS A05	MPS A06	MPS A06	MPS A06
Q2	MPS A55	MPS A55	MPS A55	MPS A55	MPS A55	MPS A55	MPS A55	MPS A56	MPS A55	MPS A56	MPS A56	MPS A56
Q3	MPS A13	MPS A13	MPS A13	MPS A13	MPS A13	MPS A13	MPS A13	MPS A13	MPS A13	MPS A13	MPS A13	MPS A13
Q4	MPS A05	MPS A05	MPS A05	MPS A05	MPS A05	MPS A05	MPS A05	MPS A05	MPS A05	MPS A06	MPS A05	MPS A06
Q5	MJE 1100	MJE 1100	MJE 1100	MJE 1100	MJE 1102	MJE 1100	MJE 6043	MJE 6044	MJE 6043	MJE 6044	MJE 6044	MJE 6044
Q6	MJE 1090	MJE 1090	MJE 1090	MJE 1090	MJE 1092	MJE 1090	MJE 6040	MJE 6041	MJE 6040	MJE 6041	MJE 6041	MJE 6041
Voltage rating on C1	35 V	40 V	40 V	50 V	40 V	50 V	45 V	60 V	50 V	65 V	60 V	75 V
Voltage rating on C2, C3	20 V	25 V	25 V	30 V	25 V	30 V	25 V	35 V	30 V	35 V	35 V	40 V
Voltage rating on C4	40 V	45 V	45 V	55 V	45 V	55 V	50 V	65 V	60 V	75 V	65 V	80 V
Min. heat sink for outputs @ 55°C ambient temperature and 10% high line voltage	9.5°C/W		7.0°C/W		5.0°C/W		6.0°C/W	5.5°C/W	4.0°C/W		3.0°C/W	

15-/20-/25-/35-/50-/60-watt AF power amplifier with AC-coupled output (courtesy Motorola Semiconductor Products Inc.).

Power Watts (RMS)	15		20		25		35		50		60	
Load Impedance (ohms)	4	8	4	8	4	8	4	8	4	8	4	8
V$_{CC}$	±16 V	±19 V	±18 V	±23 V	±19 V	±24 V	±22 V	±28 V	±25 V	±33 V	±28 V	±36 V
R4 (ohms)	1.5 k	2.2 k	2.0 k	3.3 k	2.2 k	3.3 k	3.0 k	3.9 k	3.6 k	5.6 k	3.9 k	6.2 k
R5 (ohms)	1.2 k	820	1.0 k	750	1.0 k	630	820	560	680	470	620	430
R7 (ohms)	15 k	18 k	18 k	22 k	18 k	22 k	22 k	27 k	22 k	33 k	27 k	33 k
Q1,Q2 Dual Transistors	MD 8001	MD 8001	MD 8001	MD 8001	MD 8001	MD 8001	MD 8001	MD 8001	MD 8001	MD 8002	MD 8001	MD 8002
Q3	MPS A55	MPS A55	MPS A55	MPS A55	MPS A55	MPS A55	MPS A56	MPS A56	MPS A55	MPS A56	MPS A56	MPS A56
Q4	MPS A13	MPS A13	MPS A13	MPS A13	MPS A13	MPS A13	MPS A13	MPS A13	MPS A13	MPS A13	MPS A13	MPS A13
Q5	MPS A05	MPS A05	MPS A05	MPS A05	MPS A05	MPS A05	MPS A05	MPS A06	MPS A05	MPS A06	MPS A06	MPS A06
Q6	MJE 1100	MJE 1100	MJE 1100	MJE 1100	MJE 1102	MJE 1100	MJE 6043	MJE 6044	MJE 6043	MJE 6044	MJE 6044	MJE 6044
Q7	MJE 1090	MJE 1090	MJE 1090	MJE 1090	MJE 1092	MJE 1090	MJE 6040	MJE 6041	MJE 6040	MJE 6041	MJE 6041	MJE 6041
Min. heat sink for outputs @ 55°C ambient temperature and 10% high line voltage	9.5°C/W		7.0°C/W		5.0°C/W		6.0°C/W	5.5°C/W	4.0°C/W		3.0°C/W	

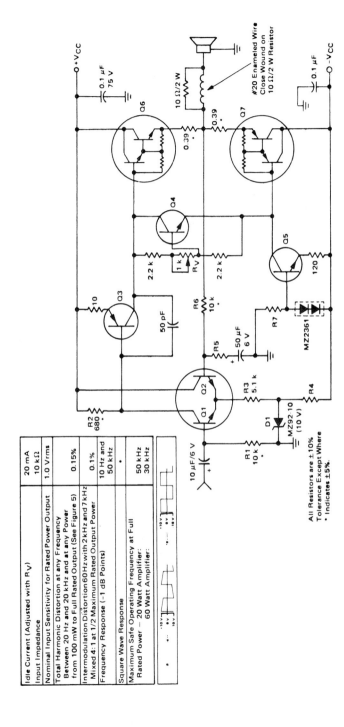

15-/20-/25-/35-/50-/60-watt AF power amplifier with DC-coupled output (courtesy Motorola Semiconductor Products Inc.).

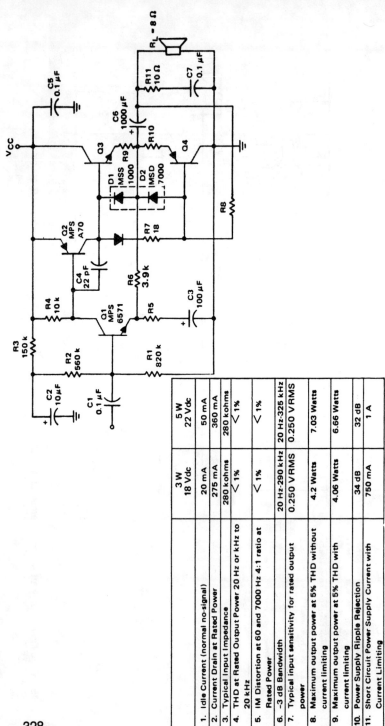

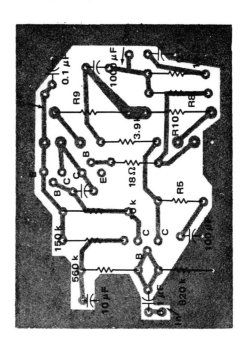

	3 W	5 W
V_{CC}	18 V	22 V
R5	180	150
R8	470	390
R9 & R10	0.82	0.56
**Q3	MPSU01 or MJE200	MPSU01 or MJE200
**Q4	MPSU51 or MJE210	MPSU51 or MJE210
Heatsink	with MPSU01/51 27.5°C/W	with MPSU01/51 16.8°C/W
	with MJE200/210 36°C/W	with MJE200/210 19.7°C/W

*Heatsink size calculation is based on a maximum ambient temperature of 50°C and a load phase angle of 60 degrees (see text for method of calculation). Heatsink is for both devices on one sink.

**Parts in same block are interchangeable.
P.C. board is for MPSU01/51, but can be changed to MJE200/210

3-watt/5-watt AF power amplifier with PNP driver (courtesy Motorola Semiconductor Products Inc.).

Component	3 W	5 W
V_{CC}	17	22
R5	120 Ω	100 Ω
R10 & R11	0.82	0.56
Q3*	MPSU01 / MJE200	MPSU01 / MJE200
Q4*	MPSU51 / MJE210	MPSU51 / MJE210
**Heatsink	MPSU01/51 27.5°C/W	MPSU01/51 16.8°C/W
**Heatsink	MJE200/210 36°C/W	MJE200/210 19.7°C/W

*Parts in same block are interchangeable. P.C. board is for MJE200/210, but can be easily changed to MPSU01/51.

**Heatsink size calculation is based on a maximum ambient temperature of 50°C and a load phase angle of 60 degrees (see text for method of calculation) Heat sink is for both devices on one sink.

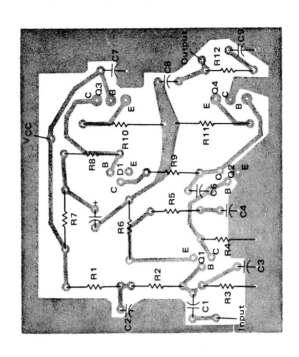

	3 W 18 Vdc	5 W 22 Vdc
1. Idle Current (nominal no-signal)	20 mA	54 mA
2. Current Drain at Rated Power	285 mA	365 mA
3. Typical Input Impedance	300 kohms	320 kohms
4. THD at Rated Output Power 20 Hz or 1 kHz 20 kHz	<1%	<1%
5. IM Distortion at 60 and 7000 Hz 4:1 ratio at Rated Power	<1%	<1%
6. −3 dB Bandwidth	20 Hz–220 kHz	20 Hz–150 kHz
7. Typical input sensitivity for rated output power	0.22 VRMS	0.23 VRMS
8. Maximum output power at 5% THD without current limiting	4.10 Watts	6.8 Watts
9. Maximum output power at 5% THD with current limiting	4.06 Watts	6.65 Watts
10. Power Supply Ripple Rejection	24 dB	36.4 dB
11. Short Circuit Power Supply Current with Current Limiting	800 mA	1 Amp

3-watt/5-watt AF power amplifier with NPN driver (courtesy Motorola Semiconductor Products Inc.).

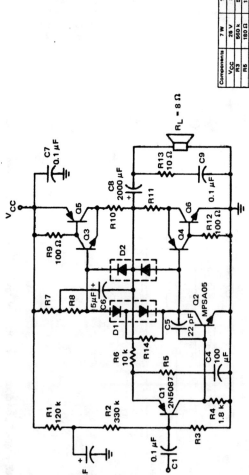

Components	7 W	10 W	15 W	20 W	25 W	35 W
V_{CC}	26 V	30 V	36 V	43 V	48 V	54 V
R3	560 k	560 k	620 k	620 k	680 k	680 k
R5	180 Ω	180 Ω	120 Ω	120 Ω	120 Ω	120 Ω
R7 & R8	3.9 k	3.9 k	3.9 k	3.9 k	3.9 k	5.6 k
R10 & R11	0.47	0.47	0.47	0.47	0.47	0.33
R14	—	—	—	—	—	470 Ω
Q3	MPSA06	MPSA05	MPSU06 MJE221	MPSU06 MJE224	MPSU06 MJE224	MPSU06 MJE224
Q4	MPSA55	MPSA55	MPSU55 MJE231	MPSU55 MJE234	MPSU55 MJE234	MPSU55 MJE234
Q5	MJE230	MJE230	MJE105	MJE105	MJE2901	MJE2901
Q6	MJE220	MJE220	MJE205	MJE205	MJE2801	MJE2801
D1 & D2	MSD7000	MSD7000	MSD7000	MSD7000	MSD7000	MSD7000
*Heatsink	18.8°C/W	9.8°C/W	6.2°C/W	4.7°C/W	3.4°C/W	2°C/W

*Heatsink size calculation is based on a maximum ambient temperature of 50°C and a load phase angle of 60 degrees (see text for methods of calculation). The heat sink is for both devices on one sink.

	7 W 26 Vdc	10 W 30 Vdc	15 W 36 Vdc	20 W 43 Vdc	25 W 48 Vdc	35 W 54 Vdc
1. Idle Current (nominal no-signal)	1.6 mA	20 mA	28 mA	65 mA	68 mA	32 mA
2. Current Drain at Rated Power	425 mA	510 mA	610 mA	720 mA	320 mA	940 mA
3. Typical Input Impedance	230 kohms	320 kohms	230 kohms	230 kohms	230 kohms	220 kohms
4. THD at Rated Output Power 20 kHz or 1 kHz to 20 kHz	<0.5%	<0.5%	<0.5%	<0.5%	<0.5%	<0.5%
5. IM Distortion at 60 and 7000 Hz 4:1 ratio at Rated Power	<1%	<1%	<1%	<1%	<1%	<1%
6. -3 dB Bandwidth	16 Hz-300 kHz	15 Hz-380 kHz	16 Hz-250 kHz	16 Hz-250 kHz	16 Hz-275 kHz	16 Hz-230 kHz
7. Typical input sensitivity for rated output power	210 mV	260 mV	200 mV	220 mV	250 mV	280 mV
8. Maximum output power at 5% THD without current limiting	8.8 Watts	12.5 Watts	18 Watts	27.38 Watts	33.6 Watts	45.4 Watts
9. Maximum output power at 5% THD with current limiting	8.85 Watts	12.5 Watts	18 Watts	26.3 Watts	30 Watts	45 Watts
10. Power Supply Ripple Rejection	44.4 dB	42 dB	40 dB	30 dB	36 dB	34 dB
11. Short Circuit Power Supply Current with Current Limiting	1.32 Amps	1.45 Amps	1.5 Amps	1.5 Amps	1.62 Amps	2.2 Amps

7-/10-/15-/20-/25-/35-watt AF power amplifier with NPN driver (courtesy Motorola Semiconductor Products Inc.).

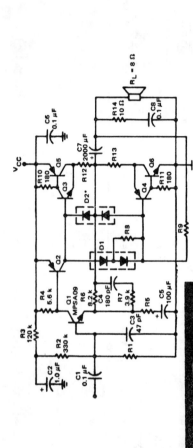

Component	7 W	10 W	15 W	20 W	25 W	35 W
V_{CC}	25 V	30 V	38 V	46 V	48 V	54 V
R1	560 k	560 k	560 k	620 k	620 k	620 k
R2	100 Ω	82 Ω	100 Ω	100 Ω	150 Ω	180 Ω
R12 & R13	Value Selected to Provide 30 mA Collector Current in Q5.					
R12 & R13	4.7 k	4.7 k	8.2 k	8.2 k	8.2 k	8.2 k
R9	0.47	0.47	0.47	0.47	0.33	0.33
Q1	MPSA06	2N6067	2N6067	2N6067	MPSA56	MPSA56
Q3	MPSA06	MPSU05	MPSU05	MPSU05	MPSU05	MPSU05
Q4	MPSA56	MPSA56	MJE220	MJE224	MJE224	MJE224
			MPSU55	MPSU55	MPSU55	MPSU55
	MJE230	MJE230	MJE230	MJE234	MJE234	MJE234
Q5	MJE220	MJE220	MJE106	MJE106	MJE106	MJE106
Q6	MJE230	MJE230	MJE206	MJE206	MJE206	MJE206
D1 & D2	MSD7000	MSD7000	MSD7000	MSD7000	MSD7000	MSD7000
			MJE2901	MJE2901	MJE2901	MJE2901
			MJE2801	MJE2801	MJE2801	MJE2801
Heatsink	18.3°C/W	9.8°C/W	6.2°C/W	4.7°C/W	3.4°C/W	2°C/W

*Heatsink size calculation is based on a maximum ambient temperature of 50°C and a load phase angle of 60 degrees (see text for method of calculation). The heatsink is for both devices on one sink.

	7 W 25 Vdc	10 W 30 Vdc	15 W 28 Vdc	20 W 46 Vdc	25 W 48 Vdc	35 W 54 Vdc
1. Idle Current (nominal no-signal)	28 mA	40 mA	20 mA	58 mA	20 mA	37 mA
2. Current Drain at Rated Power	440 mA	500 mA	670 mA	720 mA	820 mA	940 mA
3. Typical Input Impedance	230 kohms	230 kohms	210 kohms	220 kohms	220 kohms	210 kohms
4. THD at Rated Output Power 20 Hz or 1 kHz to 20 kHz	<0.5%	<0.5%	<0.5%	<0.5%	<0.5%	<0.5%
5. IM Distortion at 60 and 7000 Hz 4:1 ratio at Rated Power	<1%	<1%	<1%	<1%	<1%	<1%
6. -3 dB Bandwidth	12 Hz-250 kHz	18 Hz-110 kHz	17 Hz-55 kHz	18 Hz-150 kHz	10 Hz-160 kHz	13 Hz-60 kHz
7. Typical input sensitivity for rated output power	220 mV	270 mV	110 mV	120 mV	230 mV	270 mV
8. Maximum output power at 5% THD without current limiting	8.8 Watts	12.5 Watts	21.7 Watts	27 Watts	34 Watts	43 Watts
9. Maximum output power at 5% THD with current limiting	8.75 Watts	11.5 Watts	21.1 Watts	24.5 Watts	34 Watts	43 Watts
10. Power Supply Ripple Rejection	37 dB	26 dB	41.94 dB	33 dB	36.5 dB	38 dB
11. Short Circuit Power Supply Current with Current Limiting	1.2 Amps	1.3 Amps	1.32 Amps	1.4 Amps	2 Amps	2 Amps

7-/10-/15-/20-/25-/35-watt AF power amplifier with PNP driver (courtesy Motorola Semiconductor Products Inc.).

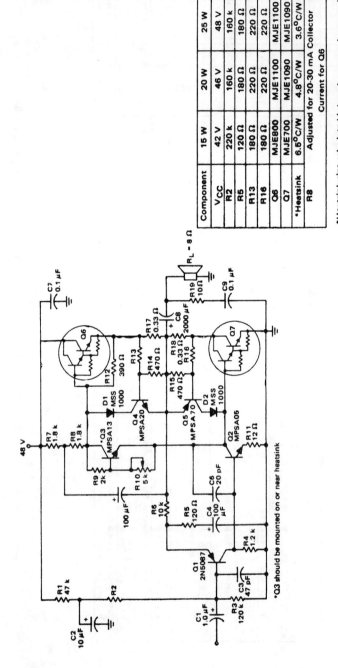

Component	15 W	20 W	25 W
V_{CC}	42 V	46 V	48 V
R2	220 k	160 k	160 k
R5	120 Ω	180 Ω	180 Ω
R13	180 Ω	220 Ω	220 Ω
R16	180 Ω	220 Ω	220 Ω
Q6	MJE800	MJE1100	MJE1100
Q7	MJE700	MJE1090	MJE1090
*Heatsink	6.5°C/W	4.8°C/W	3.6°C/W
R8	Adjusted for 20-30 mA Collector Current for Q6		

*Heatsink size calculated is based on a maximum ambient temperature of 50°C and a load phase angle of 60 degrees (see text for method of calculation). The heatsinks for both devices on one sink.

*Q3 should be mounted on or near heatsink

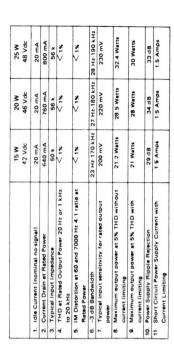

	15 W 42 Vdc	20 W 46 Vdc	25 W 48 Vdc
1. Idle Current (nominal no-signal)	20 mA	20 mA	20 mA
2. Current Drain at Rated Power	640 mA	760 mA	800 mA
3. Typical Input Impedance	50 k	56 k	56 k
4. THD at Rated Output Power 20 Hz or 1 kHz to 20 kHz	<1%	<1%	<1%
5. IM Distortion at 60 and 7000 Hz 4:1 ratio at Rated Power	<1%	<1%	<1%
6. −3 dB Bandwidth	23 Hz-170 kHz	27 Hz-180 kHz	28 Hz-190 kHz
7. Typical input sensitivity for rated output power	200 mV	220 mV	230 mV
8. Maximum output power at 5% THD without current limiting	21.2 Watts	28.5 Watts	32.4 Watts
9. Maximum output power at 5% THD with current limiting	21 Watts	28 Watts	30 Watts
10. Power Supply Ripple Rejection	29 dB	34 dB	33 dB
11. Short Circuit Power Supply Current with Current Limiting	1.5 Amps	1.5 Amps	1.5 Amps

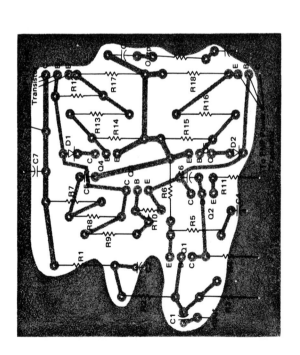

15-/20-/25-watt AF power amplifier with Darlington outputs (courtesy Motorola Semiconductor Products Inc.).

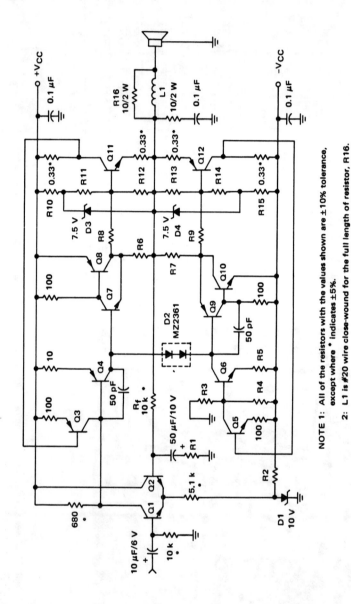

NOTE 1: All of the resistors with the values shown are ±10% tolerance, except where * indicates ±5%.

2: L1 is #20 wire close-wound for the full length of resistor, R16.

Output Power (Watts-rms)	Load Impedance (Ohms)	Output Transistors NPN (Q10)	Output Transistors PNP (Q8)	Driver Transistors NPN (Q7)	Driver Transistors PNP (Q9)	Pre-Driver Transistors NPN (Q6)	Pre-Driver Transistors PNP (Q4)	Differential Amplifier Transistors (Q1 & Q2) Single Channel	Differential Amplifier Transistors (Q1 & Q2) Dual Channel
35	4	MJ2840	MJ2940	MPSU05	MPSU55	MPSA05	MPSA55	MD8001	MFC8000
35	8	MJE2801	MJE2901	MPSU05	MPSU55	MPSA06	MPSA56	MD8001	MFC8000
50	4	2N5302	2N4399	MPSU05	MPSU55	MPSA06	MPSA56	MD8001	MFC8000
50	8	MJ2841	MJ2941	MPSU06	MPSU56	MPSA06	MPSA56	MD8002	MFC8001
60	4	2N5302	2N4399	MPSU06	MPSU56	MPSA06	MPSA56	MD8001	MFC8000
60	8	MJ2841	MJ2941	MPSU06	MPSU56	MPSA06	MPSA56	MD8002	MFC8001
75	4	MJ802	MJ4502	MPSU06	MPSU56	MPSA06	MPSA56	MD8001	MFC8000
75	8	MJ802	MJ4502	MM3007	2N5679	MM3007	MM4007	MD8003	MFC8002
100	4	MJ802	MJ4502	MPSU06	MPSU56	MPSU06	MPSU56	MD8002	MFC8001
100	8	MJ802	MJ4502	MM3007	2N5679	MM3007	MM4007	MD8003	MFC8002

The following semiconductors are used at all of the power levels:

Q11 — MPSL01
Q5 — MPSA20
Q12 — MPSL51
Q3 — MPSA70

D1 — MZ500-16 or MZ292-20 (See Note 1)
D2 — MZ2361
D3 & D4 — 1N5236B or MZ92-16A (See Note 1)

NOTE 1: For a low-cost zener diode, an emitter-base junction of a silicon transistor can be substituted. A transistor similar to the MPS6512 can be used for the 7.5 V zener. A transistor similar to the Motorola BC317 can be used for the 10 V zener diode.

35-/50-/60-/75-/100-watt AF power amplifier.

Output Power (Watts-rms)	Load Impedance (Ohms)	R1 ±5%	R2 ±10%	R3 ±5%	R4 ±5%	R5 ±5%	R6, R7 ±5%	R8, R9 ±10%	R10, R15 ±5%	R11, R14 ±5%	R12, R13 ±5%	V_{CC}
35	4	820	2.7 k	18 k	1.2 k	120	0.39	390	2.7 k	1.5 k	470	±21 V
35	8	560	3.9 k	22 k	1.2 k	180	0.47*	240	3.0 k	1.2 k	470	±27 V
50	4	680	3.3 k	22 k	1.2 k	100	0.33	360	3.3 k	1.5 k	470	±25 V
50	8	470	4.7 k	27 k	1.2 k	150	0.43*	270	3.9 k	1.2 k	470	±32 V
60	4	620	3.9 k	22 k	1.2 k	120	0.33	430	3.9 k	1.5 k	470	±27 V
60	8	430	5.6 k	33 k	1.2 k	120	0.39	300	4.7 k	1.2 k	470	±36 V
75	4	560	4.7 k	27 k	1.2 k	91	0.33	620	5.6 k	1.8 k	470	±30 V
75	8	390	6.8 k	33 k	1.2 k	150	0.39	390	6.8 k	1.5 k	470	±40 V
100	4	470	5.6 k	33 k	1.2 k	68	0.39	1.0 k	8.2 k	2.2 k	470	±34 V
100	8	330	8.2 k	39 k	1.2 k	100	0.39	510	9.1 k	1.8 k	470	±45 V

NOTE: All of the above resistor values are in ohms and are 1/2 W except for R6 and R7.
*R6 and R7 are 5 W resistors except where * indicates 2 W.

Minimum Heat Sinking Required for Safe Operation Under Shunted Load at 50°C Ambient Temperature

Output Power (Watts-rms)	Load Impedance (Ohms)	Output Transistor Heat Sink (θ_{CA}) (See Note 1)	Driver Transistor Heat Sink (θ_{CA}) (See Note 2)
35	4	4.2°C/W	None
	8	2.4°C/W	None
50	4	3.0°C/W	60°C/W
	8	2.4°C/W	60°C/W
60	4	2.5°C/W	60°C/W
	8	2.0°C/W	60°C/W
75	4	1.6°C/W	35°C/W
	8	1.6°C/W	70°C/W*
100	4	1.0°C/W	20°C/W
	8	1.0°C/W	50°C/W*

NOTE 1: All of the output transistors are in TO-3 packages with the exception of the MJE2801/2901 (35 W/8Ω), which are in the Case 90 Thermopad† plastic package.

2: All of the driver transistors are in the plastic Uniwatt† package with the exception of those marked *, which are metal cased TO-5.

†Trademark of Motorola Inc.

35-/50-/60-/75-/100-watt AF power amplifier (courtesy Motorola Semiconductor Products Inc).

Waveform Generators

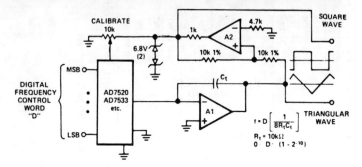

Programmable triangle/square-wave function generator. A1 and A2 are AD301 op amps. The peak-to-peak amplitude of both waveforms is approximately 15 volts (courtesy Analog Devices, Inc.).

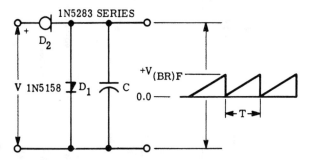

Sawtooth generator with positive-going ramp using a zener and PNPN diode (courtesy Motorola Semiconductor Products Inc.).

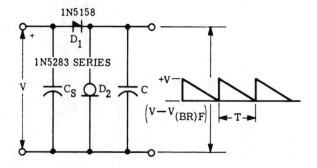

Sawtooth generator with negative-going ramp using a zener and a PNPN diode (courtesy Motorola Semiconductor Products Inc.).

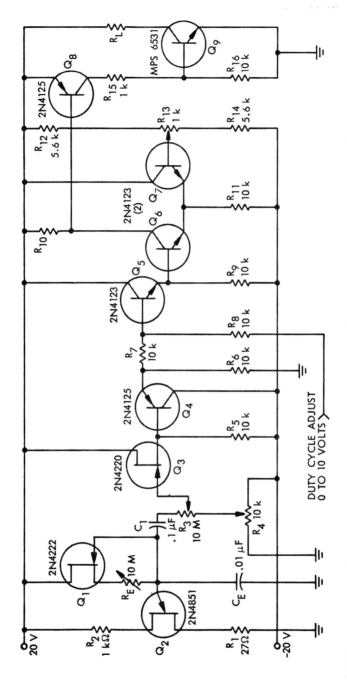

Space mark generator with voltage-controlled duty cycle. Frequency is set by R_E in the source circuit of Q_1 (courtesy Motorola Semiconductor Products Inc.).

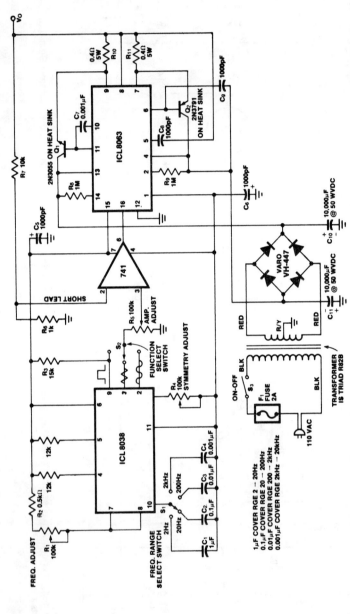

Power function generator developing sine waves, triangular waves and square waves. Operating frequency is 2 hertz to 20 kilohertz. Vo is up to ±25 volts (50 volts peak to peak) across loads as small as 10 ohms, which produces a maximum of 2.5 amperes of output current. All capacitor working voltage ratings should be 50 volts DC. All resistors should be the half-watt type unless otherwise specified. Keep the lead at pin 2 of the op amp less than 2 inches or oscillations will result. Full power output is only obtainable up to about 5 kilohertz due to the op amp (courtesy Intersil, Inc.).

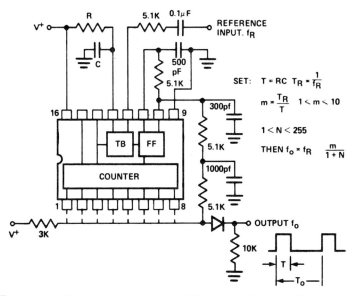

Frequency synthesizer using an external time base reference. The counter shown is an Intersil 8240 programmable timer/counter. The harmonic synchronization property of the 8240 time base can be used to generate a wide number of frequencies from a given input reference frequency (courtesy Intersil, Inc.).

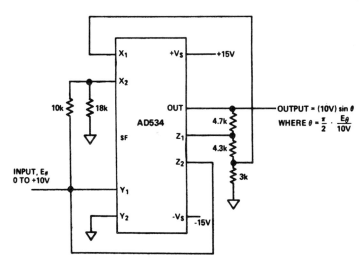

USING CLOSE TOLERANCE RESISTORS AND AD534L, ACCURACY OF FIT IS WITHIN ±0.5% AT ALL POINTS. θ IS IN RADIANS.

Sine function generation using an AD534 multiplier/divider chip (courtesy Analog Devices, Inc.).

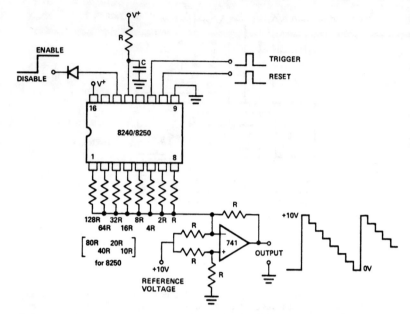

Staircase generator. The resistor array is switched to ground to generate binary or BCD weighted currents. The op amp converts these currents to an output voltage. Under reset condition the switches are off and the output is at ground. When a trigger is applied the output goes to V_{REF} and generates a negative-going staircase of 256 levels for the 8240 or 100 level for the 8250. The time duration of each step is equal to the time base period, $T = RC$. The amplitude of the staircase can be varied by changing the input reference voltage. The staircase can be stopped at any desired level by applying a disable signal to pin 14 (courtesy Intersil, Inc.).

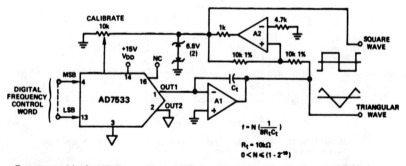

Programmable function generator with square and triangular wave outputs (courtesy Analog Devices, Inc.).

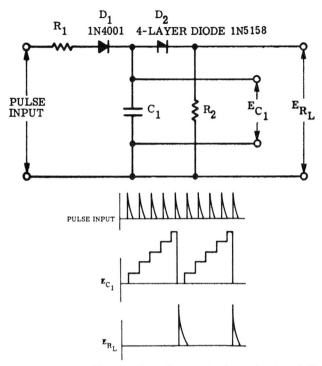

Staircase generator. The number of steps is determined as follows: $h = (1_1 R_L V_{BR})/[(V_P - V_{BR})t_p]$, where n is the number of steps, V_{BR} is the breakdown voltage of the PNPN diode, V_P is the peak pulse with and t_p is the pulse width (courtesy Motorola Semiconductor Products Inc.).

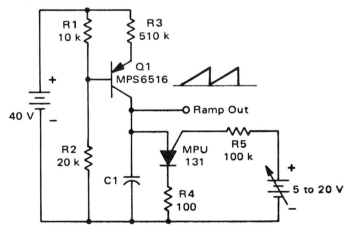

Voltage-controlled ramp generator (VCRG) using a PUT. Setting the value of C to 0.0047 μF provides a change in frequency of 3.4 ms; setting it to 0.01 μF provides a change of 5.4 ms (courtesy Motorola Semiconductor Products Inc.).

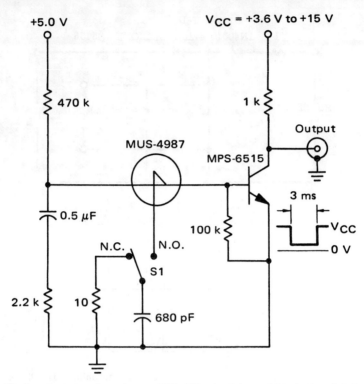

Single pulse generator using an SUS. This circuit is useful in testing digital equipment. The output is a clean debounced pulse of 3 ms duration (courtesy Motorola Semiconductor Products Inc.).

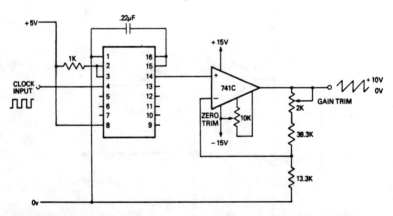

Precision ramp generator using a Datel ADC-MC8B 16-pin DIP. Pin functions of the ADC-MC8B are as follows: pin 1, ground; pin 2, logic select; pin 3, reset; pin 4, strobe; pin 5, bit 8(LSB); pin 6, bit 7; pin 7, bit 6; pin 8, +Vcc; pin 9, bit 5; pin 10, bit 4; pin 11, bit 3; pin 12, bit 2; pin 13, bit 1 (MSB); pin 14, analog output; pin 15, V_{REF} input; pin 16, V_{REF} output (courtesy Datel Systems, Inc.).

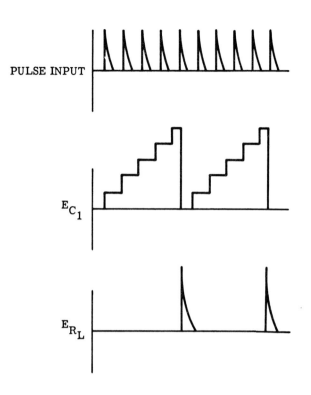
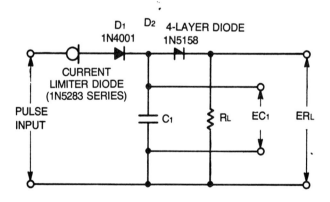

Staircase generator. The number of steps is determined as follows: $n = (C_1 V_{BR})/(I_P t_p)$, where n is the number of steps, V_{BR} is the breakdown voltage of the PNPN diode, I_P is the pinchoff voltage of the 1N5283 and t_p is the pulse width (courtesy Motorola Semiconductor Products Inc.).

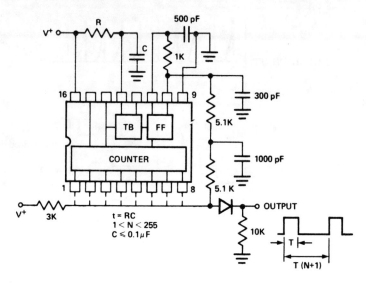

Frequency synthesizer using internal time base. The counter shown is an Intersil 8240 programmable timer/counter. The output of the circuit is a positive pulse train with a pulse width equal to T and a period equal to (N+1)T, where N is the programmable count between 1 and 255. The modulus N is the total count corresponding to the counter outputs connected to the output bus. For example if pins 1, 3 and 6 are connected to the output bus the total count is $N = 1+4+32=37$ and the period of the output waveform is equal to (N+1) T or 38T. And T=RC (courtesy Intersil, Inc.).

Oscillators

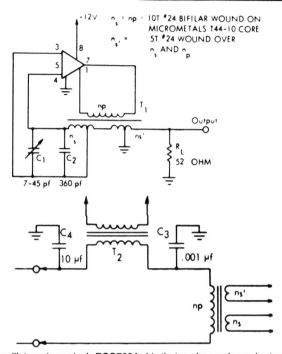

10 MHz oscillator using a single ECG703A chip that can be used as a short range transmitter. It is possible to modulate the oscillator by inserting the audio modulator shown between pin 1 and the tank circuit (courtesy GTE Sylvania Incorporated).

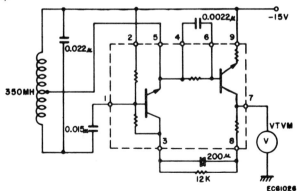

Electronic organ master oscillator using an ECG1026 thin-film hybrid module. Output is taken at pin 7, where the VTVM is shown connected to test output voltage level (courtesy GTE Sylvania Incorporated).

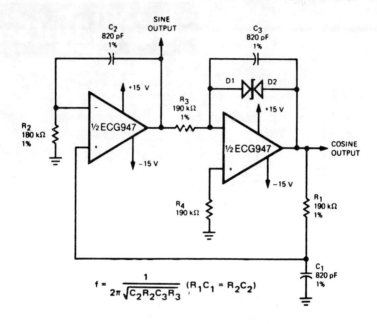

Quadrature oscillator using an ECG947 dual operational amplifier IC. The ECG947 is short-circuit proof and requires no external components for frequency compensation (courtesy GTE Sylvania Incorporated).

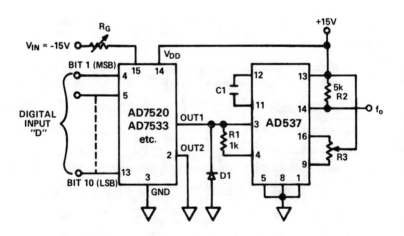

Digitally programmed oscillator. The circuit uses an AD537 monolithic V/F converter and an AD7520 multiplying D/A converter. It is a programmable square wave frequency source with excellent linearity in the range of 0 to 100 kHz (courtesy Analog Devices, Inc.).

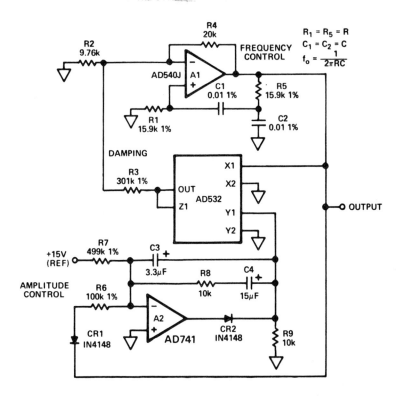

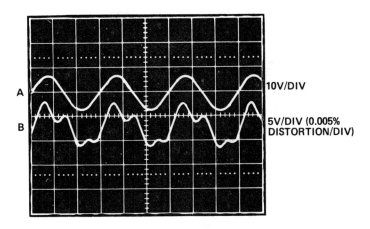

Low-distortion oscillator. A1 is connected as a noninverting amplifier and has a gain of three. As shown the oscillator output is 1 kHz, determined by R1, R5, C1 and C2. See equation in diagram to determine frequency (courtesy Analog Devices, Inc.).

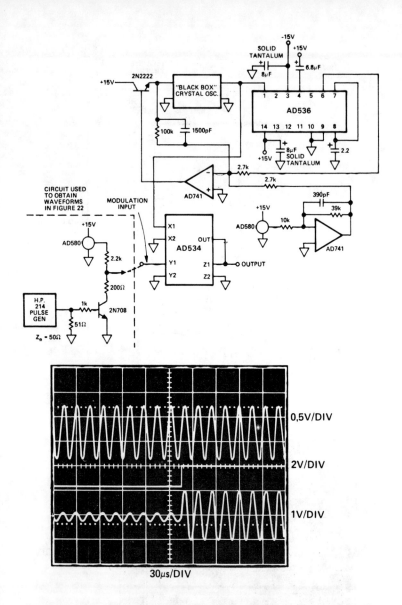

Crystal oscillator with amplitude-modulated output. The black box crystal oscillator output is converted to DC by the AD536 RMS/DC converter. The AD536 output is summed in the AD741 with the DC reference voltage obtained by inverting and amplifying the output of the AD580 band-gap reference. The AD741 drives the 2N2222 control transistor to close the feedback loop around the oscillator by adjusting its supply voltage. The AD534 serves as an amplitude modulator. The oscilloscope waveforms show the 32.768 kHz signal being switched by a fast step from 0.3 to 2.2 volts peak to peak with no ringing, overshoot, etc. (courtesy Analog Devices, Inc.).

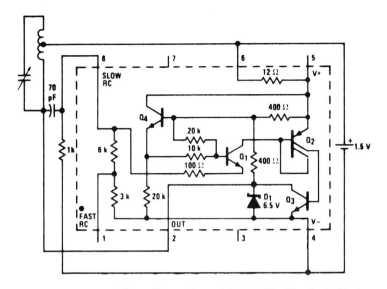

800 kHz experimental RF oscillator using an LM3909 chip. Circuitry inside the dashed lines is the LM3909. The external coil is a standard broadcast loopstick. The variable capacitor is 360 pF (courtesy National Semiconductor Corporation).

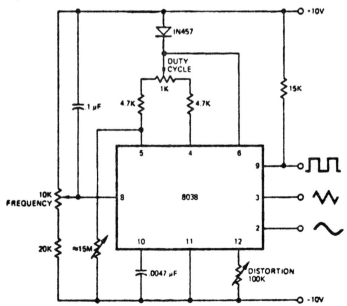

Variable audio oscillator with a range of 20 hertz to 20 kilohertz. The chip used is an Intersil 8038 waveform generator, which is a 14-pin DIP (courtesy Intersil, Inc.).

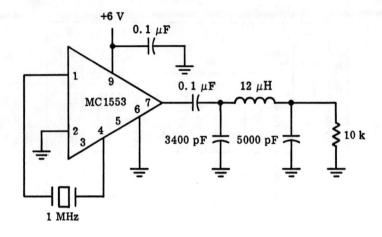

1 MHz oscillator using the MC1553 video amplifier (courtesy Motorola Semiconductor Products Inc.).

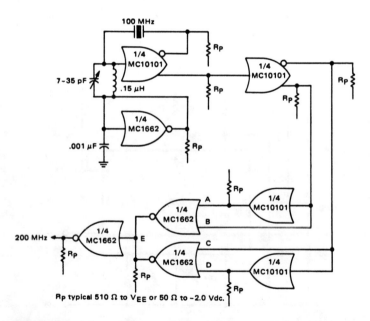

200 MHz crystal oscillator. This circuit incorporates a 100 MHz oscillator and a frequency doubler. V$_{EE}$ is −5.2 volts (courtesy Motorola Semiconductor Products Inc.).

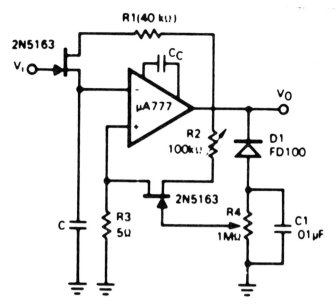

Voltage-controlled sine-wave oscillator for up to 50 kHz. For frequencies of 10 kHz to 50 kHz capacitor Cc should be 3 pF. For frequencies less than 10 kHz capacitor Cc should be 30 pF. Select capacitor C according to desired frequency, or make it variable from zero to 5 µF (courtesy Fairchild Semiconductor).

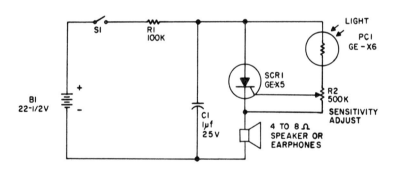

Parts List

C1 — 1-mfd, 25-volt (minimum) capacitor
PC1 — GE-X6 cadmium sulphide photoconductor
R1 — 100K-ohm, 1-watt resistor
R2 — 500K-ohm, 2-watt potentiometer
S1 — SPST switch (on R2)
SCR1 — GE-X5 silicon controlled rectifier
SPKR — 4- to 8-ohm speaker
B1 — 22-1/2-volt battery
Minibox — 6" x 5" x 4"

Watchdog light-sensitive oscillator. The basic circuit is a relaxation oscillator that sounds an alarm through a speaker or earphone. R2 is set for a bias current just below the firing level (courtesy General Electric Company).

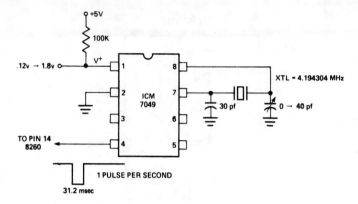

1-second reference oscillator. This circuit is ideal for clock circuits where no 60-hertz line is available (courtesy Intersil, Inc.).

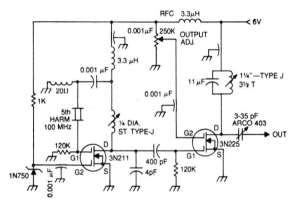

100 MHz oscillator and buffer (courtesy Texas Instruments Incorporated).

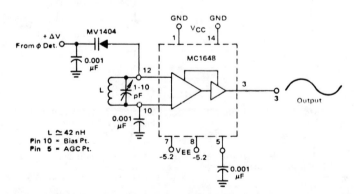

Voltage-controlled oscillator using an MC1648. This VCO is tunable from 110 MHz to 145 MHz. The control voltage range is 1.8 volts at 110 MHz to 5 volts at 145 MHz (courtesy Motorola Semiconductor Products Inc.).

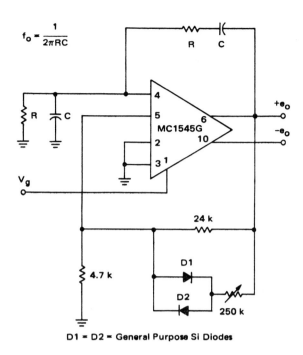

Gated oscillator using an MC1545G video amplifier chip (courtesy Motorola Semiconductor Products Inc.).

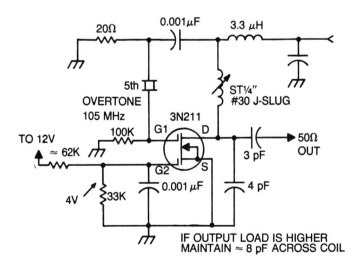

105 MHz crystal oscillator operating at the fifth overtone (courtesy Texas Instruments Incorporated).

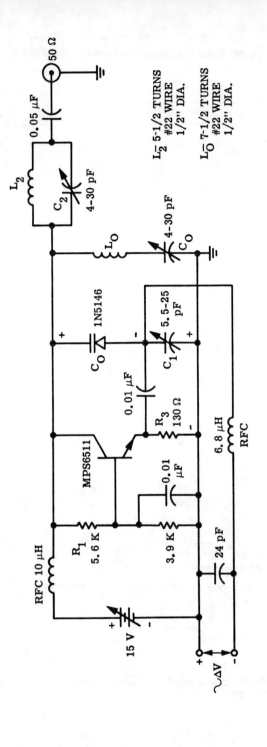

Frequency-modulated 52 MHz oscillator. The circuit uses an MPS6511 transistor especially designed for oscillators. The 1N5146 is a varactor diode rated at 33 pF for a reverse bias of −4 volts. L2 and L3 are used to tune out unwanted harmonics. Audio inputs should be limited to 200 mV or less (courtesy Motorola Semiconductor Products, Inc.).

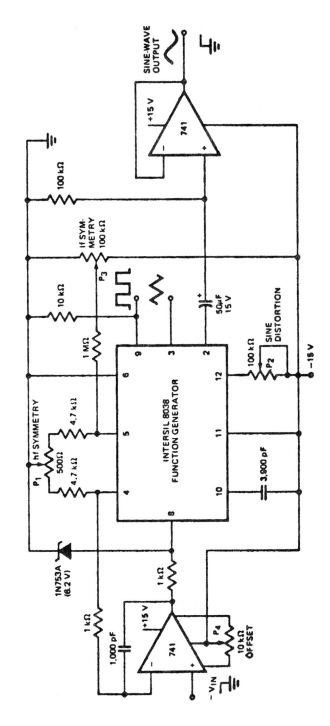

Linear voltage-controlled oscillator (courtesy Intersil, Inc.).

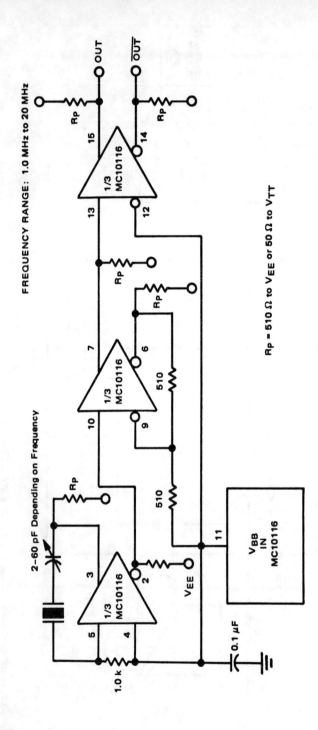

Fundamental crystal oscillator for 1 MHz to 20 MHz. V_{EE} is -5.2 volts (courtesy Motorola Semiconductor Products Inc.).

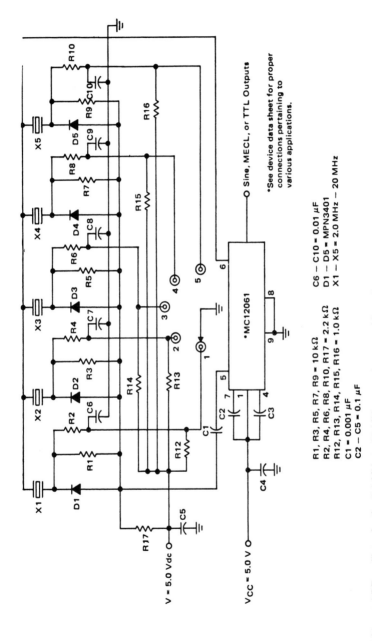

Multicrystal RF oscillator for the 2.0 MHz to 20 MHz range (courtesy Motorola Semiconductor Products Inc.).

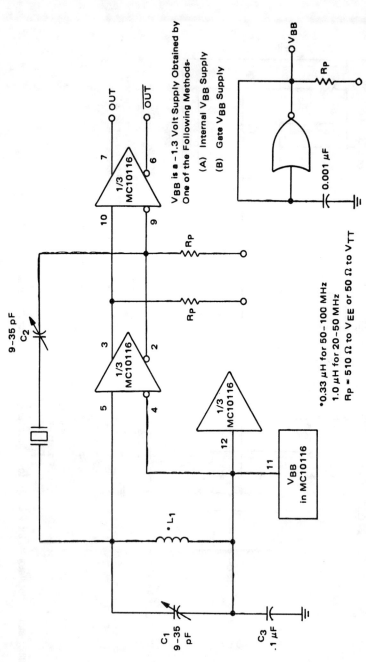

Overtone crystal oscillator with operating range of 20 MHz to 100 MHz depending on crystal selection and tank tuning. $V_{EE} = -5.2$ volts (courtesy Motorola Semiconductor Products Inc.).

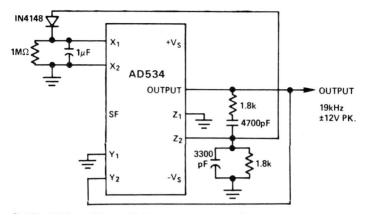

Stabilized Wien bridge oscillator (courtesy Analog Devices, Inc.).

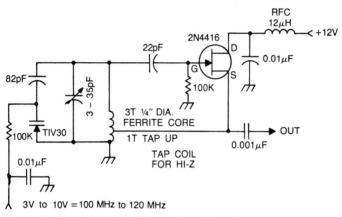

Voltage-controlled oscillator for FM operation using a 2N4416 (courtesy Texas Instruments Incorporated).

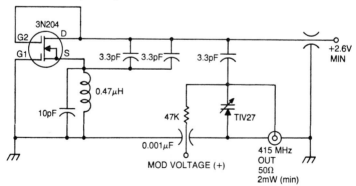

415 MHz frequency-modulated oscillator using a 3N204 dual-gate MOSFET. The 3N212 must be selected for I$_{DSS}$ greater than 20 mA (courtesy Texas Instruments Incorporated).

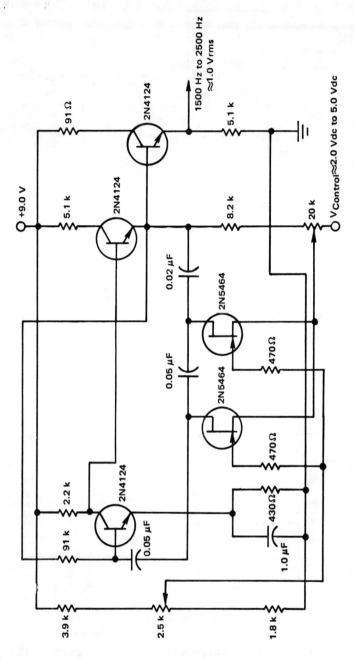

Voltage-controlled oscillator. This three-section phase-shift oscillator produces a good sine wave that is linear over the range indicated (courtesy Motorola Semiconductor Products Inc.).

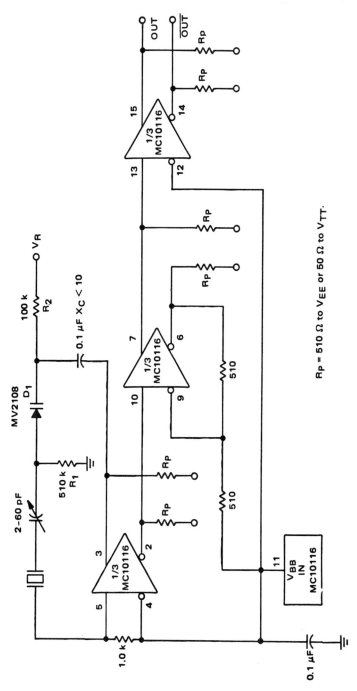

Voltage-controlled crystal oscillator. Operating range is 1 MHz to 20 MHz depending on the selected crystal and tank tuning. Tunign range is from zero to 25 volts. It is possible to make the tuning range from zero to −25 volts by reversing the varactor (courtesy Motorola Semiconductor Products Inc.).

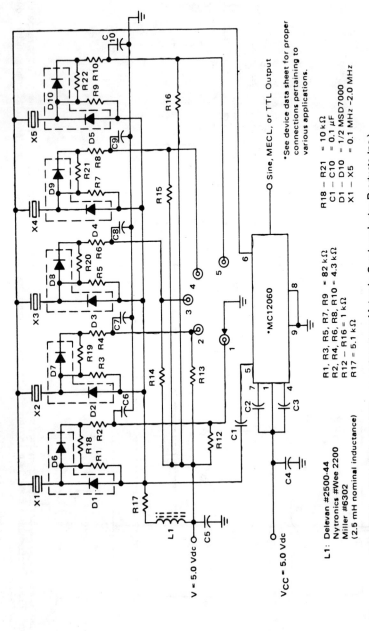

Multicrystal RF oscillator for the 100 kHz to 2.0 MHz range (courtesy Motorola Semiconductor Products Inc.).

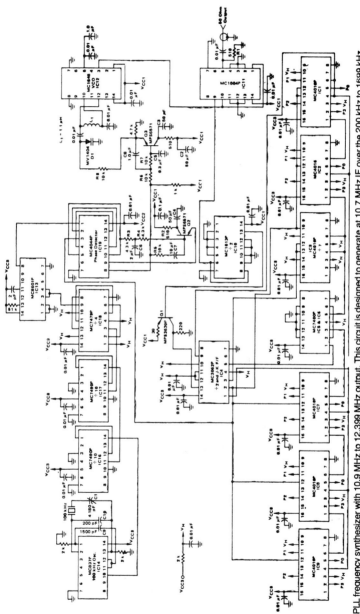

PLL frequency synthesizer with 10.9 MHz to 12.399 MHz output. This circuit is designed to generate at 10.7 MHz IF over the 200 kHz to 1699 kHz band for an ADF. Steps are 1 kHz apart (courtesy Motorola Semiconductor Products Inc.).

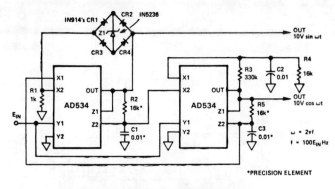

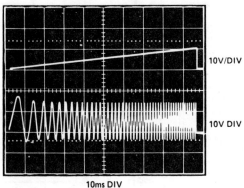

Voltage-controlled sine-wave oscillator. Two AD534 multipliers are used to form integrators with controllable time constants in a second-order differential-equation feedback loop. The waveform shows the VCO's response to a ramp input (courtesy Analog Devices, Inc.).

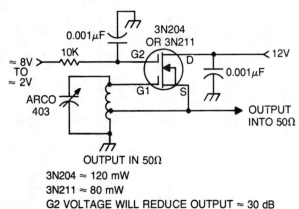

145 MHz RF oscillator using a 3N204/3N211 dual-gate MOSFET (courtesy Texas Instruments Incorporated).

Math Function Circuits

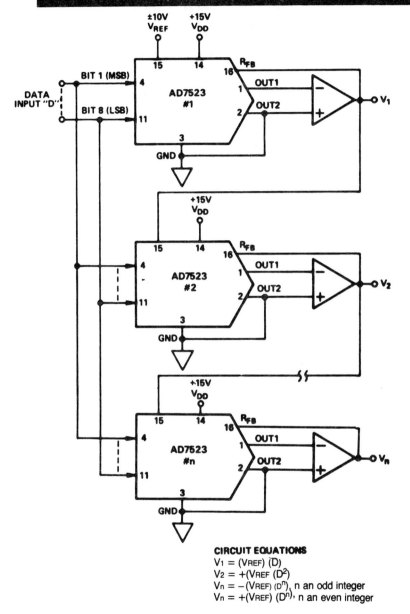

CIRCUIT EQUATIONS
$V_1 = (V_{REF})(D)$
$V_2 = +(V_{REF})(D^2)$
$V_n = -(V_{REF})(D^n)$, n an odd integer
$V_n = +(V_{REF})(D^n)$, n an even integer

Power generation circuit using *n* AD7523 8-bit multiplying D/A converters (courtesy Analog Devices, Inc.).

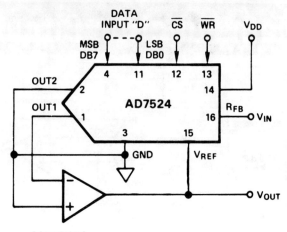

EQUATIONS

$$V_{OUT} = -\frac{V_{IN}}{D}$$

$$A_V = \frac{V_{OUT}}{V_{IN}} = -\frac{1}{D} \quad \text{where: } A_V = \text{Voltage Gain}$$

and where:

$$D = \frac{DB7}{2^1} + \frac{DB6}{2^2} + \ldots \frac{DB0}{2^8}$$

$DB_N = 1$ or 0

EXAMPLES

$D = 00000000$, $A_V = -A_{OL}$ (OP AMP)
$D = 00000001$, $A_V = -256$
$D = 10000000$, $A_V = -2$
$D = 11111111$, $A_V = -\frac{256}{255}$

Divider circuit with digitally controlled gain (courtesy Analog Devices, Inc.).

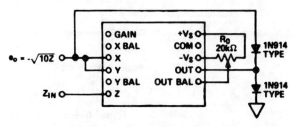

Square rooter circuit using the 435 multiplier/divider chip (courtesy Analog Devices, Inc.).

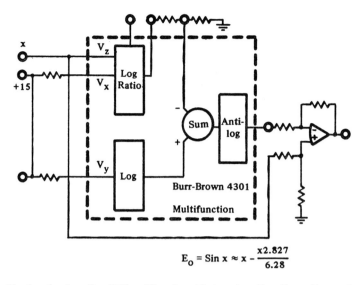

$$E_o = \sin x \approx x - \frac{x^{2.827}}{6.28}$$

Sine function from the 4301 multifunction chip (courtesy Burr-Brown Research Corporation).

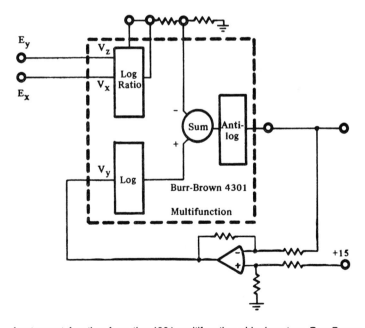

Arc tangent function from the 4301 multifunction chip (courtesy Burr-Brown Research Corporation).

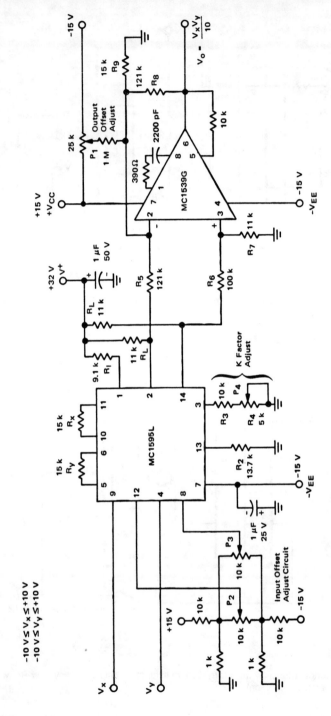

Multiplier with op amp level shift (courtesy Motorola Semiconductor Products Inc.).

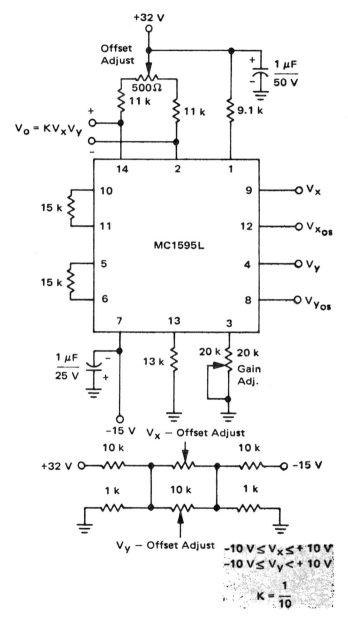

Multiplier circuit using an MC1595 (courtesy Motorola Semiconductor Products Inc.).

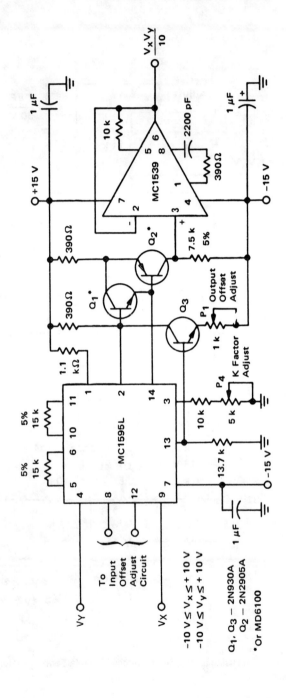

Multiplier with discrete level shift (courtesy Motorola Semiconductor Products Inc.).

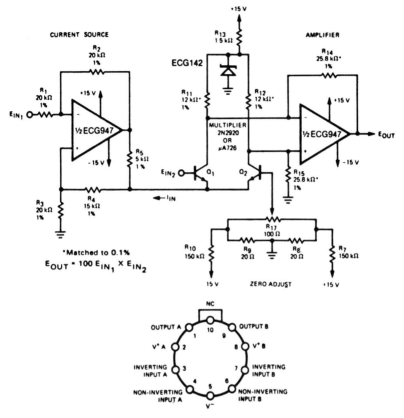

Analog multiplier using an ECG947 dual operational amplifier. The ECG947 is short-circuit protected and requires no external components for frequency compensation (courtesy GTE Sylvania Incorporated).

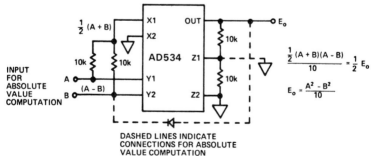

Difference-of-squares circuit. This circuit computes the difference of the squares of two input signals. It is useful in vector computations and in weighing the difference of two magnitudes to emphasize the greater nonlinearity. It can also be used to determine absolute value if A is the input, B is connected to Eo through a diode and both Z terminals are grounded (courtesy Analog Devices, Inc.).

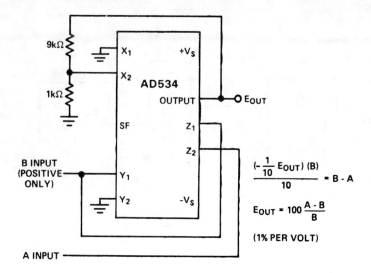

Percent of deviation ratio computer (courtesy Analog Devices, Inc.).

$$\frac{(-\frac{1}{10} E_{OUT})(B)}{10} = B - A$$

$$E_{OUT} = 100 \frac{A - B}{B}$$

(1% PER VOLT)

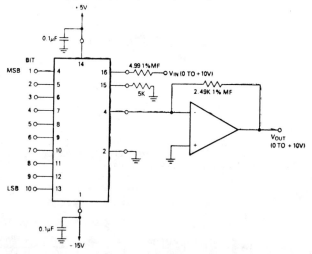

INPUT CODE		UNIPOLAR OPERATION – STRAIGHT BINARY			
MSB	LSB	0 TO +5V	0 TO +10V	0 TO −2MA	0 TO −4MA
1111	1111	+4.995V	+9.990	1.998 MA	3.996
1110	0000	+4.375	+8.750	1.750	3.500
1100	0000	+3.750	+7.500	1.500	3.000
1000	0000	+2.500	+5.000	1.000	2.000
0100	0000	+1.250	+2.500	0.500	0.100
0000	0001	+0.005	+0.010	0.002	0.004
0000	0000	0.000	0.000	0.000	0.000

One-quadrant multiplication using a DAC-IC10BC D/A converter and AM-452 op amp. See coding table. With V_{IN} connected to pin 16 the input impedance is low; with it connected to pin 15 the input impedance is high. The range is then 0 to −10 volts (courtesy Datel Systems, Inc.).

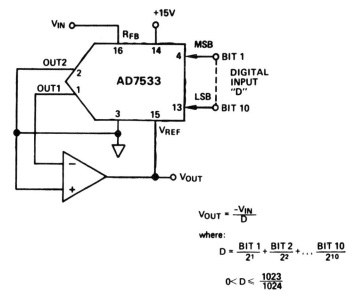

Divider with digitally controlled gain (courtesy Analog Devices, Inc.).

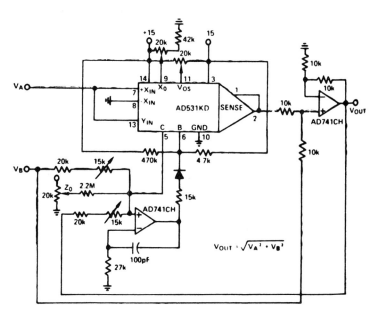

Vector computer using one AD531 multiplier/divider and two AD741 op amps. The circuit derives the square root of the sum of the squares (courtesy Analog Devices, Inc.).

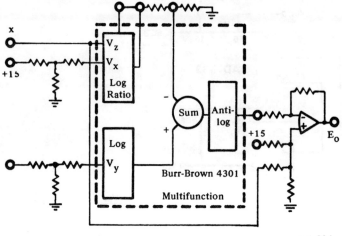

$$E_o = \cos x = 1 + 0.2325 x - \frac{x\ 1.504}{1.445}$$

Cosine function from the 4301 multifunction chip (courtesy Burr-Brown Research Corporation).

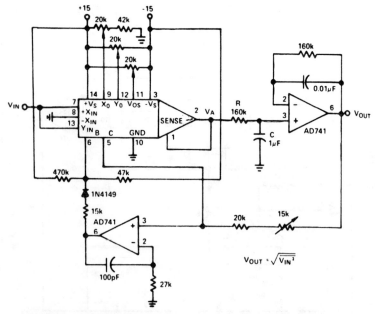

True RMS circuit using one AD531 multiplier/divider and two AD741 op amps. The AD531 is combined with a simple filter to obtain the true RMS value of an AC input signal. By scaling $V_{OUT} = 10$ volts DC for a ± 10-volt DC input this circuit can give direct RMS readings for 100 hertz to 100 kilohertz sine waves from 0.2 volt to 7.0 volts peak (courtesy Analog Devices, Inc.).

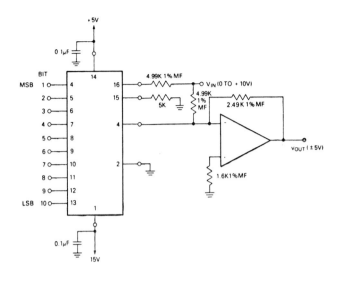

INPUT CODE	BIPOLAR OPERATION – OFFSET BINARY CODING			
MSB LSB	±5V	±10V	±1MA	±2MA
1 1 1 1 1 1 1 1 1 1	·4 990V	· 9 980V	·0 998MA	·1 996MA
1 1 1 0 0 0 0 0 0 0	·3 750	· 7 500	·0 750	·1 500
1 1 0 0 0 0 0 0 0 0	·2 500	· 5 000	·0 500	·2 000
1 0 0 0 0 0 0 0 0 0	0 000	0 000	0 000	0 000
0 1 0 0 0 0 0 0 0 0	2 500	5 000	0 500	1 000
0 0 0 0 0 0 0 0 0 1	4 990	9 980	0 998	1 996
0 0 0 0 0 0 0 0 0 0	5 000	10 000	1 000	2 000

Two-quadrant multiplier using a Datel DAC-IC10BC D/A converter. The DAC-IC10BC chip is a 16-pin DIP. V$_{IN}$ is unipolar and the digital input is bipolar with offset binary coding. V$_{OUT}$ varies over the bipolar range of ±5 volts. It is recommended that full scale I$_{REF}$ be set to 2.0 mA; the output is then monotonic as the reference current varies over 0.5 mA to 2.0 mA (courtesy Datel Systems, Inc.).

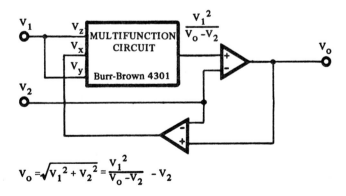

$$v_o = \sqrt{v_1^2 + v_2^2} = \frac{v_1^2}{v_o - v_2} - v_2$$

Vector magnitude function from the 4301 multifunction chip (courtesy Burr-Brown Research Corporation).

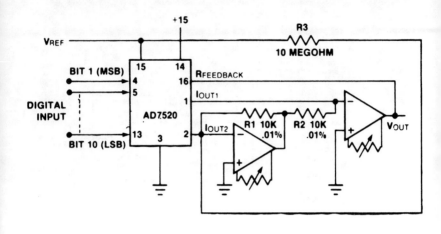

CODE TABLE — BIPOLAR (OFFSET BINARY) OPERATION

DIGITAL INPUT	ANALOG OUTPUT
1111111111	$-V_{REF}(1 - 2^{-(n-1)})$
1000000001	$-V_{REF}(2^{-(n-1)})$
1000000000	0
0111111111	$V_{REF}(2^{-(n-1)})$
0000000001	$V_{REF}(1 - 2^{-(n-1)})$
0000000000	V_{REF}

NOTE: 1. $LSB = 2^{-(n-1)} V_{REF}$ 2. n = 10(12) for 7520 (7521) 7530 (7531)

Four-quadrant multiplication with the AD7520 in bipolar operation. The Intersil AD7520 is an 18-pin multiplying D/A converter. A logic one input at any digital input forces the corresponding ladder switch to steer the bit current to the I_{OUT1} bus. A logic zero input forces the bit current to the I_{OUT2} bus. For any code the I_{OUT1} and I_{OUT2} bus currents are complements of one another. The current amplifier at I_{OUT2} changes the polarity of I_{OUT2} current and the transconductance amplifier at I_{OUT1} output sums the two currents. This configuration doubles the output range but halves the resolution of the D/A converter. For the offset adjustment set V_{REF} to approximately +10 volts and connect all digital inputs to logic one. Adjust the I_{OUT2} amplifier offset zero pot for 0V ± 1 mV at I_{OUT2} amplifier output. Connect the MSB to logic one and all other bits to logic zero. Adjust I_{OUT1} offset zero pot for 0V ± mV at V_{OUT}. For the gain adjustment connect all digital inputs to V_{DD}. Monitor V_{OUT} for a $-V_{REF}(1 - 2^{-(n-1)})$ volt reading, where n is equal to 10. To increase V_{OUT} connect a series resistor, 0 to 500 ohms, in the I_{OUT1} amplifier feedback loop. To decrease V_{OUT} connect a series resistor, 0 to 500 ohms, between the reference voltage and pin 15 (courtesy Intersil, Inc.).

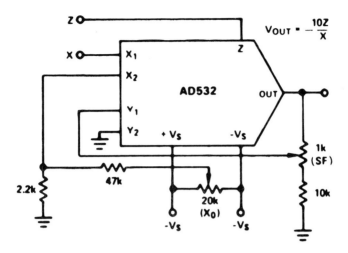

Divider circuit using an AD532 multiplier/divider chip. The AD532 is available as a 10-pin TO-100 or as a 14-pin DIP (courtesy Analog Devices, Inc.).

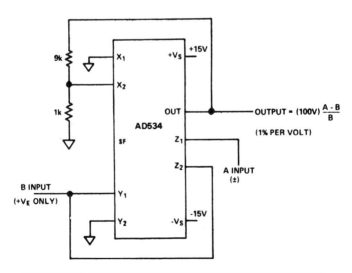

OTHER SCALES, FROM 10% PER VOLT TO 0.1% PER VOLT CAN BE OBTAINED BY ALTERING THE FEEDBACK RATIO.

Percentage computer using an AD534 multiplier/divider chip (courtesy Analog Devices, Inc.).

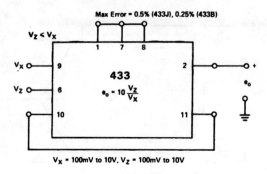

Divider circuit using the 433 multiplier/divider chip (courtesy Analog Devices, Inc.).

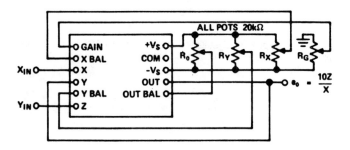

Divider circuit using the 435 multiplier/divider chip (courtesy Analog Devices, Inc.).

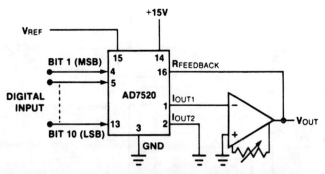

Two-quadrant multiplication with the AD7520 in unipolar binary operation. The Intersil AD7520 is an 18-pin multiplying D/A converter. To adjust the zero offset connect all digital input to ground, and adjust the zero trimmer on the op amp for 0V ± 1 mV at VOUT. To adjust the gain connect all digital inputs of the AD7520 to VDD. Monitor VOUT for a $-V_{REF}(1 - 2^{-n})$ reading, where n is equal to 10. To decrease VOUT connect a series resistor, 0 to 500 ohms, between the reference voltage and pin 15. To increase VOUT connect a series resistor, 0 to 500 ohms, in the IOUT1 amplifier feedback loop (courtesy Intersil, Inc.).

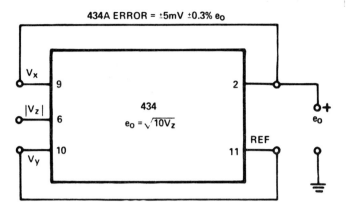

Square rooter circuit using the 434 multiplier/divider chip (courtesy Analog Devices, Inc.).

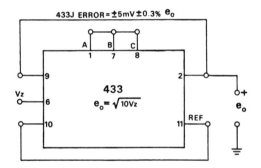

Square rooter using the 433 multiplier/divider chip (courtesy Analog Devices, Inc.).

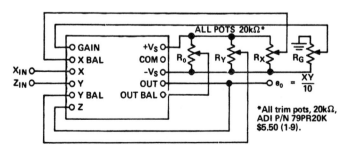

Multiplier circuit using the 435 multiplier/divider chip (courtesy Analog Devices, Inc.).

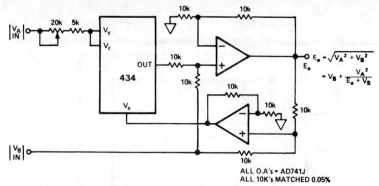

Square root of the sum of the squares circuit. This circuit performs vector computations (courtesy Analog Devices, Inc.).

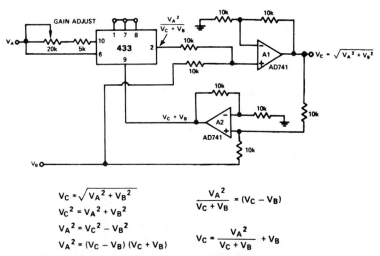

$$V_C = \sqrt{V_A^2 + V_B^2}$$
$$V_C^2 = V_A^2 + V_B^2$$
$$V_A^2 = V_C^2 - V_B^2$$
$$V_A^2 = (V_C - V_B)(V_C + V_B)$$

$$\frac{V_A^2}{V_C + V_B} = (V_C - V_B)$$

$$V_C = \frac{V_A^2}{V_C + V_B} + V_B$$

Vector computation circuit for $V_C^2 = V_A^2 + V_B^2$ using the 433 multiplier/divider and two AD741 op amps (courtesy Analog Devices, Inc.).

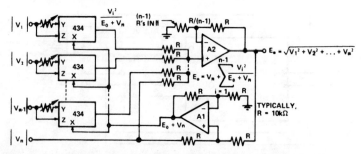

Vector computation circuit for n variables using three 434 multiplier/dividers and two AD741 op amps (courtesy Analog Devices, Inc.).

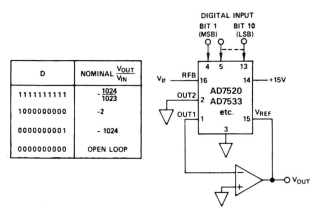

Variable gain amplifier (two-quadrant divider). Examples of gains at various codes are given in the table (courtesy Analog Devices, Inc.).

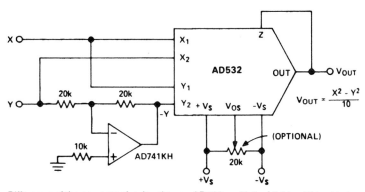

Difference of the squares circuit using an AD532 multiplier/divider. This chip is available as a 10-pin TO-100 or as a 14-pin DIP (courtesy Analog Devices, Inc.).

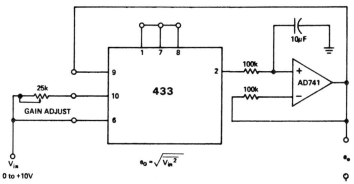

True RMS circuit using the 433 multiplier/divider chip and an AD741 op amp (courtesy Analog Devices, Inc.).

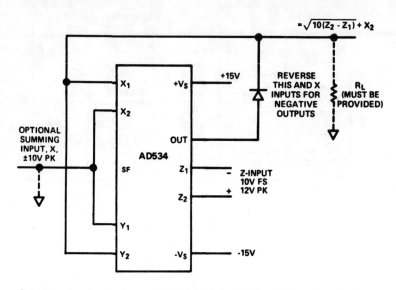

Square rooter circuit using an AD534 multiplier/divider chip (courtesy Analog Devices, Inc.).

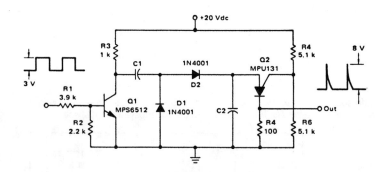

Low-frequency divider circuit using a PUT (courtesy Motorola Semiconductor Products Inc.).

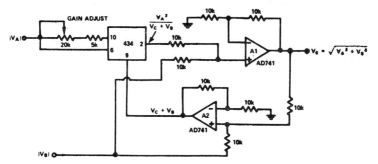

Vector computation circuit for $V_C^2 = V_A^2 + V_B^2$ using the 434 multiplier/divider chip and two AD741 op amps (courtesy Analog Devices, Inc.).

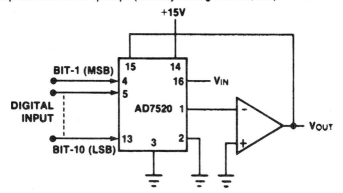

A/D divider using the AD7520 18-pin DIP. With all bits off the amplifier saturates since division by zero isn't defined. With the LSB (bit 10) on the gain is 1023. With all bits on the gain is 1 (courtesy Intersil, Inc.).

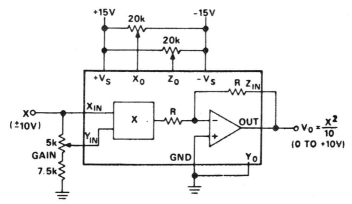

Squarer circuit using an AD533 multiplier/divider chip. With X at zero volts adjust Zo for a zero-volt DC output. With X equal to +10 volts DC adjust gain for +10 volts DC output. Reverse the polarity of X input and adjust Xo to reduce the output error to half its original value and adjust gain to take out the remaining error. Check the output offset with the input grounded. If nonzero repeat the above procedure (courtesy Analog Devices, Inc.).

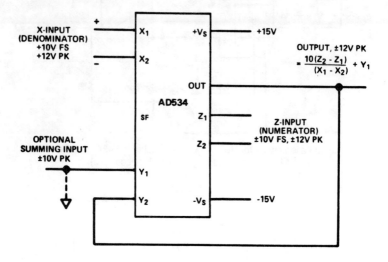

Divider circuit using an AD534 multiplier/divider chip (courtesy Analog Devices, Inc.).

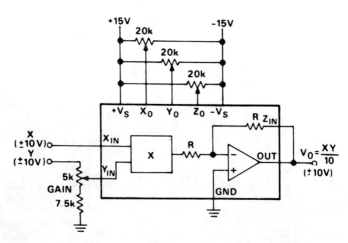

Multiplier circuit using an AD533. With X and Y at zero volts adjust Z_O for a zero-volt DC output. With Y at 20 volts peak to peak 50 hertz and X at zero volts adjust X_O for minimum AC output. With X equal to 20 volts peak to peak at 50 hertz and Y at zero volts adjust Y_O for minimum AC output. Adjust Z_O again for zero volts DC output. With X equal to 10 volts DC and Y equal to 20 volts peak to peak at 50 hertz adjust gain for output equal to Y_{IN} (courtesy Analog Devices, Inc.).

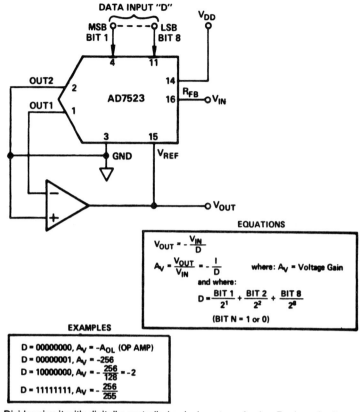

Divider circuit with digitally controlled gain (courtesy Analog Devices, Inc.).

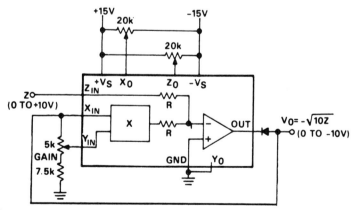

Square rooter using an AD533 multiplier/divider chip. With Z equal to +0.1 volt DC adjust Zo for a −1.0 volt DC output. With Z equal to +10.0 volts DC adjust gain for −10.0 volts DC output. With Z equal to +2.0 volts DC adjust Xo for a −4.47 ±0.1 volt DC output (courtesy Analog Devices, Inc.).

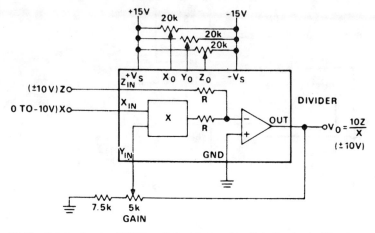

Divider circuit using an AD533 multiplier/divider chip. Set all pots at midrange. With Z equal to zero volts, trim Zo to hold the output constant as X is varied from −10 volts DC to −1 volt DC. With Z equal to zero volts and X equal to −10 volts DC trim Yo for zero volts DC. With Z equal to X or −X trim Xo for the minimum worst-case variations as X is varied from −10 volts DC to −1 volt DC. With Z equal to X or −X trim the gain for the closest average approach to ±10 volts DC output as X is varied from −10 volts DC to −3 volts DC (courtesy Analog Devices, Inc.).

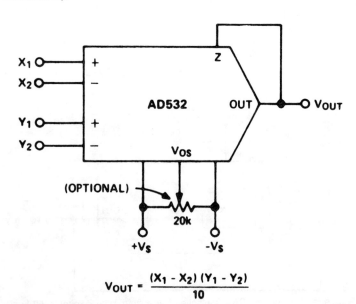

$$V_{OUT} = \frac{(X_1 - X_2)(Y_1 - Y_2)}{10}$$

Multiplier circuit using an AD532. Inputs can be fed differentially to the X- and Y-inputs or single-endedly by simply grounding the unused input (courtesy Analog Devices, Inc.).

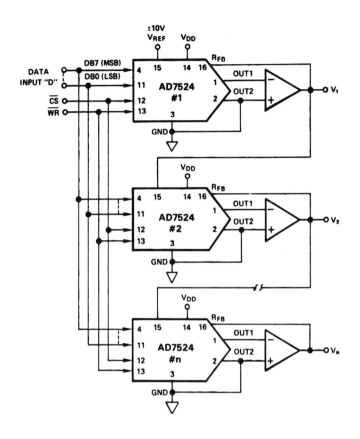

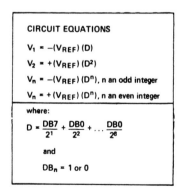

Power generation circuit using n number of AD7524 8-bit D/A converters (courtesy Analog Devices, Inc.).

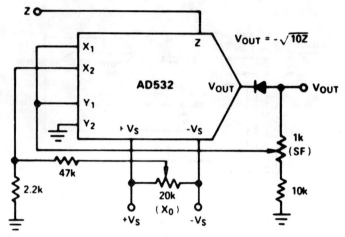

Square rooter circuit using an AD532 multiplier/divider chip. The AD532 is available as a 10-pin TO-100 or as a 14-pin DIP (courtesy Analog Devices, Inc.).

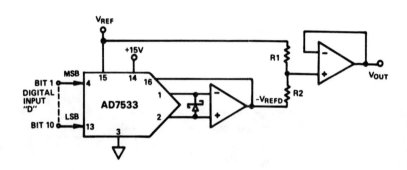

$$V_{OUT} = V_{REF} \left[\left(\frac{R_2}{R_1 + R_2} \right) - \left(\frac{R_1 D}{R_1 + R_2} \right) \right]$$

where:

$$D = \frac{BIT\,1}{2^1} + \frac{BIT\,2}{2^2} + \ldots \frac{BIT\,10}{2^{10}}$$

$$\left(0 < D < \frac{1023}{1024} \right)$$

Modified scale factor and offset circuit (courtesy Analog Devices, Inc.).

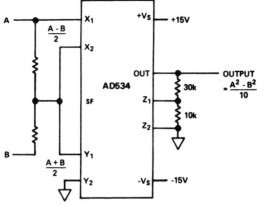

Difference of squares circuit using an AD534 multiplier/divider chip (courtesy Analog Devices, Inc.).

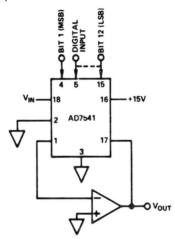

A/D divider (courtesy Analog Devices, Inc.).

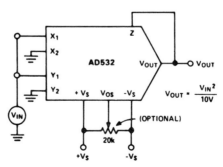

Squarer circuit using an AD532 multiplier/divider. This chip is available as a 10-pin TO-100 or a 14-pin DIP (courtesy Analog Devices, Inc.).

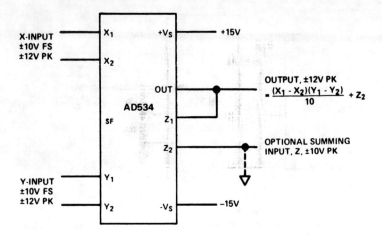

Multiplier circuit using an AD534 multiplier/divider chip (courtesy Analog Devices, Inc.).

Power-Controlling Circuits

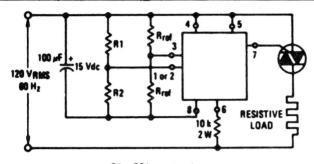

R1 or R2 is an external sensor

Basic triac trigger circuit utilizing the zero crossing detector and the input comparator to control the gate of the triac.

Triac control circuit using an ECG776 zero voltage switch IC. Resistor R2 must be the external sensor for the internal short and open protection to be operative. Select the triac from the ECG5600 series for the particular application (courtesy GTE Sylvania Incorporated).

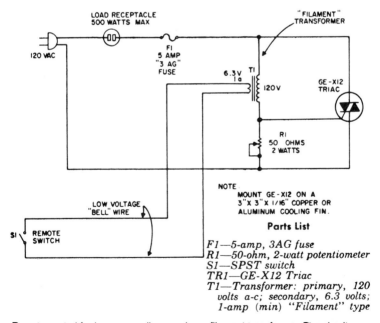

Parts List

F1—5-amp, 3AG fuse
R1—50-ohm, 2-watt potentiometer
S1—SPST switch
TR1—GE-X12 Triac
T1—Transformer: primary, 120 volts a-c; secondary, 6.3 volts; 1-amp (min) "Filament" type

Remote control for lamp or appliance using a filament transformer. The circuit can handle up to 500 watts. R1 is adjusted for the highest resistance that will not trigger the triac with S1 open (courtesy General Electric Company).

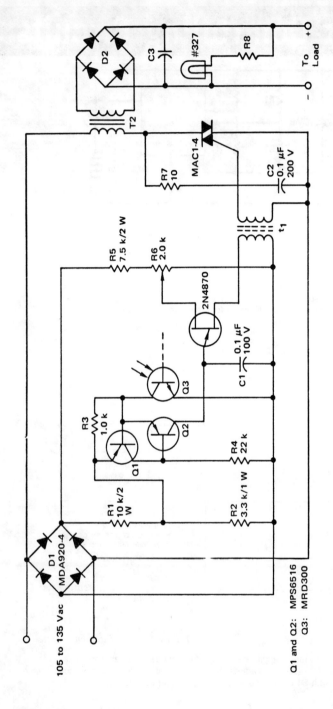

RMS regulator for a DC power supply using a triac, phototransistor and UJT (courtesy Motorola Semiconductor Products Inc.).

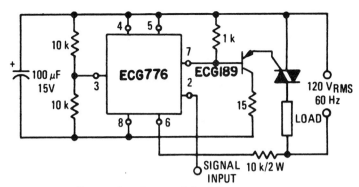

Zero crossing triac control circuit for gate current requirements greater than 50 mA.

Triac control circuit employing an ECG776 zero voltage switch with current boost utilizing an AC supply. The circuit is for applications requiring gate currents greater than 50 mA. Select the triac from the ECG5600 series for the particular application (courtesy GTE Sylvania Incorporated).

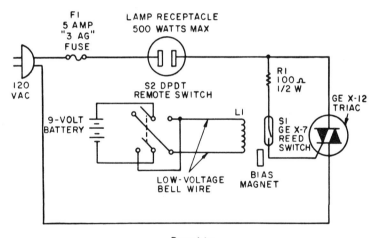

Parts List

Triac — GE-X12
F1 — 5-amp fuse
L1 — 1000 turns of No. 36 enameled copper wire on 1/4" diam. x 2" coil form (this form supplied with reed switch)
R1 — 100-ohm, 1/2-watt resistor
S1 — GE-X7 reed switch (includes coil form and bias magnet)
Receptacle — Amphenol type 61F socket, or equivalent
Misc. — Line cord, fuse holder, 9-volt transistor battery and clip, grommet, and heat sink.
S2 — DPDT switch with center OFF position (spring loaded or telephone type optional)

Remote control for lamp or appliance using a reed switch and coil. The circuit will handle up to 500 watts (courtesy General Electric Company).

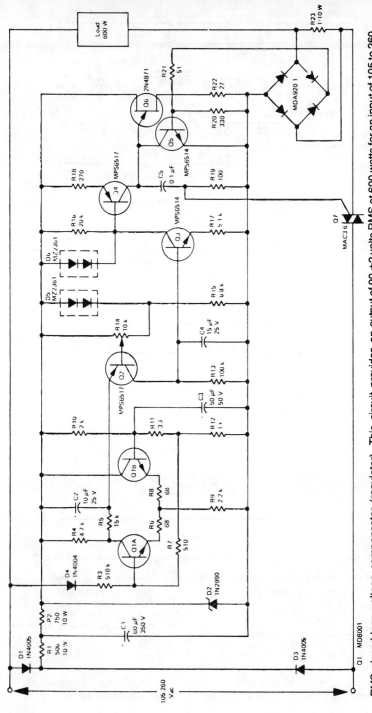

RMS closed-loop voltage compensator (regulator). This circuit provides an output of 90 ±2 volts RMS at 600 watts for an input of 105 to 260 volts (courtesy Motorola Semiconductor Products Inc.).

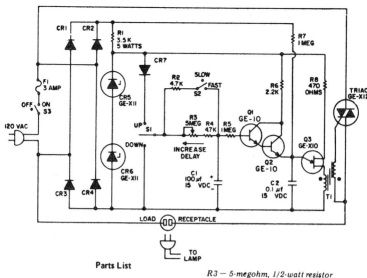

Parts List

C1 — 100-mfd, 15-volt electrolytic capacitor
C2 — 0.1-mfd, 15-volt capacitor
CR1 thru CR4 — GE-504A rectifier diode
CR5, CR6 — GE-X11 zener diode
CR7 — GE-504A rectifier diode
F1 — 3-ampere fuse
Q1, Q2 — GE-10 transistor
Q3 — GE-X10 unijunction transistor
R1 — 3500-ohm, 5-watt resistor
R2, R4 — 4700-ohm, 1/2-watt resistor
R3 — 5-megohm, 1/2-watt resistor
R5, R7 — 1-megohm, 1/2-watt resistor
R6 — 2200-ohm, 1/2-watt resistor
R8 — 470-ohm, 1/2-watt resistor
S1 — SPDT toggle switch
S2, S3 — SPST toggle switch
Triac — GE-X12
T1 — Pulse transformer — available from GE distributors as ETRS-4898, or from General Electric Co., Dept. B, 3800 N. Milwaukee Ave., Chicago, Ill. 60641.
Minibox — 4" x 2-1/2" x 2-1/2"

Time-dependent light dimmer. The circuit allows bright lights to slowly fade over a period of 15 to 20 minutes and permits loads up to 500 watts (courtesy General Electric Company).

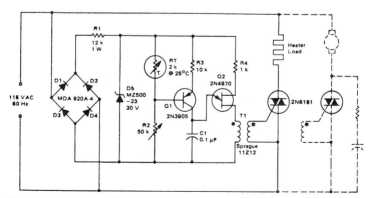

Triac temperature-sensitive heater control. This circuit can be modified as shown to control a motor with a constant load. As shown the circuit is for heating applications but can be used for cooling by interchanging R_T and R2 (courtesy Motorola Semiconductor Products Inc.).

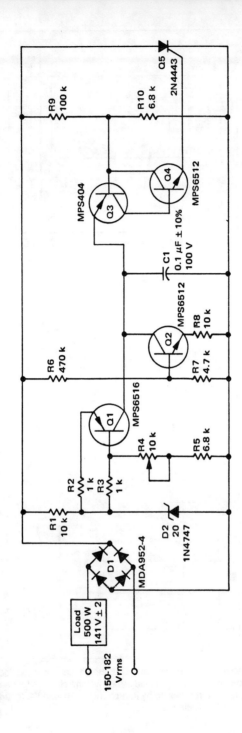

RMS open-loop voltage compensator (regulator) for application requiring large conduction angles. It provides an output 141 ±2 volts RMS at 500 watts for an input of 150 to 182 volts RMS (courtesy Motorola Semiconductor Products Inc.).

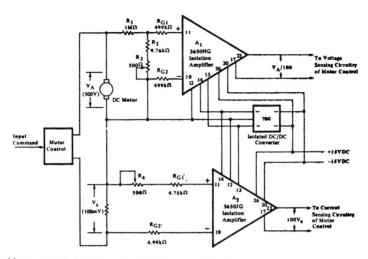

Motor control circuitry using 3650 optical isolation amplifiers (courtesy Burr-Brown Research Corporation).

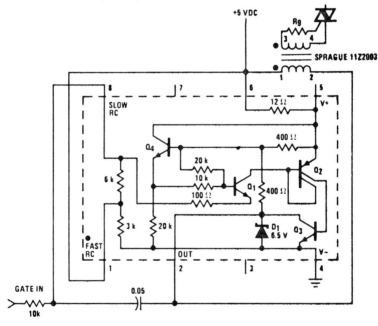

Triac trigger using an LM3909 chip. This circuit operates from a 5-volt logic supply and provides gate trigger pulses of up to 200 mA. The LM3909 provides a 10 μs pulse at about a 7 kHz rate. This is not the synchronized zero crossing type since the first trigger could occur at any time; however, the repetition rate is such that after the first cycle, a triac is triggered within 8 volts of zero with a resistive load and a 115-volt AC line. The Sprague transformer provides 2-to-1 stepup (courtesy National Semiconductor Corporation).

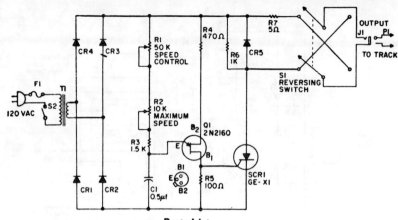

Parts List

- C1 — 0.5-mfd, 200-volt capacitor
- CR1 thru CR5 — GE-504A rectifier diode
- F1 — 1/2-amp fuse and holder
- J1 — Output jack
- P1 — Output plug to track connections
- Q1 — G-E Type 2N2160 unijunction transistor
- R1 — 50000-ohm, 2-watt potentiometer with SPST switch
- R2 — 10000-ohm, 2-watt potentiometer
- R3 — 1500-ohm, 1/2-watt resistor
- R4 — 470-ohm, 1/2-watt resistor
- R5 — 100-ohm, 1/2-watt resistor
- R6 — 10000-ohm, 1/2-watt resistor
- R7 — 5-ohm, 20-watt resistor or two 10-ohm, 10-watt resistors in parallel
- S1 — DPDT switch
- S2 — SPST switch (on R1)
- SCR1 — GE-X1 silicon controlled rectifier
- T1 — Transformer: primary, 120 volts a-c; secondary, 25 volts a-c (Stancor P-6469, or equivalent)
- Minibox — Aluminum, 6" x 5" x 4"

SCR model railroading control. Bridge circuit CR1 through CR4 supplies pulsating DC to firing circuit Q1, R1 through R5 and C1, which phase controls SCR1. The SCR is in series with the train power and thereby controls the amount of current it receives (courtesy General Electric Company).

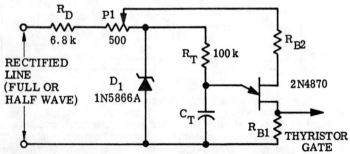

Line-voltage compensation circuit using UJT trigger for a thyrister gate (courtesy Motorola Semiconductor Products Inc.).

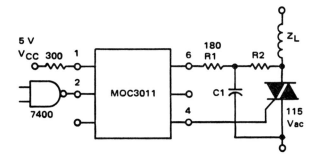

NOTE: Circuit supplies 25 mA drive to gate of triac at V_{in} = 25 V and $T_A \leqslant 70°C$.

TRIAC		
I_{GT}	R2	C
15 mA	2400	0.1
30 mA	1200	0.2
50 mA	800	0.3

Logic to inductive load interface using an MOC3011 optically coupled triac driver (courtesy Motorola Semiconductor Products Inc.).

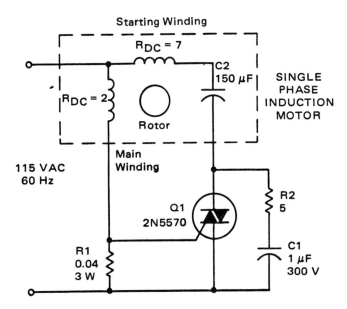

Triac static motor-starting switch for 0.5 HP 115-volt AC single-phase induction motors (courtesy Motorola Semiconductor Products Inc.).

Parts List

C1 — 5.0-mfd, 10-volt miniature electrolytic capacitor
CR1 — GE-504A rectifier diode, or 1N5059, or equivalent
J1 — Socket, Amphenol Type 61F, or equivalent
L1 — 120-volt, 6-watt lamp (night lite type)
Q1 — GE-X19 phototransistor
Q2 — GE-10 or 2N5172 transistor
R1 — 5-ohm, 5-watt resistor
R2 — 47,000-ohm, 1/2-watt resistor
R3 — 22,000-ohm, 1/2-watt resistor
R4 — 1000-ohm, 1/2-watt resistor
R5 — 10,000-ohm, 1/2-watt resistor
R6 — 220,000-ohm, 1/2-watt resistor
R7 — 100,000-ohm, 1/2-watt resistor
R8 — 47,000-ohm, 1/2-watt resistor
SCR-GEMR-5
Line Cord
Touch Switch — see text
Minibox — 5" x 2-1/2" x 2-1/2", Bud CR-2104A, or equivalent

Touch switch for loads up to 180 watts. This circuit is good for turning on and off lights, TVs, stereos, etc. Q1 is a phototransistor (courtesy General Electric Company).

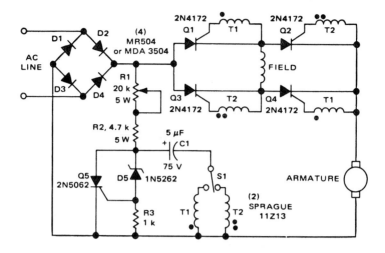

Direction and speed control for series-wound motors (courtesy Motorola Semiconductor Products Inc.).

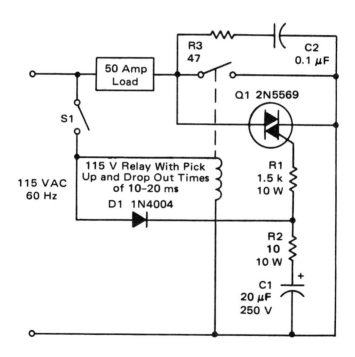

Triac relay-contact protection circuit (courtesy Motorola Semiconductor Products Inc.).

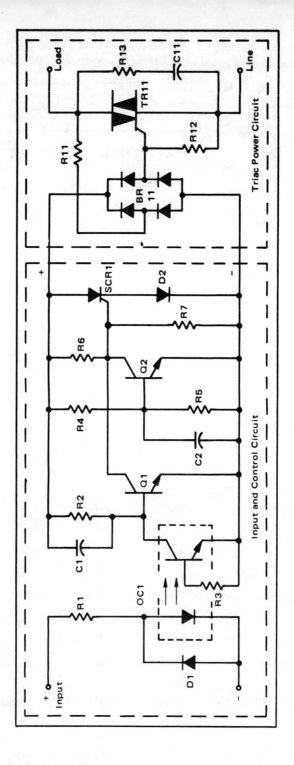

Triac Power Circuit Parts List

Voltage	120 Vrms				240 Vrms			
RMS Current Amperes	8.0	12	25	40	8.0	12	25	40
BR11	MDA102	MDA102	MDA102	MDA102	MDA104	MDA104	MDA104	MDA104
C11, μF (10%, line voltage ac rated)	0.047	0.047	0.1	0.1	0.047	0.047	0.1	0.1
R11 (10%, 1 W)	39	39	39	39	39	39	39	39
R12 (10%, 1/2 W)	18	18	18	18	18	18	18	18
R13 (10%, 1/2 W)	620	620	330	330	620	620	330	330
TR11 Plastic	2N6342	2N6342A	–	–	2N6343	2N6343A	–	–
TR11 Metal	MAC40799	MAC40800	2N6163	MAC4688	–	MAC40800	2N6164	MAC4689

Control Circuit Parts List

	LINE VOLTAGE	
Part	120 Vrms	240 Vrms
C1	220 pF, 20%, 200 Vdc	100 pF, 20%, 400 Vdc
C2	0.022 μF, 20%, 50 Vdc	0.022 μF, 20%, 50 Vdc
D1	1N4001	1N4001
D2	1N4001	1N4001
OC1	MOC1005	MOC1005
Q1	MPS5172	MPS5172
Q2	MPS5172	MPS5172
R1	1 kΩ, 10%, 1 W	1 kΩ, 10%, 1 W
R2	47 kΩ, 5%, 1/2 W	100 kΩ, 5%, 1 W
R3	1 MΩ, 10%, 1/4 W	1 MΩ, 10%, 1/4 W
R4	110 kΩ, 5%, 1/2 W	220 kΩ, 5%, 1/2 W
R5	15 kΩ, 5%, 1/4 W	15 kΩ, 5%, 1/4 W
R6	33 kΩ, 10%, 1/2 W	68 kΩ, 10%, 1 W
R7	10 kΩ, 10%, 1/4 W	10 kΩ, 10%, 1/4 W
SCR1	2N5064	2N6240

Triac solid-state relay circuit for AC power control. The input circuit will function over the range of 3 to 33 volts (courtesy Motorola Semiconductor Products Inc.).

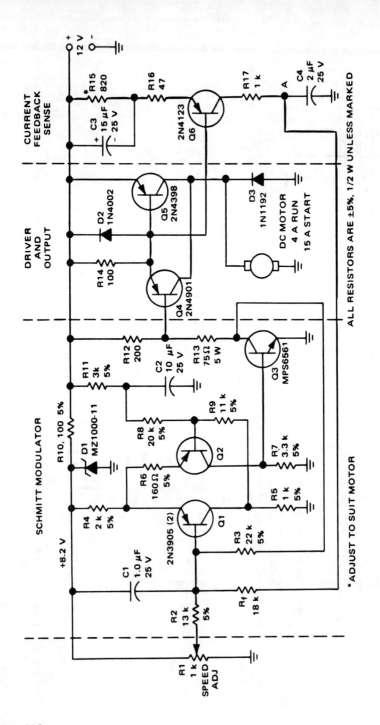

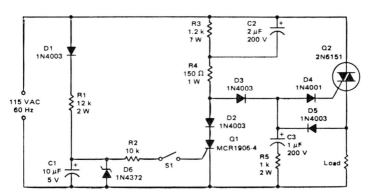

Triac zero-point switch for resistive loads (courtesy Motorola Semiconductor Products Inc.).

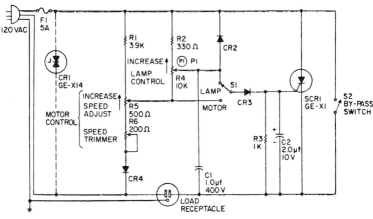

C1 — 1-mfd, 400-volt capacitor
C2 — 2-mfd, 10-volt capacitor
CR1 — GE-X14 thyrector diode (optional transient voltage suppressor)
CR2 thru CR4 — GE-504A rectifier diode
F1 — 5-amp fuse and holder
R1 — 3900-ohm, 2-watt resistor
R2 — 330-ohm, 1-watt resistor
R3 — 1000-ohm, 1-watt resistor
R4 — 10K-ohms, 2-watt potentiometer

Parts List

R5 — 500-ohm, 2-watt potentiometer
R6 — 200-ohm, 2-watt potentiometer
S1 — SPDT toggle switch
S2 — SPST switch (on speed and lamp adjust potentiometers)
SCR1 — GE-X1 silicon controlled rectifier mounted on 3" x 3" x 1/16" copper cooling fin
Minibox — 6" x 5" x 4"

Half-wave variable AC control for small motors and lamps. In the lamp position the SCR is controlled by P1. In the motor position and with S2 open the circuit incorporates a feedback feature that maintains a constant motor speed as the load changes. Do not use this circuit for controlling fluorescent lamps, transformers, AC motors with capacitor start, induction motors, shaded pole motors and the like. The motor controlled must have a commutator as found in DC or AC-DC universal motors (courtesy General Electric Company).

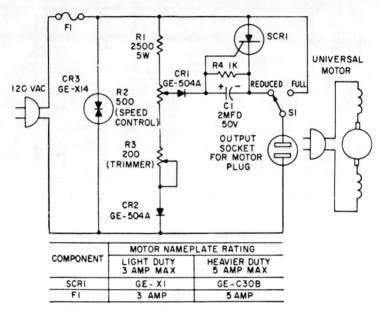

Parts List

C1 — 2-mfd, 50-volt electrolytic capacitor
CR1, CR2 — GE-504A rectifier diode
CR3 — GE-X14 thyrector diode
R1 — 2500-ohm, 5-watt resistor
R2 — 500-ohm, 2-watt potentiometer
R3 — 200-ohm, 1-watt potentiometer
R4 — 1000-ohm, 1/2-watt resistor
F1 — See schematic diagram
S1 — SPDT toggle switch
SCR1 — See schematic diagram
Minibox — 4" x 2-1/4" x 2-1/4"
Line cord and grommet
Socket, AC output

Plug-in speed control for tools and appliances. Do not use this control on motors without commutators (courtesy General Electric Company).

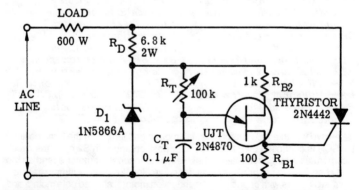

Thyrister half-wave control circuit with UJT trigger designed for a 600-ohm load (courtesy Motorola Semiconductor Products Inc.).

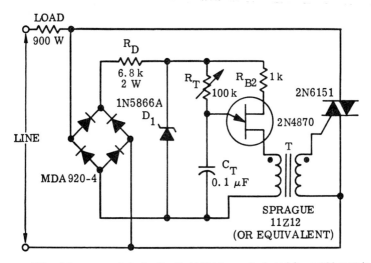

Triac full-wave control circuit with UJT trigger designed for a 900-watt load (courtesy Motorola Semiconductor Products Inc.).

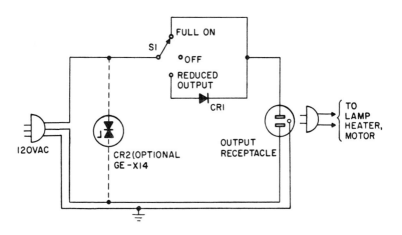

Parts List

$CR1$ — GE-504A rectifier diode for 130 watts output
 — GE-X4 rectifier diode for higher output

$CR2$ — GE-X14 thyrector diode (optional transient voltage suppressor)

$S1$ — SPDT, 3-amp, 125-volt a-c switch with center "off" position

High-low AC switch for light dimming or small motor two-speed control (courtesy General Electric Company).

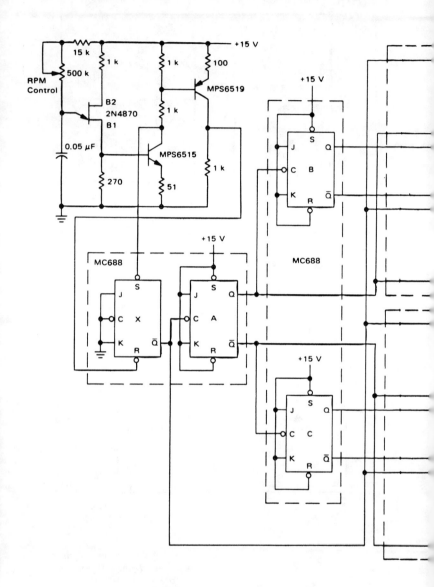

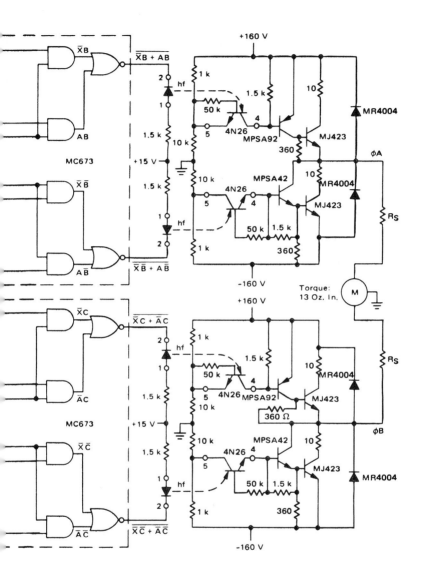

Variable speed control for induction motors (courtesy Motorola Semiconductor Products Inc.).

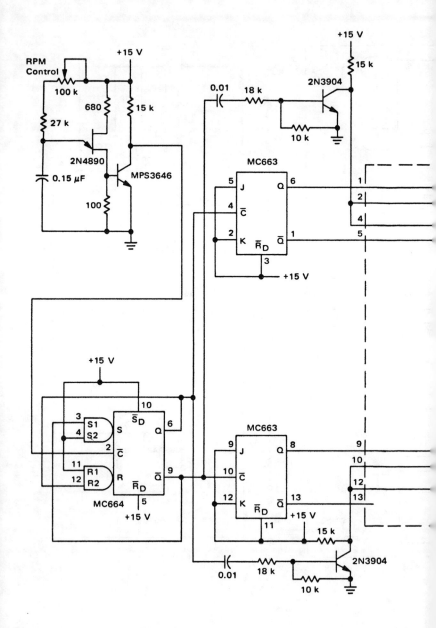

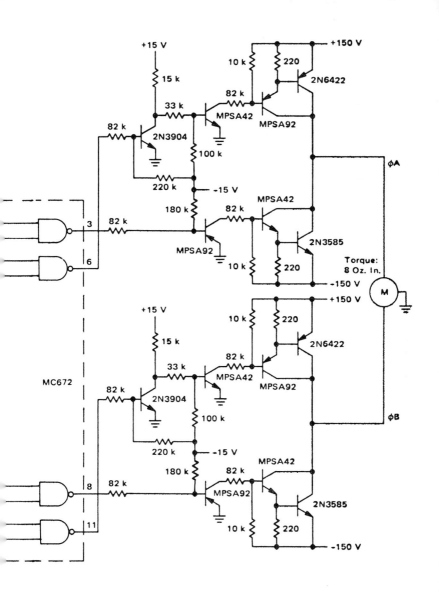

MHTL Logic: Pin 14 V_{CC}
Pin 7 Gnd

Variable speed control for induction motors (courtesy Motorola Semiconductor Products Inc.).

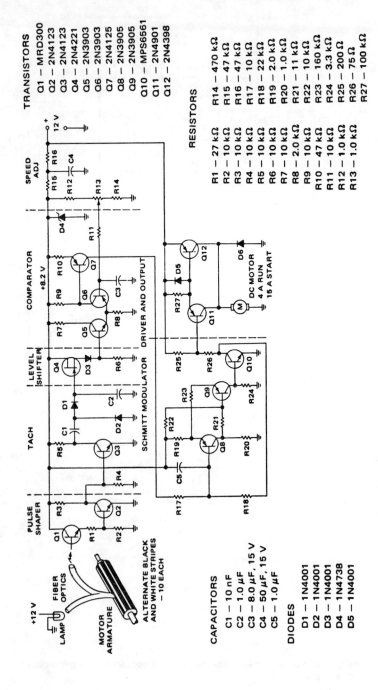

Regulated DC motor control with feedback from optical sensor (courtesy Motorola Semiconductor Products Inc.).

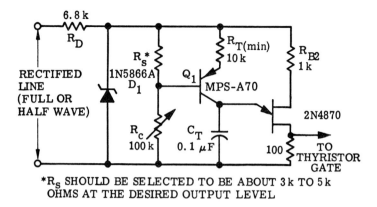

*R_S SHOULD BE SELECTED TO BE ABOUT 3k TO 5k OHMS AT THE DESIRED OUTPUT LEVEL

Feedback control circuit to trigger a thyrister (courtesy Motorola Semiconductor Products Inc.).

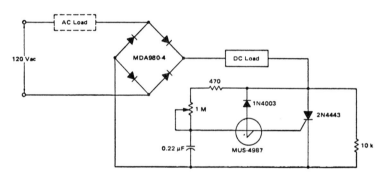

SCR full-range power controller incorporating an SUS (courtesy Motorola Semiconductor Products Inc.).

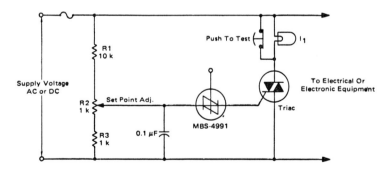

Electronic crowbar circuit using an SBS and a triac. This circuit protects equipment by placing a short-circuit across the line, thereby blowing the fuse. It works on DC circuits as well as AC types since the SBS and triac are both bilateral devices (courtesy Motorola Semiconductor Products Inc.).

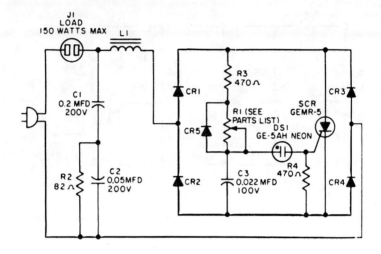

Parts List

C1 — 0.2-mfd, 200-volt capacitor
C2 — 0.05-mfd, 200-volt capacitor
C3 — 0.022-mfd, 100-volt capacitor
CR1 thru CR5 — GE-504A rectifier diode
DS1 — GE-5AH neon glow lamp
L1 — RFI choke; 65 turns of No. 18 varnished wire on 1/4" ferrite rod such as antenna loopstick rod (see Theory section)
J1 — Power socket; Amphenol type 61F, or equivalent

R1 — 100K two-watt potentiometer for fan-motor load or 250K for high-intensity lamp load
R2 — 82-ohm, 1/2-watt resistor
R3 — R4 — 470-ohm, 1/2-watt resistors
SCR — GEMR-5 silicon controlled rectifier
Minibox — 3-1/4" x 2-1/8" x 1-5/8"
Misc. — Line cord, vectorboard, push-in terminals

High-intensity lamp dimmer or fan control. This circuit is for use with high-intensity lamps with a built-in transformer only; do not use with 120-volt standard household incandescent or fluorescent lamps. Fan motors should not draw more than 1.5 amperes (courtesy General Electric Company).

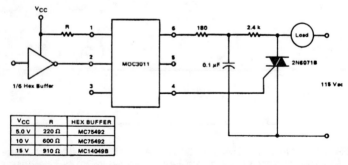

V_{CC}	R	HEX BUFFER
5.0 V	220 Ω	MC75492
10 V	600 Ω	MC75492
15 V	910 Ω	MC14049B

MOS to AC load interface using an MOC3011 optically coupled triac driver (courtesy Motorola Semiconductor Products Inc.).

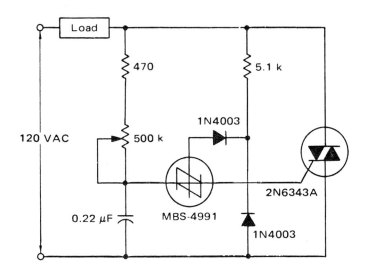

Hysteresis-free power controller using an SBS and a triac (courtesy Motorola Semiconductor Products Inc.).

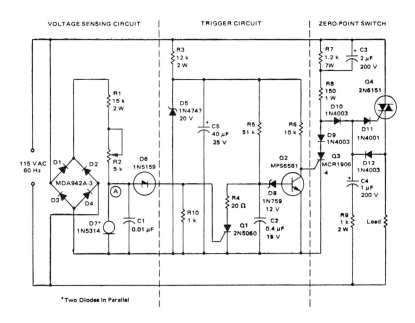

Triac overvoltage protection circuit with automatic reset. If the voltage at point A exceeds 11 volts during any half-cycle D6 fires and turns on SCR Q1, removing power from the load (courtesy Motorola Semiconductor Products Inc.).

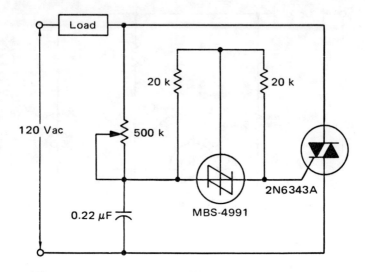

Low-cost light dimmer using an SBS and triac. Shunting the SBS with two 20K resistors minimizes the flash-on effect (courtesy Motorola Semiconductor Products Inc.).

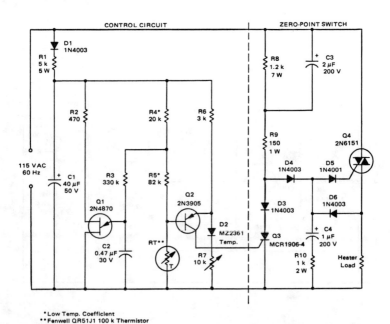

*Low Temp. Coefficient
**Fenwell QR51J1 100 k Thermistor

Triac heater temperature control circuit (courtesy Motorola Semiconductor Products Inc.).

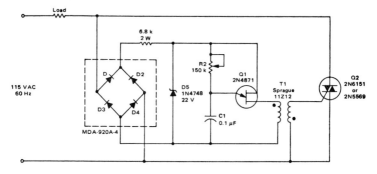

800-watt triac light dimmer (courtesy Motorola Semiconductor Products Inc.).

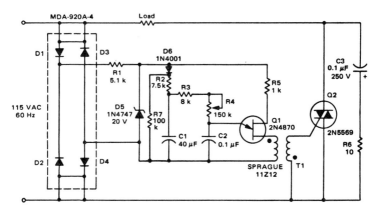

800-watt soft-start triac light dimmer (courtesy Motorola Semiconductor Products Inc.).

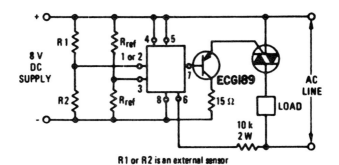

R1 or R2 is an external sensor

Triac control circuit with current boost using an ECG776 zero voltage switch. Resistor R2 must be the external sensor for the internal short and open protection to be operative. Notice that the circuit utilizes a DC supply. Select the triac for the particular application from the ECG5600 series (courtesy GTE Sylvania Incorporated).

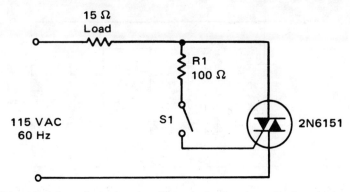

Triac AC static contactor (courtesy Motorola Semiconductor Products Inc.).

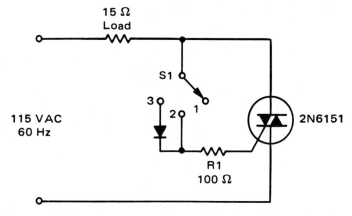

Three-position static switch. In position 1 the switch is off; in position 2 the switch supplies full-wave power to the load; in position 3 the diode conducts only on half-cycles and the load is supplied with half-wave power (courtesy Motorola Semiconductor Products Inc.).

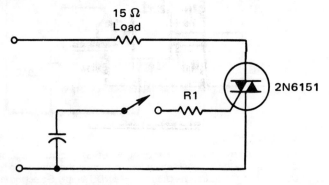

AC-controlled triac switch. R1 is 100 ohms (courtesy Motorola Semiconductor Products Inc.).

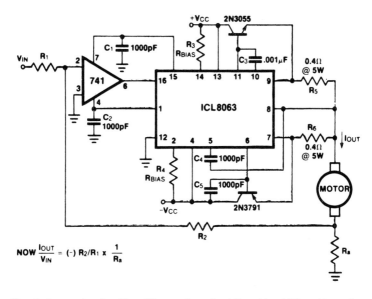

Constant-current motor drive. All capacitors should be at least 50 working volts and resistors the half-watt type, except those valued at 0.4 ohm and R_a. Use heat sinks for the power transistors with thermal compound. This circuit can be used where there is some likelihood of stalling or lockup. If the motor locks the current drive remains constant and the system does not destroy itself (courtesy Intersil, Inc.).

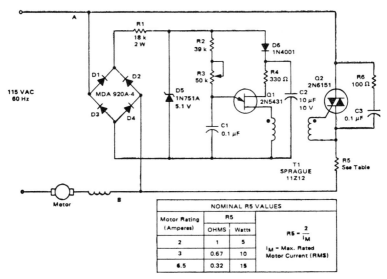

Triac motor speed control with feedback (courtesy Motorola Semiconductor Products Inc.).

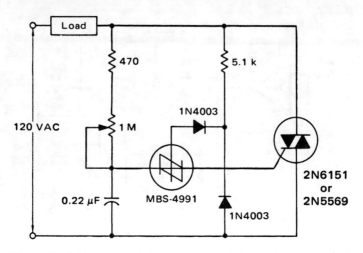

800-watt triac light dimmer with silicon bilateral switch, SBS (courtesy Motorola Semiconductor Products Inc.).

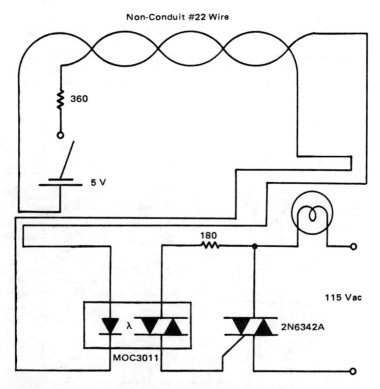

Remote control of AC load using an MOC3011 optically coupled triac driver (courtesy Motorola Semiconductor Products Inc.).

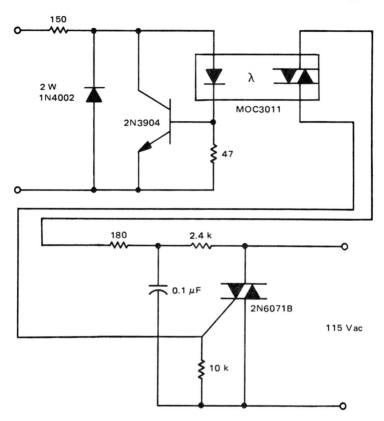

Solid-state relay circuit with input protection of the MOC3011 triac driver. The input voltage to the protection circuit can be 3 to 30 volts DC (courtesy Motorola Semiconductor Products Inc.).

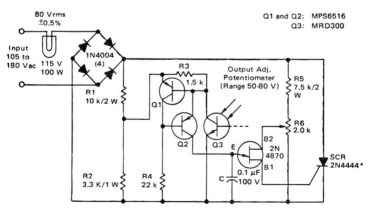

Projection lamp voltage regulator using a phototransistor, SCR and UJT (courtesy Motorola Semiconductor Products Inc.).

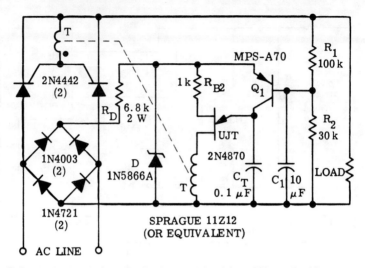

Full-wave average voltage feedback control circuit for a 600-watt load (courtesy Motorola Semiconductor Products Inc.).

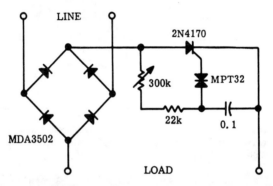

Simple DC power control circuit using an SCR (courtesy Motorola Semiconductor Products Inc.).

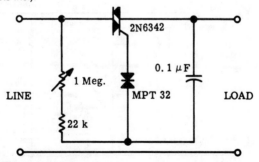

Simple full-wave power control circuit using a triac (courtesy Motorola Semiconductor Products Inc.).

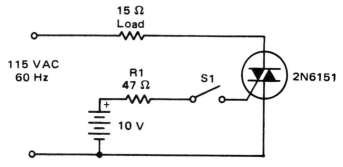

Low-voltage-controlled triac switch (courtesy Motorola Semiconductor Products Inc.).

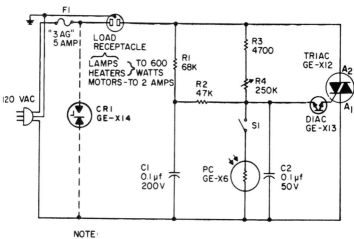

NOTE: MOUNT GE-X12 ON 3" X 3" X 1/16" COPPER OR ALUMINUM COOLING FIN

Parts List

C1 — 0.1-mfd, 200-volt capacitor
C2 — 0.1-mfd, 50-volt capacitor
CR1 — GE-X14 thyrector diode
  (optional transient voltage suppressor)
PC — GE-X6 cadmium sulfide photoconductor
R1 — 68K-ohms, 1/2-watt resistor
R2 — 47K-ohms, 1/2-watt resistor
R3 — 4700-ohm, 1/2-watt resistor
R4 — 250K-ohm, 2-watt potentiometer
Triac — GE-X12
Diac — GE-X13

Full-wave variable AC control for motors and lamps using a diac and triac. This circuit gives full symmetrical control from 0 to 100% over the AC load. For fan or blower operation it may be desirable to place a 100-ohm 1-watt resistor in series with a 0.1 μF capacitor directly across the triac to improve performance. Best operation is achieved when the circuit is adjusted during the brightest part of the day. R4 is adjusted so that the lamp plugged into the outlet is just off. In this manner as it gets darker the light will come on (courtesy General Electric Company).

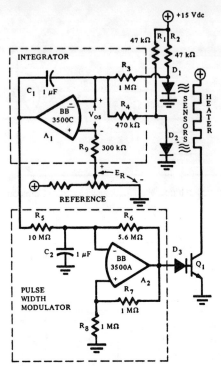

Proportional controller for heater using two op amps. The integrating error amplifier holds the heater current pulse width at a sustaining value at equilibrium (courtesy Burr-Brown Research Corporation).

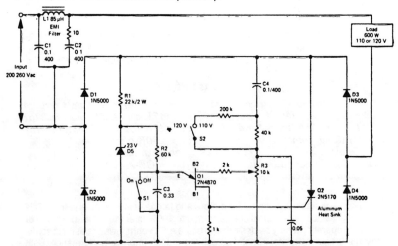

RMS open-loop voltage compensator (regulator) for small conduction angles. This circuit provides an output of 110/115 ± 2.5 volts at 600 watts for an input of 200 to 260 volts. This circuit is suitable for applications requiring a conduction angle of less than 90° (courtesy Motorola Semiconductor Products Inc.).

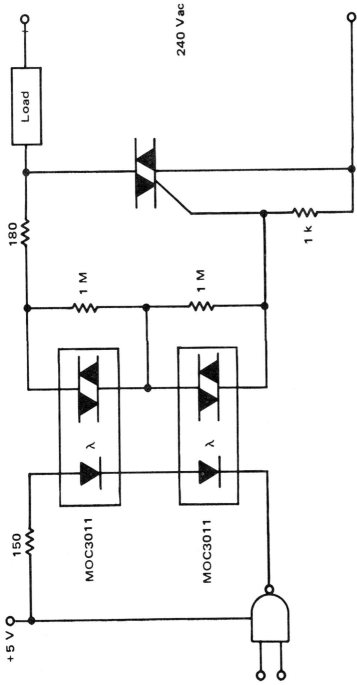

240-volt triac control circuit driven by two MOC3011 optically coupled triac drivers (courtesy Motorola Semiconductor Products Inc.).

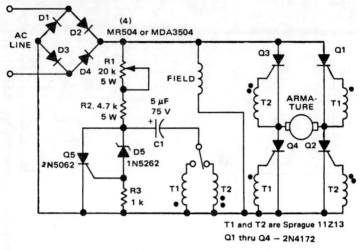

Direction and speed control for shunt-wound motors (courtesy Motorola Semiconductor Products Inc.).

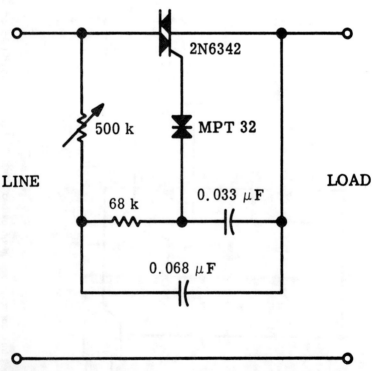

Full range AC power control circuit using a triac (courtesy Motorola Semiconductor Products Inc.).

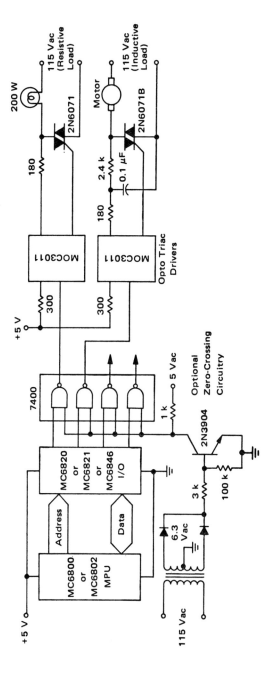

M6800 microprocessor to 115-volt AC load interface using optically coupled MOC3011 triac drivers (courtesy Motorola Semiconductor Products Inc.).

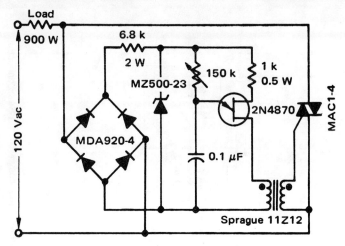

Full-wave trigger circuit for a 900-watt load using a triac and UJT (courtesy Motorola Semiconductor Products Inc.).

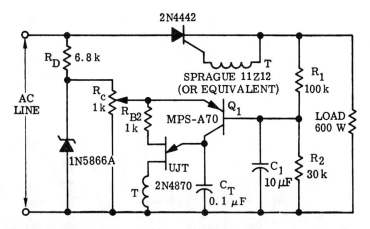

Half-wave average voltage control circuit for a 600-watt load (courtesy Motorola Semiconductor Products Inc.).

Computer-Related Circuits

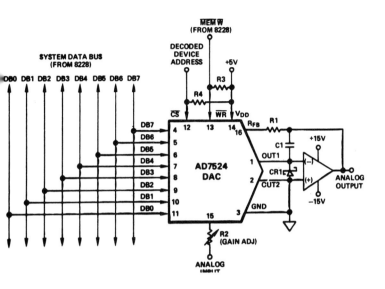

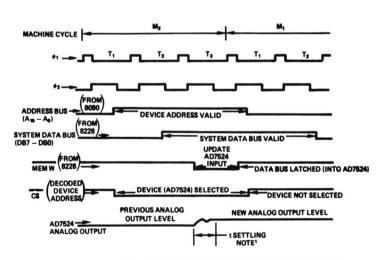

NOTE: ¹SETTLING TIME IS DEPENDENT PRIMARILY UPON OUTPUT AMPLIFIER SLEWING AND SETTLING CHARACTERISTICS. WAVEFORM SHOWN IS NOT REPRESENTATIVE OF ANY SPECIFIC AMPLIFIER.

Memory mapped output device interfaced with the MCS-80 (8080A) CPU system (courtesy Analog Devices, Inc.).

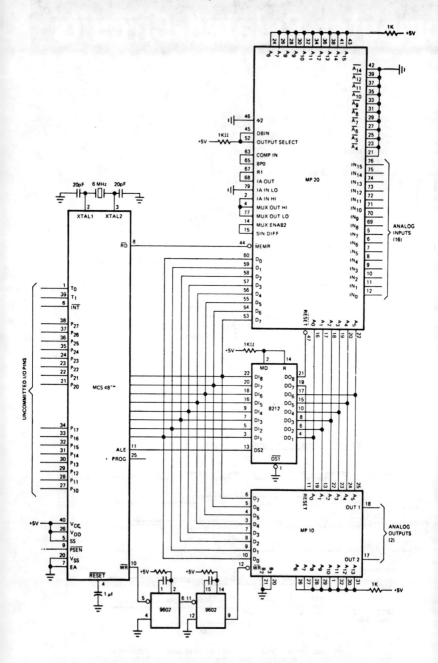

MCS-48 based analog processor. This circuit shows the MP-10 D/A converter and the MP-20 A/D converter connected to the 8748 processor (courtesy Intel Corporation).

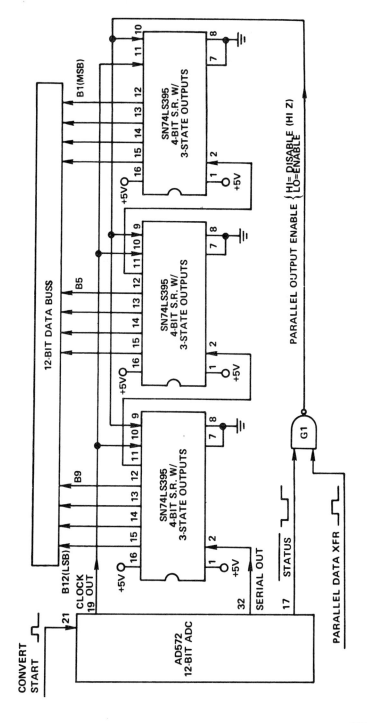

Serial data transfer circuit into shift registers with parallel output to data bus (courtesy Analog Devices, Inc.).

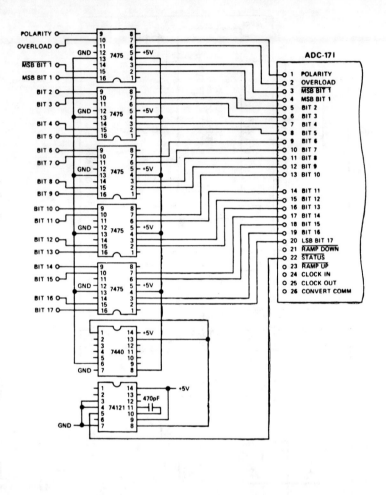
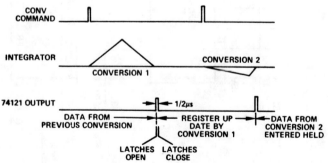

Output storage register using an ADC171 dual-slope integrating A/D converter (courtesy Analog Devices, Inc.).

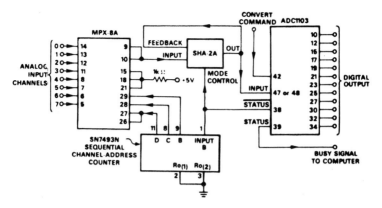

Sequentially addressed 8-channel data acquisition subsystem capable of acquiring data to 12-bit accuracy at a 220 kHz throughput rate. The MPX-8A is a multiplexer, the SHA-2A a sample-and-hold amplifier and the ADC1103 a high-speed A/D converter (courtesy Analog Devices, Inc.).

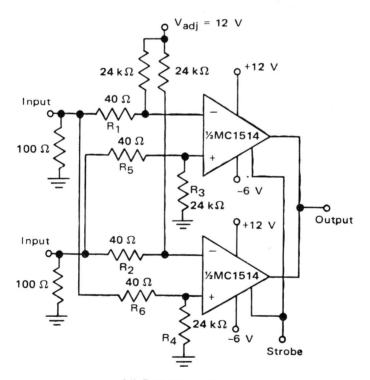

All Resistors in Ohms

Memory core sense amplifier using an MC1514 (courtesy Motorola Semiconductor Products Inc.).

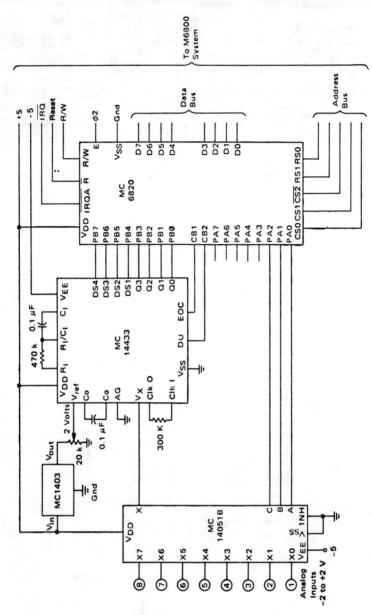

Eight-channel data acquisition circuit for the M6800 (courtesy Motorola Semiconductor Products Inc.).

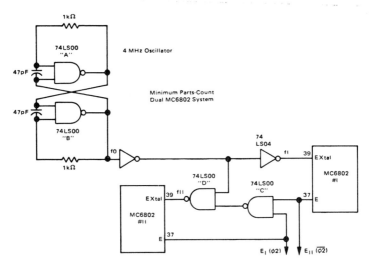

Synchronizing two M6802 microprocessors on one bus (courtesy Motorola Semiconductor Products Inc.).

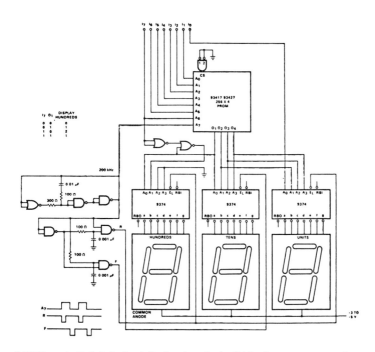

8-bit binary to 3-digit decimal display decoder for 8-bit microprocessor systems with 256 by 4 PROM, three 7-segment decoder/drivers with 9374 input latches and two gates (courtesy Fairchild Semiconductor).

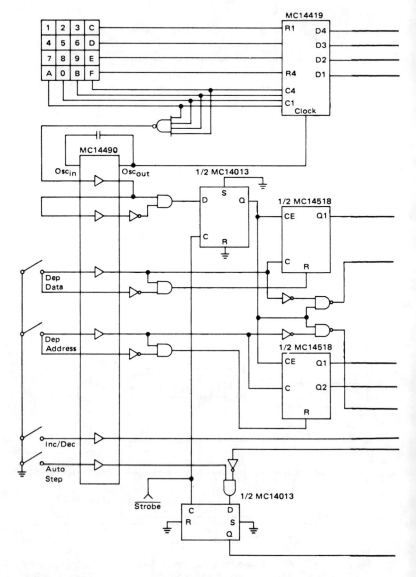

CMOS keyboard data entry system (courtesy Motorola Semiconductor Products Inc.).

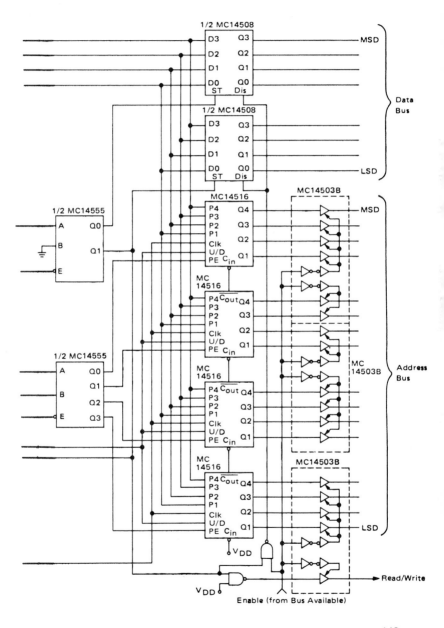

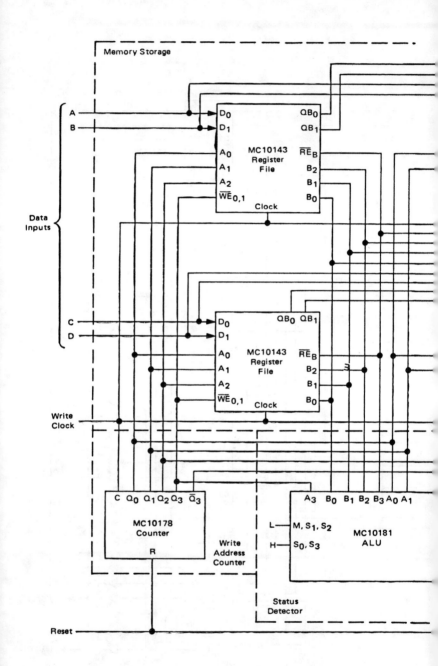

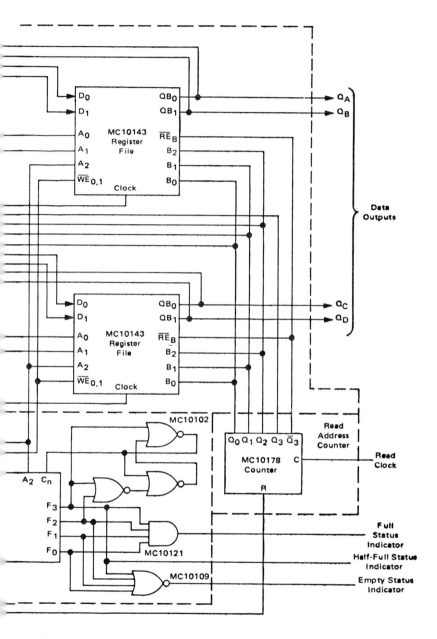

High-speed FIFO memory using the MC10143 register file (courtesy Motorola Semiconductor Products Inc.).

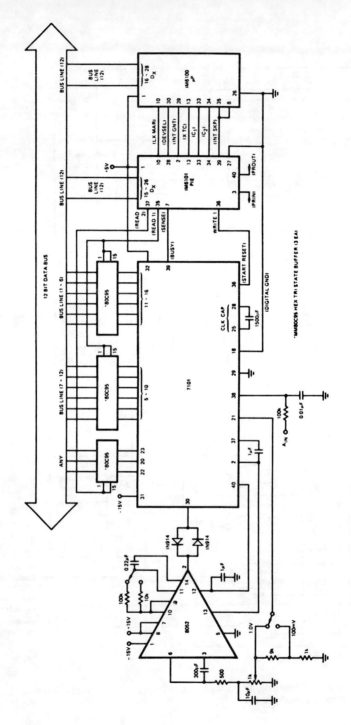

3½-digit parallel data acquisition system using the 8052/7101 A/D pair (courtesy Intersil, Inc.).

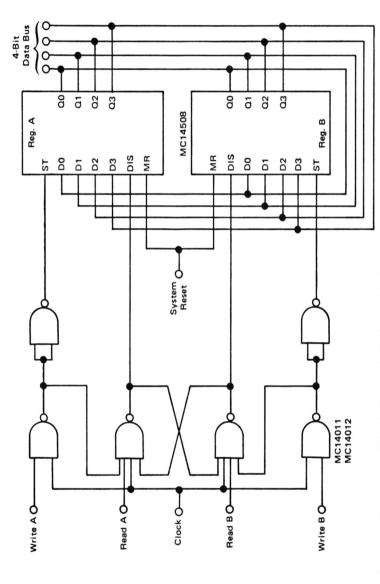

Dual 4-bit storage register (courtesy Motorola Semiconductor Products Inc.).

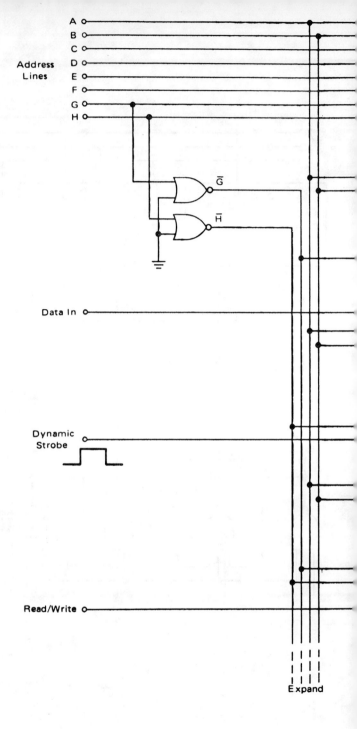

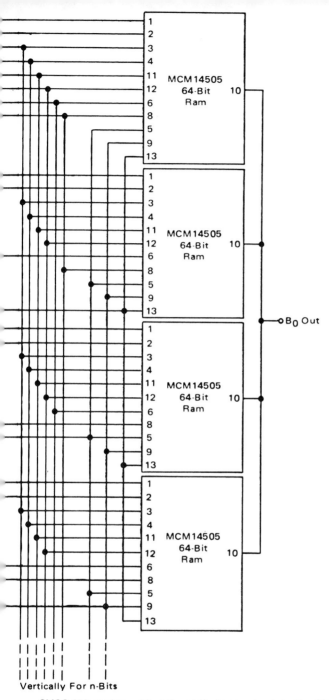

CMOS 256-word by n-bit static read/write memory (courtesy Motorola Semiconductor Products Inc.).

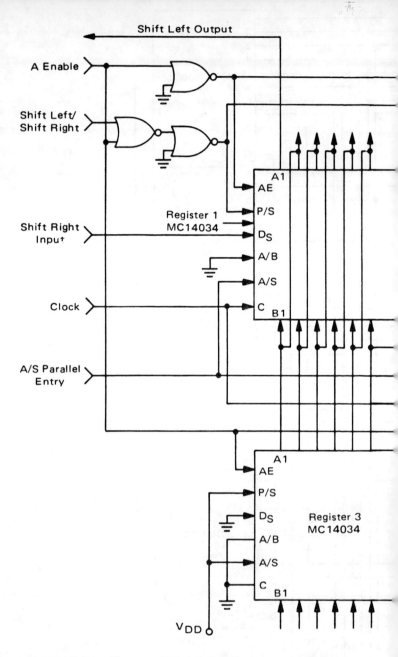

A "High" ("Low") on the Shift Left/Shift Right input allows serial data on the Shift Left Input (Shift Right Input) to enter the register on the positive transition of the clock signal. A "high" on the "A" Enable Input disables the "A" parallel data lines on Reg. 1 and 2 and enables the "A" data lines on registers 3 and 4 and allows parallel data

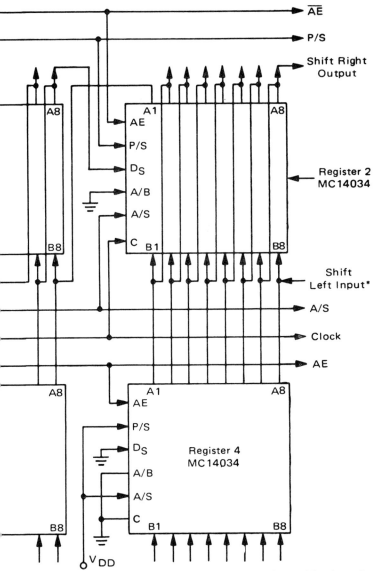

into registers 1 and 2. Other logic schemes may be used in place of registers 3 and 4 for parallel loading.

When parallel inputs are not used Reg. 3 and 4 and associated logic are not required.

*Shift left input must be disabled during parallel entry.

Shift-right/shift-left register with parallel inputs (courtesy Motorola Semiconductor Products Inc.).

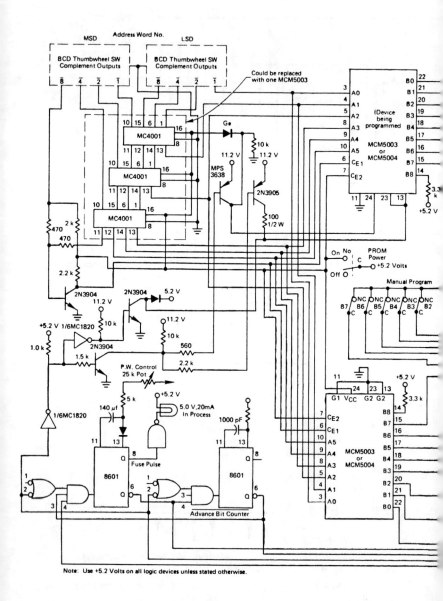

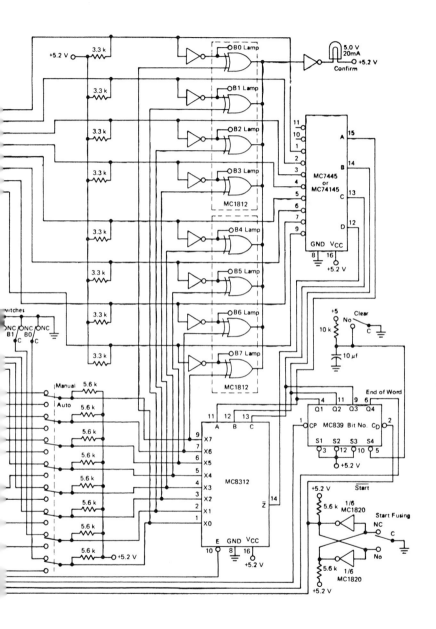

Manual/automatic PROM programmer (courtesy Motorola Semiconductor Products Inc.).

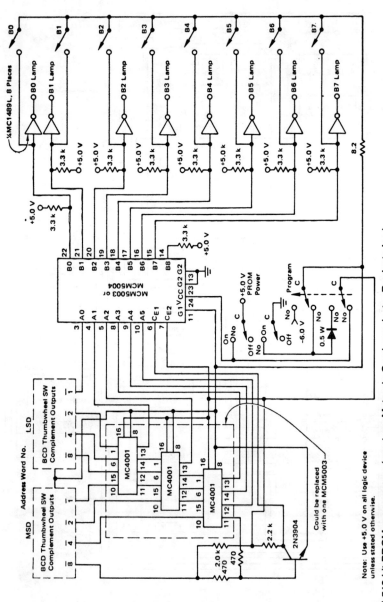

Manual 512-bit PROM programmer (courtesy Motorola Semiconductor Products Inc.).

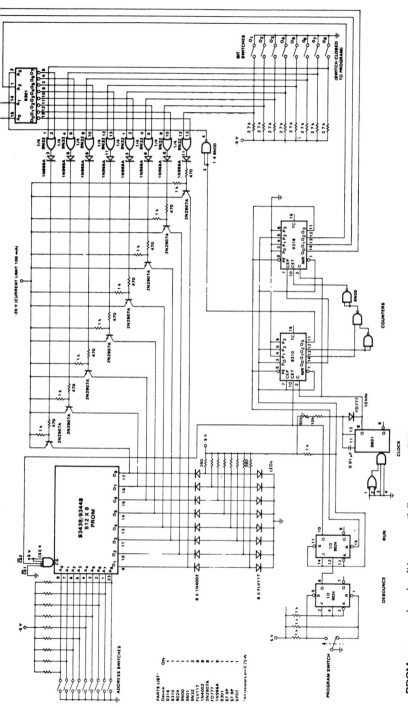

PROM programming circuit to sequentially program all bits of a given word for up to an 8-output PROM (courtesy Fairchild Semiconductor).

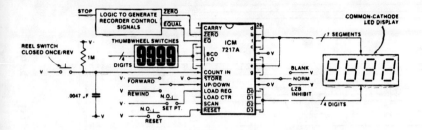

Tape recorder position indicator/controller using an ICM7217A 28-pin DIP. This circuit employs the up/down feature of the ICM7217 to keep track of the tape position. The not-equal and not-zero outputs can be used to control the recorder. To make the recorder stop at a particular point on the tape, the register can be set with the stop point and the equal output used to stop the recorder, either on fast forward, play or rewind. To have the recorder stop before the tape comes free of the reel on rewind a leader should be used. Resetting the counter at the starting point of the tape a few feet from the end of the leader allows the zero output to be used to stop the recorder on rewind. The 1M resistor and 0.0047 μF capacitor on the count input provides a time constant of about 5 ms to debounce the reel switch (courtesy Intersil, Inc.).

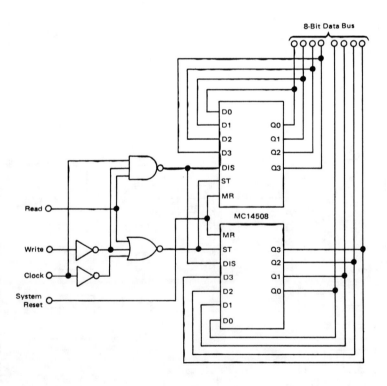

8-bit storage register (courtesy Motorola Semiconductor Products Inc.).

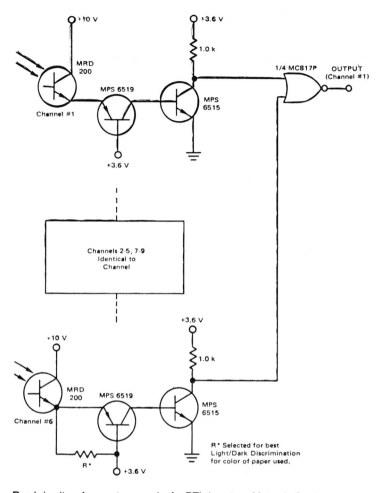

Read circuitry of paper-tape reader for RTL (courtesy Motorola Semiconductor Products Inc.).

Timers & Counters

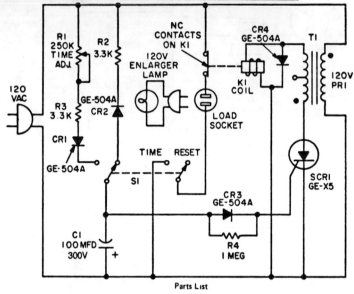

Parts List

C1 — 100-mfd, 300-volt electrolytic capacitor
CR1 thru *CR4* — GE-504A rectifier diode
K1 — 12-volt a-c relay (Potter and Brumfield-No. MR5A, or equivalent)
R1 — 250K-ohm, 2-watt potentiometer
R2, R3 — 3.3K-ohm, 1/2-watt resistor
R4 — 1-megohm, 1/2-watt resistor
S1 — DPDT toggle switch
SCR1 — GE-X5 silicon controlled rectifier
T1 — Filament transformer: primary, 120-volts a-c; secondary, 12.6-volts center tapped (Triad F25X, or equivalent)
Line cord, vectorboard, minibox etc.

Enlarger phototimer (courtesy General Electric Company).

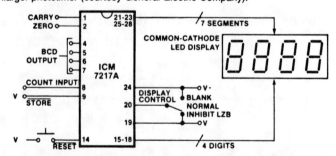

4-digit unit counter with BCD output using an Intersil ICM7217A 28-pin DIP. All that is required is an ICM7217, a power supply and a 4-digit display. Add a momentary switch for reset and an SPDT center-off switch to blank the display or view leading zeros. One more SPDT switch gives you up/down capabilities. With an ICM7217A and a common-cathode calculator-type display this is the least expensive digital counter/display system you can make (courtesy Intersil, Inc.).

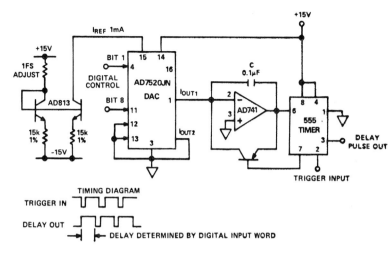

Digitally controlled time delay. This circuit generates a time delay controlled by the digital input of the AD7520. The current fed into the integrator's summing point at OUT 1 is determined by the D/A converter's input. The output of the integrator remains low until the trigger input of the 555 timer goes low. At that time the timer's output goes high, unclamping the output and permitting the integrator to change upward until it reaches ⅔V$_{DD}$, at which time the timer's output goes low and the integrator is reset to zero. The time delay is D/D1$_{REF}$ × 10V, where D is the fractional binary equivalent of the digital input word (courtesy Analog Devices, Inc.).

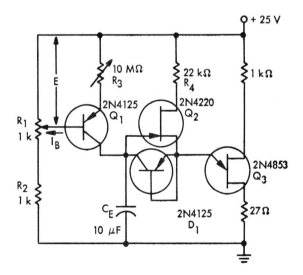

Long duration time delay circuit. Time delays of up to 10 hours are possible with this circuit (courtesy Motorola Semiconductor Products Inc.).

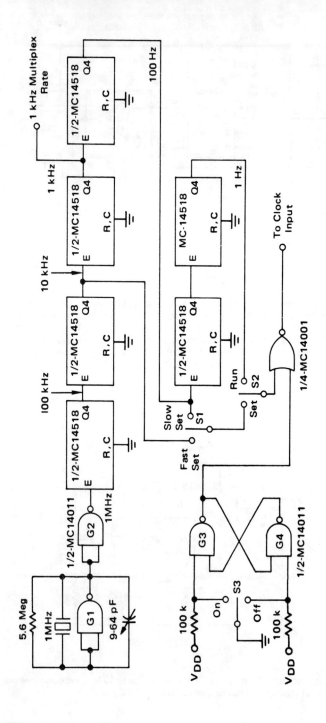

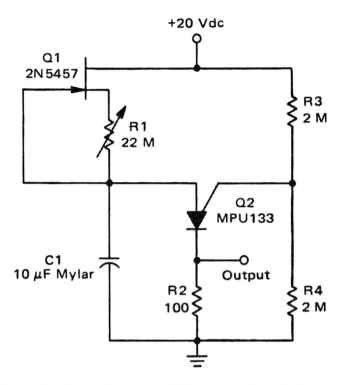

20-minute long-duration timer using a PUT (courtesy Motorola Semiconductor Products Inc.).

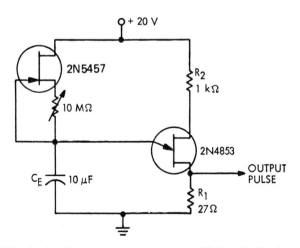

Time delay circuit with constant-current charging of timing circuit using a UJT and JFET. Constant currents of less than 1 µA can easily be obtained that result in time delays up to 10 minutes.

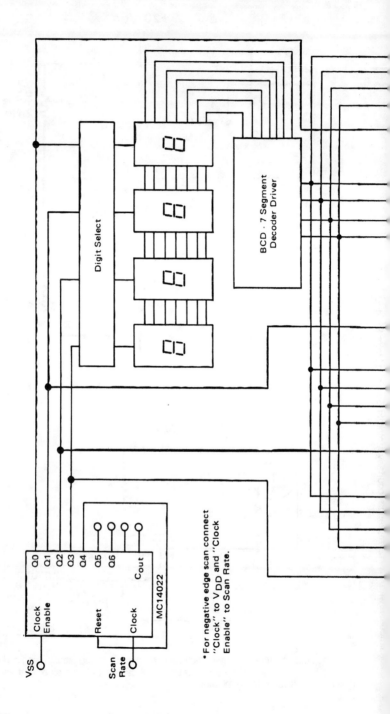

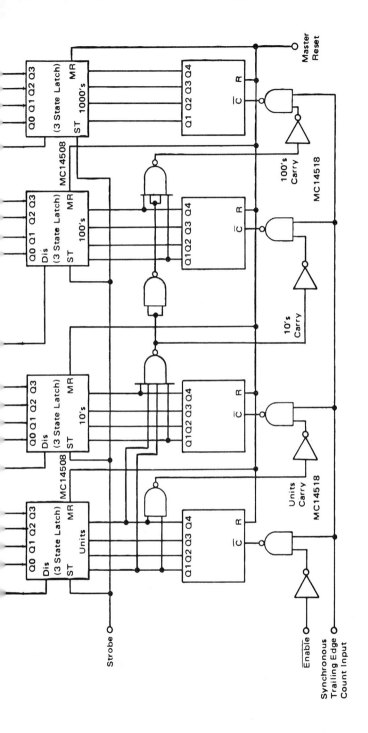

4-decade synchronous counter with data display multiplexing (courtesy Motorola Semiconductor Products Inc.).

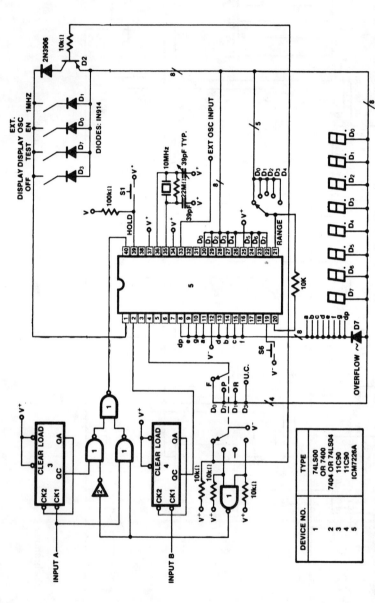

100 MHz multifunction counter using the Intersil ICM7226A 40-pin DIP. This circuit uses a divide-by-10 prescaler in the frequency counter mode (courtesy Intersil, Inc.).

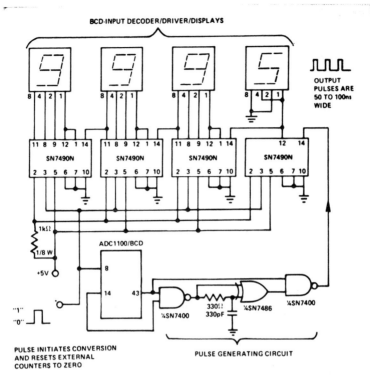

Count-by-5 counter using an ADC1100 dual-slope A/D converter (courtesy Analog Devices, Inc.).

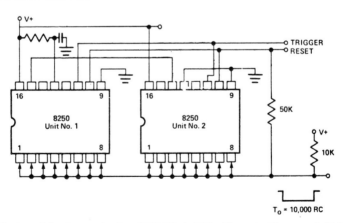

Ultralong delay timer using two 8250 16-pin DIPs. The output is normally high when reset, and goes low upon application of a trigger input. It stays low for a duration of $(100)^2$ or 10,000 cycles. Total timing cycle for the two 8250s can be programmed from To = 1RC to To = 9999RC in 10,000 discrete steps by selectively shorting any combination of pins 1 through 8 from both units to the output bus (courtesy Intersil, Inc.).

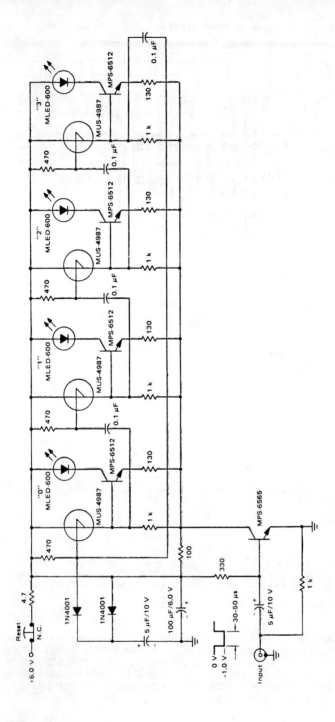

Four-stage ring counter using SUS devices (courtesy Motorola Semiconductor Products Inc.).

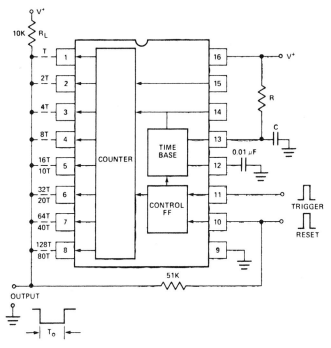

Precision programmable timer using the Intersil 8240/8250 16-pin DIP. The timer is used in the monostable mode. The output is normally high and goes low to trigger an input. Components R and C at pin 13 determine the time cycle. The timing duration, T_o is equal to NRC, where R and C are component values and N is a number between 1 and 255. Integer N is determined by the combination of pins 1 through 8 connected to the output bus. The 8250 can be programmed for numbers between 1 and 99. The numbers shown on the schematic adjacent to pin 1 through 8 are the respective N integers, with the upper ones for the 8240 and the lower ones for the 8250 (courtesy Intersil, Inc.).

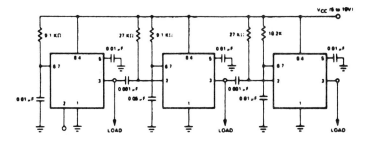

Time sequencer using ECG955M timer/oscillator chips. The first timer is started by momentarily connecting pin 2 to ground. It runs for 10 ms, then triggers the second timer. The second one runs for 50 ms, at which time it triggers the third. Note that the timing resistors and capacitors can be programmed and that each circuit could trigger several other timers (courtesy GTE Sylvania Incorporated).

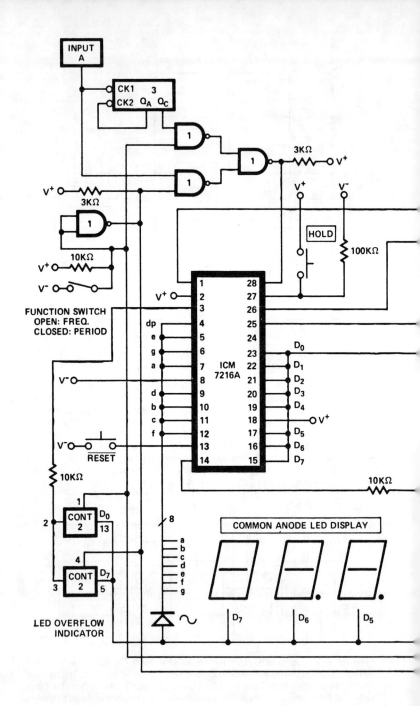

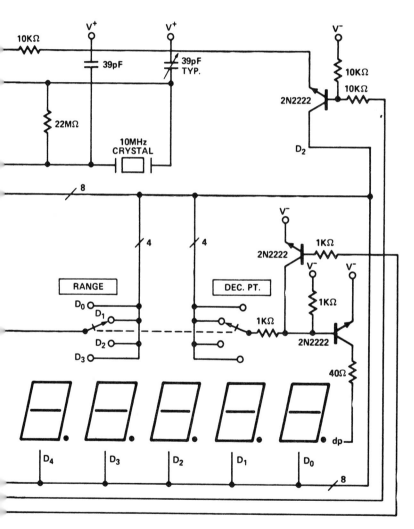

100 MHz frequency and 2 MHz period counter using an Intersil ICM7216A 28-pin DIP (courtesy Intersil, Inc.).

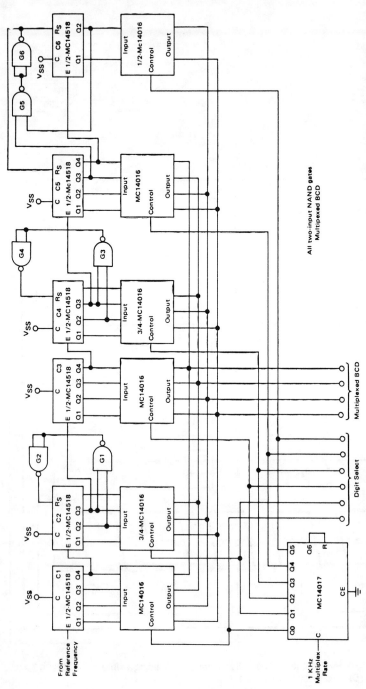

24-hour clock without the 1-hertz reference circuitry. It provides seconds, minutes and hours (courtesy Motorola Semiconductor Products Inc.).

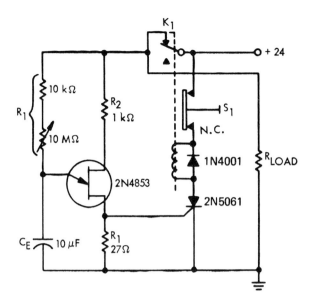

Time delay circuit using a UJT. Maximum time delay is set by the 10M pot. Time delay can be set from less than a second to approximately 2.5 minutes (courtesy Motorola Semiconductor Products Inc.).

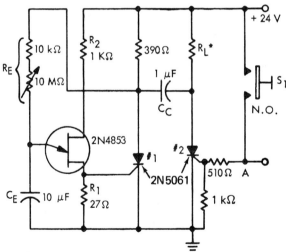

*Value of R_L must be low enough to allow hold current to flow in the SCR.

Time delay circuit using a UJT and two SCRs. Time delay is determined by setting of 10M pot from less than a second to approximately 2.5 minutes (courtesy Motorola Semiconductor Products Inc.).

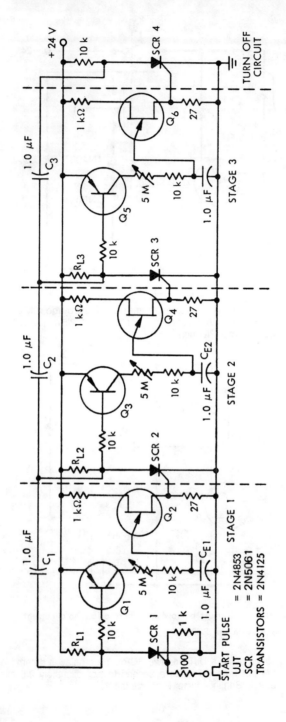

Sequential UJT-SCR timer circuit. This circuit is useful in washing machine circuits where different time cycles are required (courtesy Motorola Semiconductor Products Inc.).

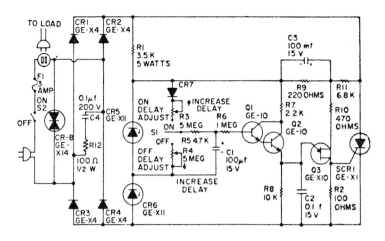

Parts List

C1, C3 — 100-mfd. 12-volt electrolytic capacitor
C2 — 0.1-mfd, 12-volt capacitor
C4 — 0.1-mfd, 200-volt capacitor
CR1, CR2, CR3 CR4 — GE-X4 rectifier diodes
CR5, CR6 — GE-X11 zener diode
CR7 — GE-504A rectifier diode
CR8 — GE-X14 thyrector diode
F1 — 3-amp. fuse
R1 — 3500-ohm, 5-watt resistor
R2, R12 — 100-ohm, 1/2-watt resistor
R3, R4 — 5-megohm potentiometers
R5 — 4700-ohm, 1/2-watt resistor
R6 — 1-megohm, 1/2-watt resistor
R7 — 2200-ohm, 1/2-watt resistor
R8 — 10000-ohm, 1/2-watt resistor
R9 — 220-ohm, 1/2-watt resistor
R10 — 470-ohm, 1/2-watt resistor
R11 — 6800-ohm, 1/2-watt resistor
Q1, Q2 — GE-10 transistor
Q3 — GE-X10 unijunction transistor
S1 — SPDT switch
S2 — SPST switch
SCR1 — GE-X1 silicon controlled rectifier
Minibox — 5" x 4" x 3"

Long time delay power switch for porch or garage. The circuit can handle loads up to 500 watts (courtesy General Electric Company).

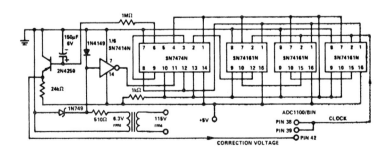

PLL clock for an ADC1100/BIN dual-slope A/D converter (courtesy Analog Devices, Inc.).

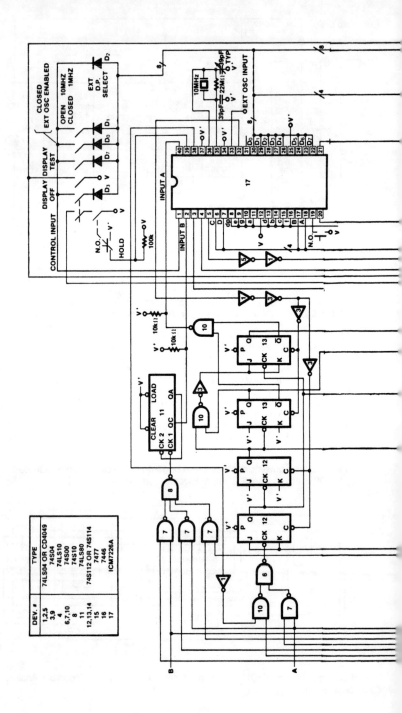

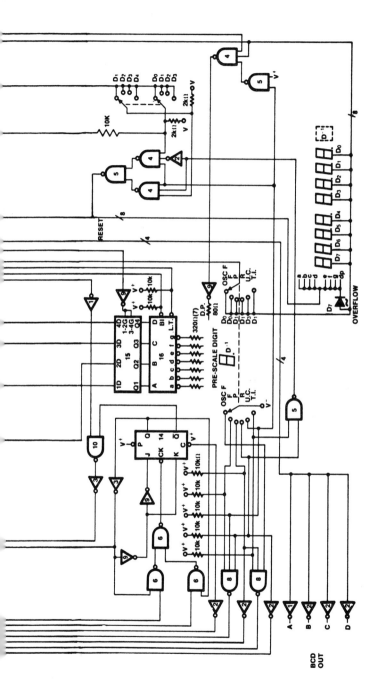

9-digit universal counter using the ICM7226A 40-pin DIP (courtesy Intersil, Inc.).

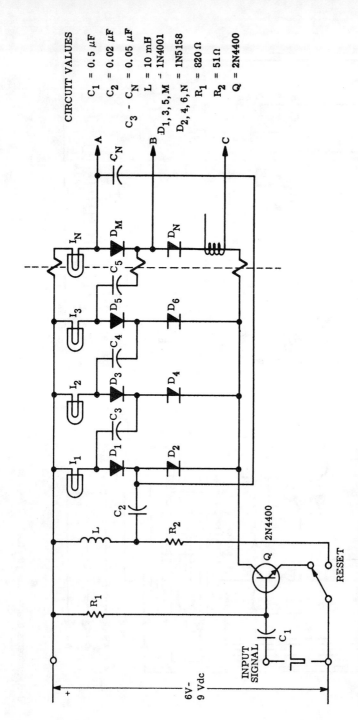

Ring counter that transfers power sequentially to a series of *n* loads (courtesy Motorola Semiconductor Products Inc.).

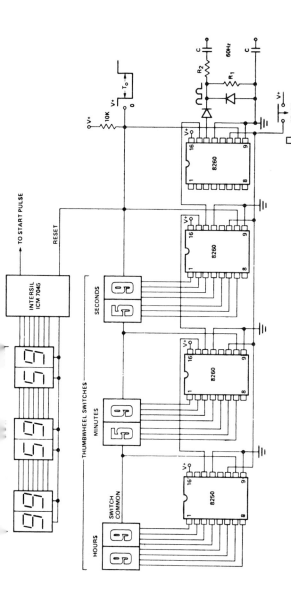

Programmable 100-hour timer with display. The 8260 on the right uses the carryout gate to generate a 1-second clock from the 60-hertz line. The diodes on the time base input rectify the input signal and alternately clamp and release the internal pull-up resistor at pin 14. The input network depends on the amplitude of the 60-hertz signal available. The internal oscillators are disabled with a 1K resistor at pin 13. The second and third 8260s are programmable with thumbwheel switches up to 59 seconds and 59 minutes. The carryout of each divider drives the next counter. The 8250 was chosen to give the maximum count of 99 hours. All reset pins are tied together and back to the 10K output pull-up resistor at the thumbwheels. The timing cycle begins by pressing the pushbutton. The 7045 provides a counter chip pluse direct drive to the 7-segment LEDs. Timing resolution can be increased to hundredths of a second by substituting 8250s for the initial stages and using the 60-hertz line to generate a 100-hertz clock (courtesy Intersil, Inc.).

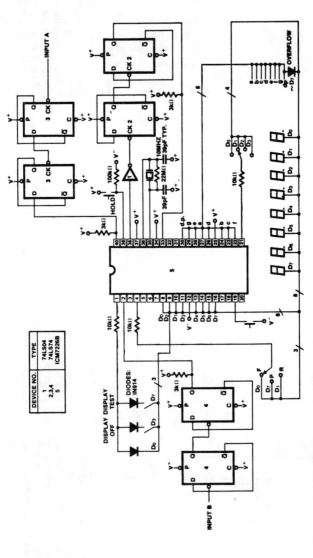

Notes: 1) If a 2.5MHz crystal is used then diode D1 and I.C's 1 and 2 can be eliminated.

40 MHz frequency and period counter. To obtain the frequency and period it is necessary to divide the 10 MHz oscillator frequency down to 2.5 MHz. In doing this the time between measurements is 800 ms and the display multiplex rate is 125 hertz (courtesy Intersil, Inc.).

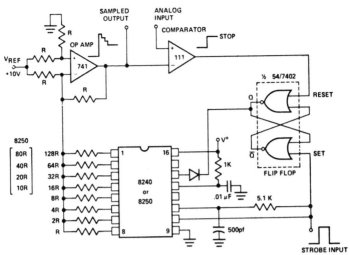

Digital sample-and-hold circuit. When a strobe input is applied the 8240/8250 is first reset, then triggered through the small RC network at pin 11, which delays the strobe signal. The strobe also sets the flip-flop, which in turn enables the counter via pin 14. The op amp goes to the high state and begins to count down at a rate set by the counter time base. When the op amp output reaches the analog input to be sampled the comparator switches, resetting the flip-flop and stops the count. The op amp output will accurately hold the sampled value until the next strobe pulse is applied. If the time base here is used the maximum acquisition time would be 256 (8240, 100 for the 8250) times 0.01 ms, or 2.6 ms (courtesy Intersil, Inc.).

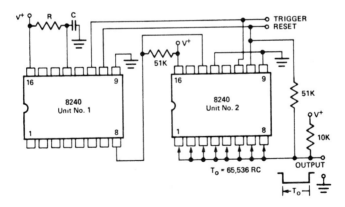

Ultralong delay timer using two 8240 16-pin DIPs. The time base of unit 2 is disabled. The output is normally high when the system is reset. Upon application of a trigger the output goes low and stays that way for a duration of $(256)^2$ or 65,536 cycles of the time base oscillator. Total timing cycle of the circuit is programmed from $T_o = 256RC$ to $T_o = 65,536 RC$ in 256 discrete steps by selecting any combination of the counter outputs of unit 2 (courtesy Intersil, Inc.).

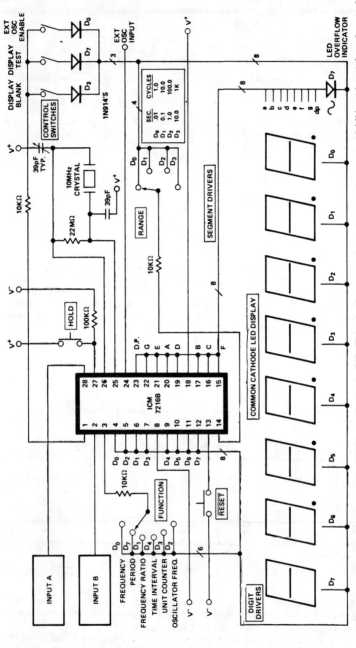

10 MHz universal counter using the Intersil ICM7216 B 28-pin DIP. This circuit can use input frequencies up to 10 MHz at input A and up to 2 MHz at input B (courtesy Intersil, Inc.).

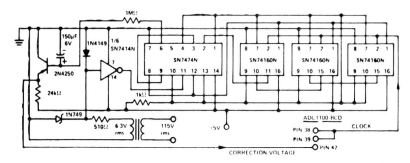

PLL clock for a ADC1100/BCD dual-slope A/D converter (courtesy Analog Devices, Inc.).

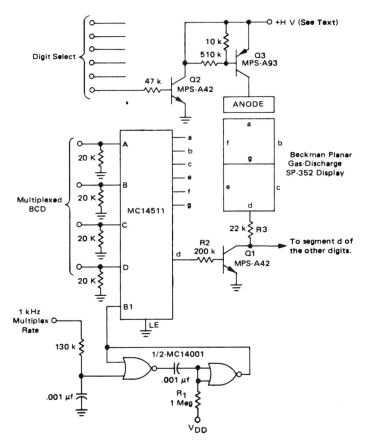

Multiplexed BCD-to-7-segment decoder/driver/displays for a 24-hour clock. Supply voltage of 160 to 200 volts is required to power the display (courtesy Motorola Semiconductor Products Inc.).

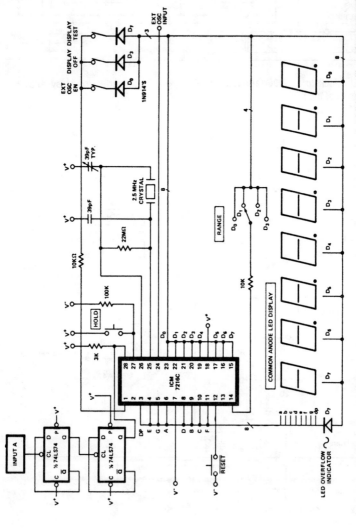

40 MHz frequency counter using the Intersil ICM7216C 28-pin DIP. To measure the correct value the 2.5 MHz oscillator frequency is divided by four as well as the input frequency (courtesy Intersil, Inc.).

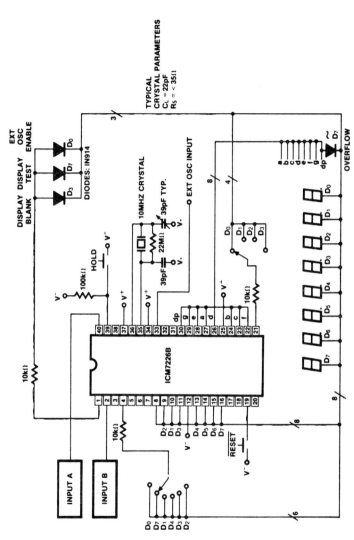

10MHz universal counter using an Intersil ICM7226B 40-pin DIP. This circuit measures frequencies at inptu A up to 10 MHz and at input B up to 2 MHz (courtesy Intersil, Inc.).

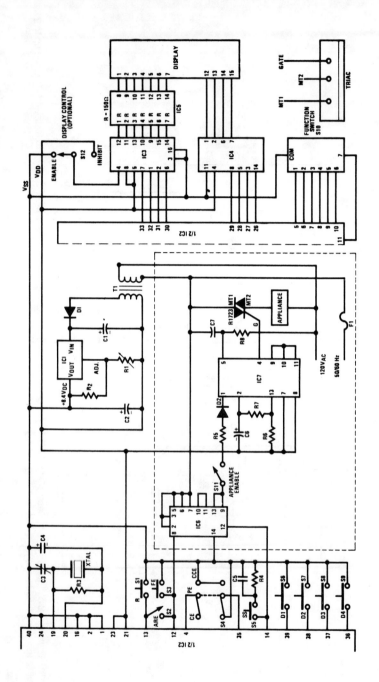

PARTS LIST

R1	5kΩ trimpot	Triac	HEP R1723
R2	240Ω, ¼W, 5% resistor	F1	1A fast or normal blow fuse
R3	20MΩ, ¼W, 5% resistor	XTAL	32.8kHz crystal (32.768kHz can be substituted. Timer will lose about 35 sec in 11 hr 20 min of use.)
R4	1MΩ, ¼W, 5% resistor		
R5	100kΩ, ¼W, 5% resistor		
R6	5.1kΩ, ¼W, 5% resistor	S1, S3, S5	SPST, NO, momentary pushbutton switches; part of flex-circuit switch assembly.
R7	4.7kΩ, ¼W, 5% resistor		
R8	10kΩ, 1W, 5% resistor	S2	SPST slide switch; part of flex-circuit switch assembly.
C1	470 - 1000mF, 25 V capacitor	S4	DPDT, center OFF toggle switch
C2	10mF, 25WV$_{DC}$ solid tantalum capacitor	S6 - S9	SPST, NO, momentary pushbutton switches
C3	6 - 25pF variable capacitor. Sprague QT1-18 4 - 30pF may be used.	S10	7 - 12 position rotary switch – Centralab PS-101 or Alcoswitch MRC-1-10.
C4	25 - 27pF, disc ceramic capacitor	S11	SPST toggle switch
C5	0.01mF disc ceramic capacitor	S12	SPDT toggle switch (optional)
C6	100mF, 25WV$_{DC}$ capacitor	Display	National Semiconductor NSB5411 4-digit multiplexed display.
C7	0.05mF, 200WV$_{DC}$ capacitor		
D1, D2	IN4003	Heat Sink	TO-220 heat sink. Two needed.
T1	10 - 16.5 V$_{AC}$ @ 300mA transformer	Misc.	16 display mounting pins (strip of 16 pins); 1 case; Clock/Instrument (available from James Electronics); 1 flex-circuit; 1 flex-circuit insulator; 2 Tinnerman nuts, #6; fuseholder; appliance control box, # LMB C.R.-234; 115V$_{AC}$ chassis mounting socket; miniature jacks; phone cable (shielded); IC sockets.
IC1	LM317T voltage regulator		
IC2	MM5865 universal timer		
IC3	CD14511 decoder/driver/latch		
IC4	DS8877 or DS75492 digit driver		
IC5	RA07 - 150 resistor array		
IC6	74C02 quad 2-input NOR gate		
IC7	CA3059 zero voltage switch		

Switch Abbreviations

Abbreviation	Switch	Functions
ARE	Automatic Reset Enable	6
C	Comparator	1-7
D1	LSD Programming	6, 7
D2	Digit 2 Programming	6, 7
D3	Digit 3 Programming	6, 7
D4	MSD Programming	6, 7
F	Function	1-7
FE	Final Event	1-5
LC	Latch Control	3, 4
R	Reset	1-7
SS	Start/Stop	1-7

4-digit 7-function stopwatch/timer. See parts list for component part numbers and values. This circuit is based on the MM5865 universal timer IC. As drawn the display resolution is 1 second. A SPST switch can be included between pin 16 of IC2 and V$_{SS}$ to provide a resolution of 0.01 second and 1 second. Another option is the display control switch which may be used to inhibit the display (courtesy National Semiconductor Corporation).

Parts List

- C1 — 100-mfd, 200-volt electrolytic capacitor
- C2 — 0.1-mfd, 50-volt paper capacitor
- C3 — 100-mfd, 12-volt electrolytic capacitor
- CR1 thru CR5 — GE-504A rectifier diode
- R1 — 3.3K-ohm, 1/2-watt resistor
- R2 — 1K-ohm, 2-watt potentiometer
- R3 — 150K-ohm, 2-watt potentiometer
- R4 — 1 megohm, 1/2-watt resistor
- R5 — 3.3K-ohm, 2-watt resistor
- R6 — 150-ohm, 1/2-watt resistor
- S1 — SPDT toggle switch
- SCR1 — GEMR-5 silicon controlled rectifier
- SCR2 — GE-X1 silicon controlled rectifier
- Line cord, load socket, minibox and heatsink (1-1/2" x 2-1/2" x 1/8" aluminum or copper)

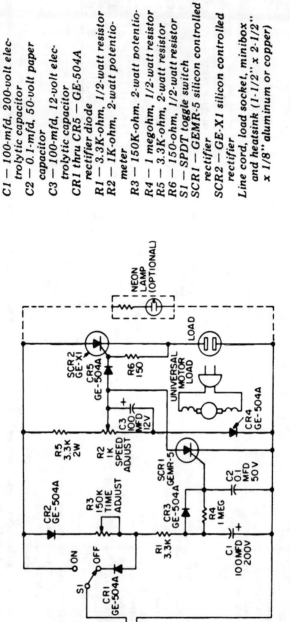

Universal motor control with built-in self-timer. Use this circuit only with motors having commutators. If heavy motor loads are anticipated, use a larger-rated C30B SCR in place of the GE-X1 for SCR2. To increase the time delay increase the capacitance of C1 (courtesy General Electric Company).

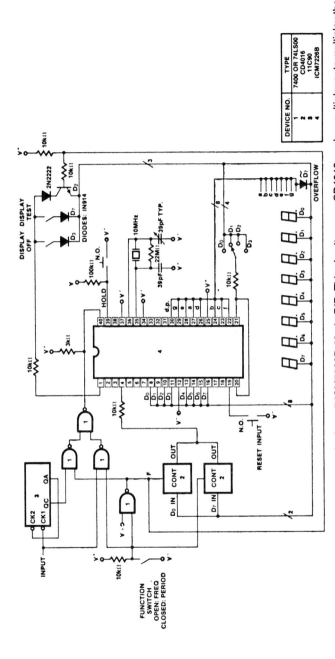

100 MHz frequency and period counter using the Intersil ICM7226B 40-pin DIP. This circuit uses a CD4016 analog multiplexer to multiplex the digital outputs back to the function input. Since the CD4016 is a digitally controlled analog transmission gate no level shifting of the digital output is required. CD4051s or CD4052s could also be used to select the proper inputs for the multiplexed input on the ICM7226 from 2- or 3-bit digital inputs. These analog multiplexers could also be used in systems in which the mode of operation is controlled by a microprocessor rather than directly by front panel switches (courtesy Intersil, Inc.).

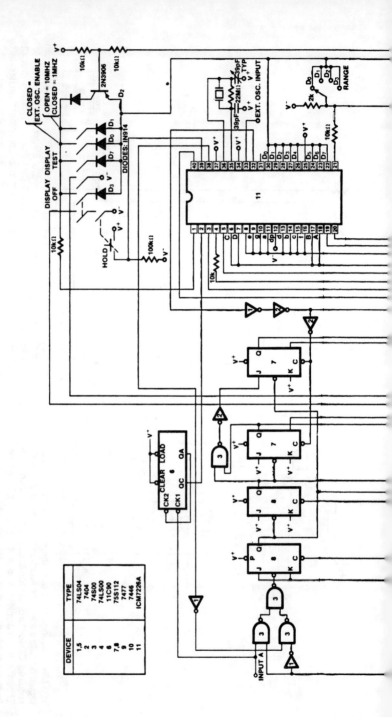

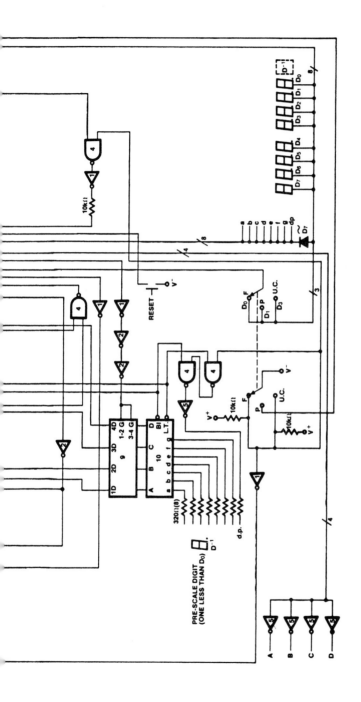

9-digit multifunction counter using the Intersil ICM7226A 40-pin DIP (courtesy Intersil, Inc.).

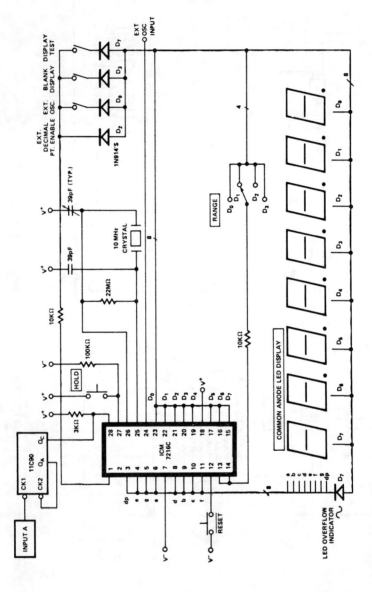

100 MHz frequency counter using the Intersil ICM7216C 28-pin DIP (courtesy Intersil, Inc.).

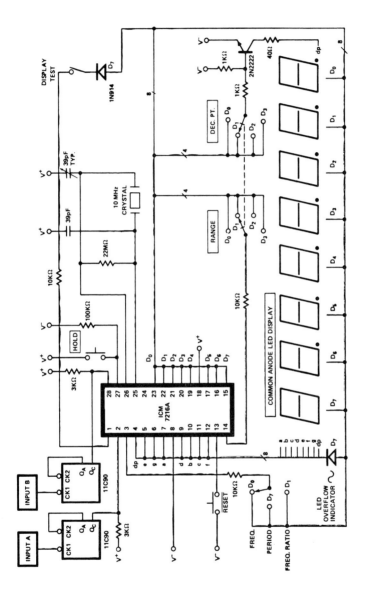

100 MHz multifunction counter using an Intersil ICM7216A 28-pin DIP (courtesy Intersil, Inc.).

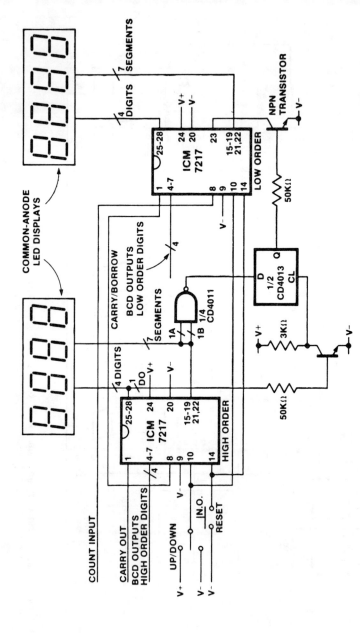

8-digit up/down counter using two cascaded ICM7217 28-pin DIPs. The NAND gate detects whether a digit is active since one of the two segments not-a or not-b is active on any unblanked number. The flip-flop is clocked by the LSB of the higher-order counter so if this digit is unblanked the Q output of the flip-flop goes high and turns on the NPN transistor, inhibiting leading zero blanking on the lower-order counter (courtesy Intersil, Inc.).

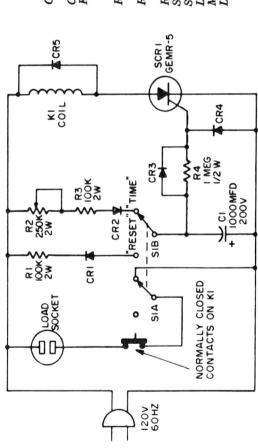

Time-delayed relay for up to 1 minute (courtesy General Electric Company).

Parts List

C1 — 1000-mfd, 200-volt electrolytic capacitor
CR1 – CR5 — GE 504A diodes
K1 — 115-volt, a-c, relay (Potter and Brumfield KA11AY, or equivalent)
R1, R3 — 100K-ohms, 2-watt resistor
R2 — 250,000-ohms, 2-watt potentiometer
R4 — 1 megohm, 1/2-watt resistor
S1 — DPDT toggle switch
SCR1 — GEMR-5
Load Socket
Minibox — 5" x 4" x 3"
Line cord and grommet

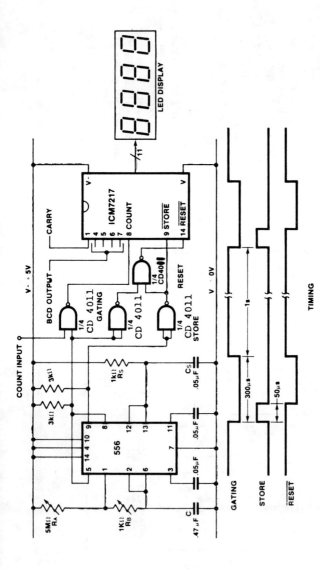

Inexpensive frequency counter/tachometer. This circuit uses a 556 dual timer to generate the gating, not-store and not-reset signals for a ICM7217 counter. One timer is an astable multivibrator using R_A, R_B and C to provide an output that is positive for approximately 1 second and negative for approximately 300 to 500 μs to serve as the gating signal. The system is calibrated by using a 5M pot for R_A as a coarse control and a 1K pot for R_B as a fine control. The other timer is a one-shot multivibrator triggered by the negative-going edge of the gating. This output at pin 9 is inverted to serve as the store pulse and to hold not-reset high. When the one-shot times out and the store goes high, not-reset goes low, resetting the counter. The one-shot pulse width is approximately 50 μs with the components shown. When fine trimming with R_B care should be taken to keep the gating low time at least twice as long as the one-shot pulse width (courtesy Intersil, Inc.).

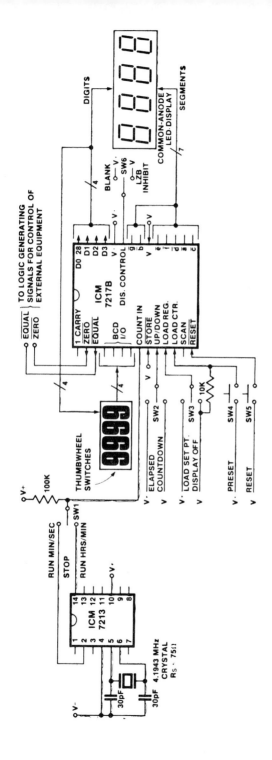

Precision elapsed time/countdown timer using an Intersil ICM7217B 28-pin DIP. This circuit also uses an ICM7213 precision 1-minute/1-second time-base generator to provide a 4.1943 MHz crystal oscillator and divider, generating pulses counted by the ICM7217B, which is a 5959-maximum-count device. The thumbwheel switches allow a starting time to be entered into the counter for a preset countdown type timer, and allow the register to be set for compare functions. For instance, to make a 24-hour clock with BCD output the register can be reset with 2400 and the equals output used to reset the counter (courtesy Intersil, Inc.).

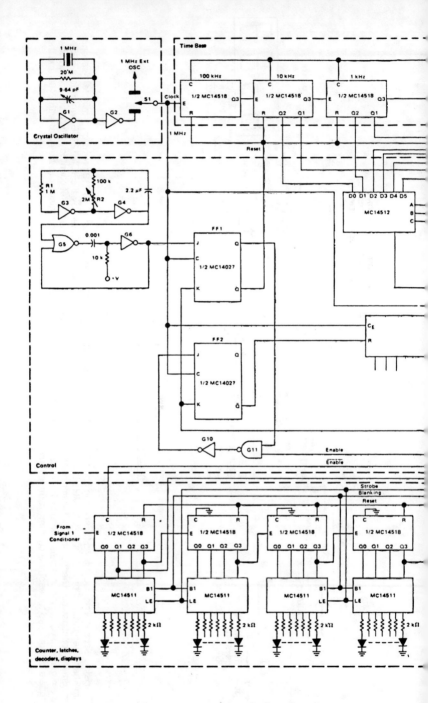

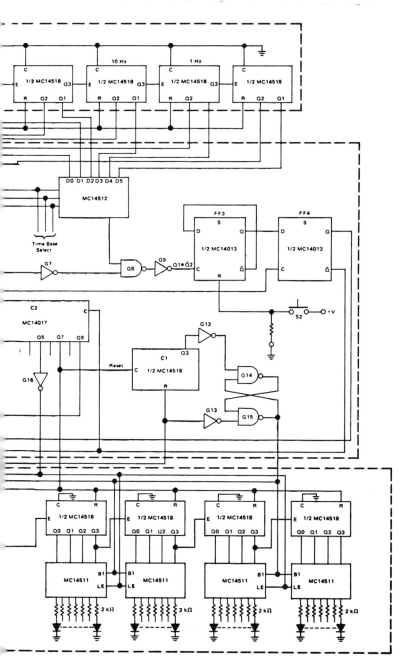

Battery powered (12-volt) 5 MHz frequency counter using McMOS logic (courtesy Motorola Semiconductor Products Inc.).

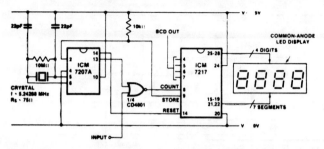

Precision frequency counter/tachometer. The ICM7207A provides a 1-second gating window and the store and reset signals. The display reads hertz directly. With pin 11 of the ICM7207A connected to V+ the gating time will be 0.1 second, which gives tens of hertz in the LSB position. For shorter gating times a 6.5536 MHz crystal may be used, giving a 0.01-second gating with pin 11 connected to V+ and a 0.1-second gating with pin 11 open. To implement the 4-digit tachometer the ICM7207A with a 1-second gating should be used. In order to get the display to read directly in RPM the rotational frequency must be multiplied by 60. This can be done electronically using a PLL or mechanically using a disc rotating with the object with appropriate hole drilled for light from an LED to a photodevice (courtesy Intersil, Inc.).

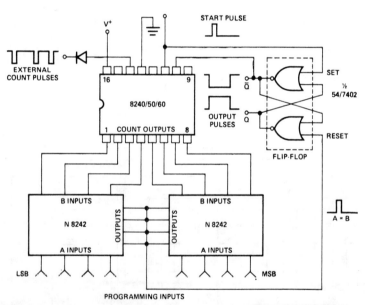

Electronically programmed counter using the Intersil 8240/50/60 16-pin DIP. Two quad exclusive-NOR circuits with open-collector outputs are wired together to form an inexpensive digital comparator. A start pulse triggers the 8240/50/60 counter and sets the output flip-flop high. The digital comparator output goes high momentarily when A is equal to B. This resets the flip-flop, which in turn resets the counter. For extended temperature range or higher speed operation individual pull-up resistor may be required on the counter outputs (courtesy Intersil, Inc.).

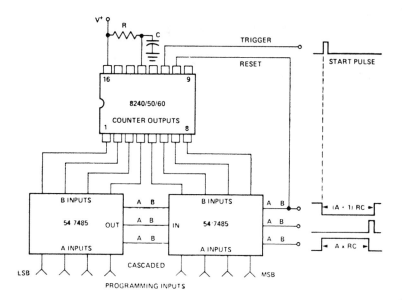

Electronically programmed timer using an Intersil 8240/50/60 16-pin DIP. This circuit uses a standard 54/74 series TTL 4-bit magnitude comparator to compare the digitally programmed input with the 8240/50/60 counter outputs. The greater, less-than and equal waveforms provide several outputs to choose from. An external start pulse triggers the timer and the A<B output is used as a reset (courtesy Intersil, Inc.).

499

Sensing Circuits

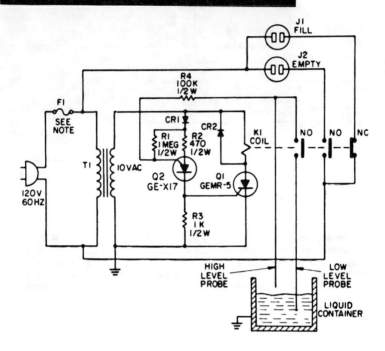

NOTE: SELECT FUSE RATING FOR SPECIFIC LOAD (3AG TYPE)
NO - NORMALLY OPEN CONTACT
NC - NORMALLY CLOSED CONTACT

Parts List

CR1, CR2 — GE-504A diodes
J1, J2 — Cinch a-c receptacles, or equivalent
K1 — 12-volt a-c relay with DPDT contacts (Potter and Brumfield KA11AY, or equivalent)
Q1 — GEMR-5 or GE-X5 silicon controlled rectifier
Q2 — GE-X17 programmable unijunction transistor

R1 — 1 megohm, 1/2-watt resistor
R2 — 470-ohm, 1/2-watt resistor
R3 — 1000-ohm, 1/2-watt resistor
R4 — 100,000-ohm, 1/2-watt resistor
T1 — Power transformer, 120-VAC primary and 10-VAC secondary (Triad F-90X, or equivalent)
Minibox — See text
Line cord and fuse holder

Automatic liquid-level control for sump pumps, water storage tanks, swimming pools, animal drinking troughs, etc. This device can operate equipment to keep the liquid level between two prescribed points (courtesy General Electric Company).

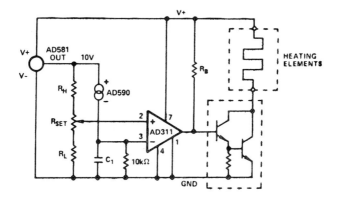

Electronic thermostat (courtesy Analog Devices, Inc.).

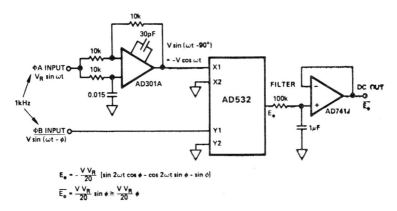

$$E_o = -\frac{V V_R}{20} [\sin 2\omega t \cos \phi - \cos 2\omega t \sin \phi - \sin \phi]$$

$$\overline{E_o} = \frac{V V_R}{20} \sin \phi \cong \frac{V V_R}{20} \phi$$

Phasemeter for sinusoidal signals. This circuit provides a good linear measure of phase for small angles up to 14° and a sinusoidal measure of angles between +90° and −90° (courtesy Analog Devices, Inc.).

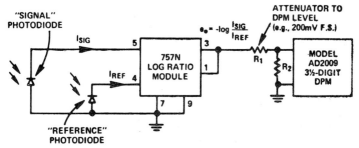

Light intensity absorbance circuit using the 757N log ratio module and the AD2009 3½-digit DPM (courtesy Analog Devices, Inc.).

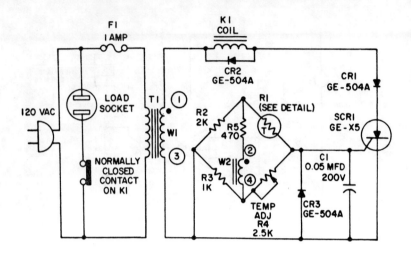

Parts List

C1 — 0.05-mfd, 200-volt capacitor
CR1 thru CR3 — GE-504A rectifier diodes
F1 — 1-amp fuse
J1 — Open-circuit phono jack
K1 — Relay, DPDT, 5-amp contacts with 6-volt d-c GPD coil (Potter & Brumfield GP11, or equivalent)
P1 Standard phono plug
R1 — GE Type 1D303 thermistor,* 0.3-inch dia, 1000 ohms at approximately 70 degrees F
R2, R3 — 1000-ohm, 2-watt resistor
R4 — 2500-ohm, 4-watt, wire wound potentiometer
R5 — 470-ohm, 2-watt resistor
SCR1 — GE-X5 silicon controlled rectifier
T1 — Power transformer: 120-volt a-c primary; two 12.6-volt secondaries W1 and W2 (UTC F10, or equivalent)
Load socket
Minibox — Aluminum, 6" x 5" x 4"

Temperature-operated relay for soldering iron, furnace, etc. This circuit will control the temperature at thermistor R1 within 1° over the temperature range of 20°F to 150°F. For other temperature ranges change R1; it should have about 1000 ohms resistance in the center of the desired control range. R1 can be substituted with other types of sensing devices (e.g., photo resistor). By using K1 to operate a large contactor heavier loads can be controlled. As is, K1 is rated for 5 amperes (courtesy General Electric Company).

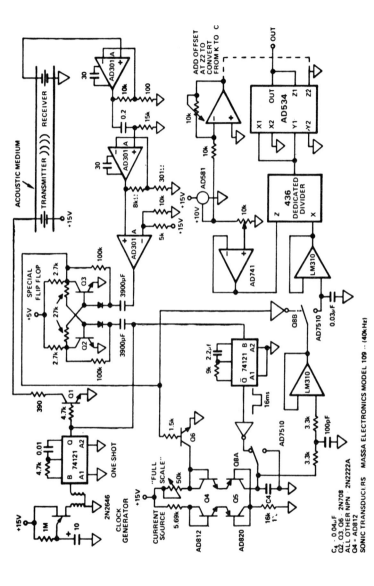

Acoustic thermometer. This circuit relies on the principle that the speed of sound varies predictably with temperature in a known medium. The sensor used is in effect a thermally dependent delay line. A 40 kHz clock is used to drive a transmitting transducer (courtesy Analog Devices, Inc.).

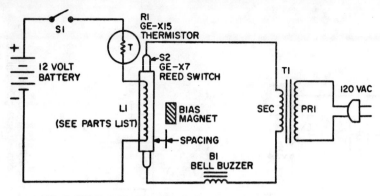

NOTE: BATTERY CIRCUIT CURRENT APPROXIMATELY 10 MILLIAMPERES. SELECT SUITABLE BATTERY

Parts List

Battery — 12 volts
L1 — 7000 turns of No. 38 wire (440 ohms). This is General Electric C-1 coil available at G-E distributors
R1 — GE-X15 thermistor
S1 — SPST toggle switch
S2 — GE-X7 magnetic reed switch (magnet supplied with reed switch)
B1 — Standard bell or buzzer
T1 — Standard 12/24-volt a-c bell transformer

Thermistor-operated temperature alarm that can ring a bell, turn on a light or ring a buzzer. Any temperature between 115° and 165°F can be detected. To calibrate adjust the following spacings of the magnet from the coil for the indicated temperatures: 1.5 inches for 117°F, 1 inch for 129°F, 0.75 inch for 138°F, 0.625 inch for 149°F and 0.5 inch for 164°F. Then immerse R1 in the medium heated to the desired point to sound the alarm. Turn on S1 and adjust the bias magnet to sound the alarm. Secure the magnet (courtesy General Electric Company).

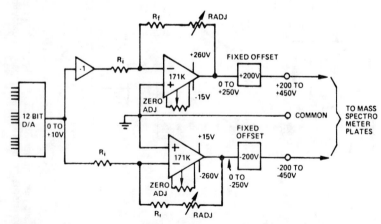

Programmable mass spectrometer voltage source (courtesy Analog Devices, Inc.).

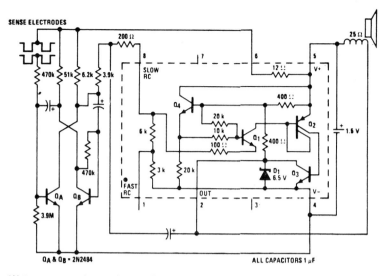

Water seepage alarm using an LM3909 chip. Circuitry inside dashed lines is the LM3909. Standby battery current drain is 100 μA. The multivibrator rate is about 1 hertz when it starts and increases as more moisture passes across the sensor. The sensor should be part of the box housing the circuitry. The sensor consists of two 6-inch strips of metal about an eighth-inch apart. Stainless steel or copper will work well. Battery life for a single alkaline penlight cell is about 3 months (courtesy National Semiconductor Corporation.)

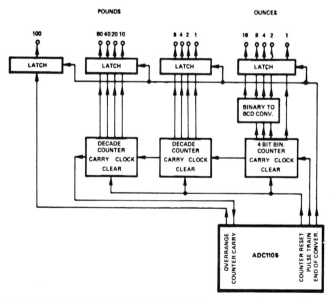

Simple weighing system with 199-pound 15-ounce maximum using an ADC1105 (courtesy Analog Devices, Inc.).

505

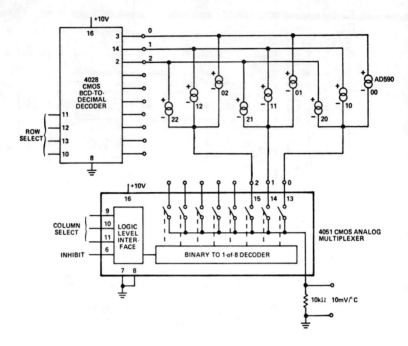

Matrix temperature multiplexer (courtesy Analog Devices, Inc.).

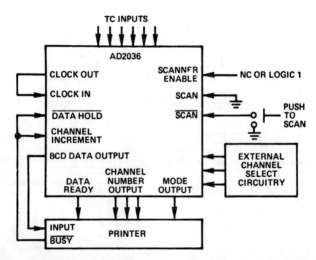

6-channel scanning digital thermometer with printer output (courtesy Analog Devices, Inc.).

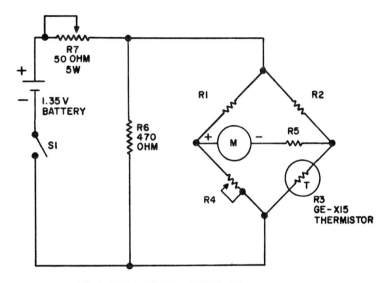

TL SEE PARTS LIST FOR VALUES OF
R1, R2, R4 AND R5

Parts List

High temperature range (+32 to +122 F)
R1 — 1000-ohm, 1/4-watt resistor
R2 — 1000-ohm, 1/4-watt resistor
R3 — GE-X15 thermistor
R4 — 5000-ohm, 5-watt potentiometer
R5 — 9500-ohm, 1/4-watt resistor
R6 — 470-ohm, 1/4-watt resistor
R7 — 50-ohm, 5-watt potentiometer
S1 — SPST toggle switch
M — 50-microampere d-c meter (G-E Type DW-91), or equivalent
Battery — 1.35-volt mercury cell

Low temperature range (-40 to +32 F)
R1 — 7300-ohm, 1/4-watt resistor
R2 — 7300-ohm, 1/4-watt resistor
R3 — GE-X15 thermistor
R4 — 50,000-ohm, 5-watt potentiometer
R5 — 4850-ohm, 1/4-watt resistor
R6 — 470-ohm, 1/4-watt resistor
R7 — 50-ohm, 5-watt potentiometer
S1 — SPST toggle switch
M — 50-microampere d-c meter (G-E Type DW-91), or equivalent
Battery — 1.35-volt mercury cell

Thermistor thermometer that can be calibrated for two temperature ranges. If another meter is used in place of the one specified (1500 ohms) the resistance of M plus R5 should equal 5350 ohms. To calibrate for the 32°F to 122°F scale immerse R3 in crushed ice and adjust R4 for zero current; mark the location of R4's knob pointer. Replace R3 with the low-resistance test resistor supplied, and adjust R7 for full-scale deflection; mark R7's position. Save the low-resistance test resistor. This completes the 32°F to 122°F calibration. To calibrate for −40°F to +32°F replace R3 with the high-resistance test resistor supplied, and adjust R4 for zero current; mark R4's location. Replace the rest resistor with R3. Immerse R3 in crushed ice, and adjust R7 for full-scale deflection; mark R7's location. This completes the −40° to +32°F calibration (courtesy General Electric Company).

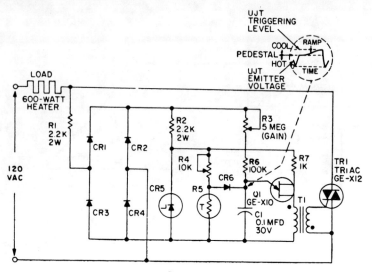

Parts List

C1 — 0.1-mfd, 30-volt capacitor
CR1 thru CR4 — GE-504A rectifier diode
CR5 — Two GE-X11 zener diodes in series
CR6 — GE-504A rectifier diode
Q1 — GE-X10 unijunction transistor
R1, R2 — 2200-ohm, 2-watt resistor
R3 — 5-megohm, 2-watt potentiometer
R4 — 10K-ohm, wirewound, potentiometer
R5 — ETRS-4942 thermistor*, approximately 5000 ohms at operating temperature
R6 — 100K-ohm, 1/2-watt resistor
R7 — 1K-ohm, 1/2-watt resistor
T1 — Pulse transformer — available from GE distributors as ETRS-4898, or from General Electric Co., Dept. B, 3800 N. Milwaukee Ave., Chicago, Ill. 60641.
TR1 — GE-X12 triac

Precision temperature regulator for regulating temperature of ovens, hot plates, fluids, air and gases. It can control up to 600 watts and has adjustable gain and temperature with built-in protection against transient voltages (courtesy General Electric Company).

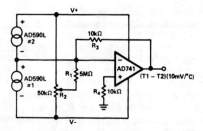

Temperature differential measurement circuit (courtesy Analog Devices, Inc.).

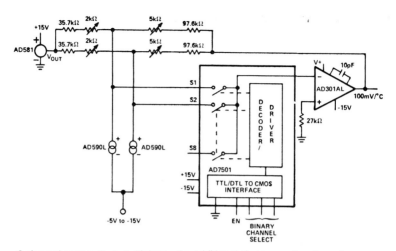

8-channel temperature multiplexer. An additional six temperature transducers should be added; only two are shown (courtesy Analog Devices, Inc.).

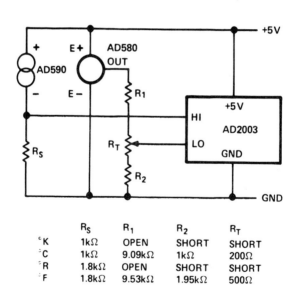

	R_S	R_1	R_2	R_T
°K	1kΩ	OPEN	SHORT	SHORT
°C	1kΩ	9.09kΩ	1kΩ	200Ω
°R	1.8kΩ	OPEN	SHORT	SHORT
°F	1.8kΩ	9.53kΩ	1.95kΩ	500Ω

Electronic thermometer for all temperature scales (courtesy Analog Devices, Inc.).

Multiplexers

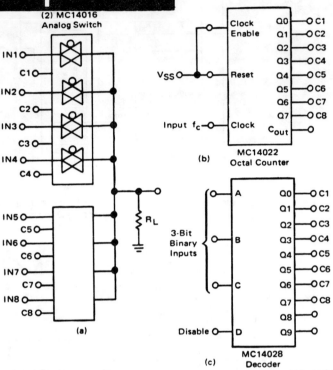

8-channel analog multiplexer with two methods of selection. Two MC14016s as shown in (a) can be connected for an 8-channel multiplexer. The MC14022 octal counter in (b) is used for time division multiplexing (stepping switch) and each channel is on for one period of the scanning clock. The MC14028 decoder is used for control in applications requiring selection with a 3-bit binary code (courtesy Motorola Semiconductor Products Inc.).

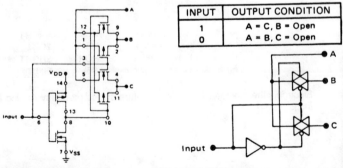

2-input analog multiplexer using an MC14007 dual pair plus inverter (courtesy Motorola Semiconductor Products Inc.).

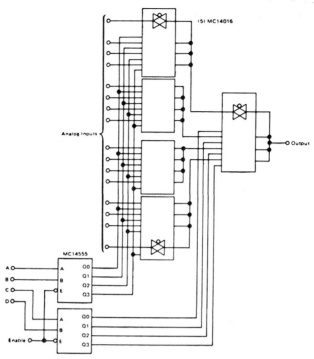

2-level 16-channel multiplexer (courtesy Motorola Semiconductor Products Inc.).

4-CHANNEL SEQUENCING MUX

Truth Table (IH5052)

ENABLE	MUX SEQUENCE RATE	SEQUENCER OUTPUT		SWITCH STATES (– DENOTES OFF)			
		2^0	2^1	SW1	SW2	SW3	SW4
0	0	0	0	–	–	–	–
1	0	0	0	ON	–	–	–
1	1 pulse	1	0	–	ON	–	–
1	2 pulses	0	1	–	–	–	–
1	3 pulses	1	1	–	–	–	ON
1	4 pulses	0	0	ON	–	–	–

Four-channel sequencing multiplexer. The analog switch package is an Intersil IH5052 16-pin CMOS DIP (courtesy Intersil, Inc.).

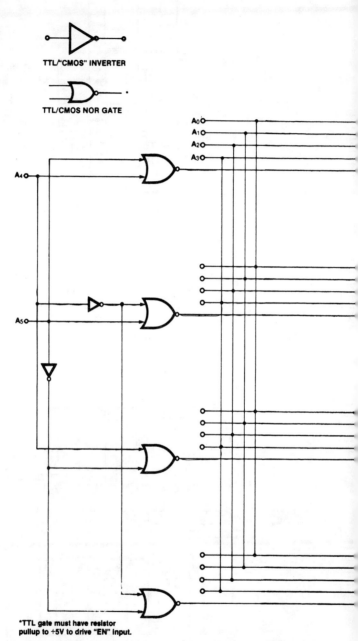

1-out-of-64 multiplexer using four IH6116s and one IH5053 (courtesy Intersil, Inc.).

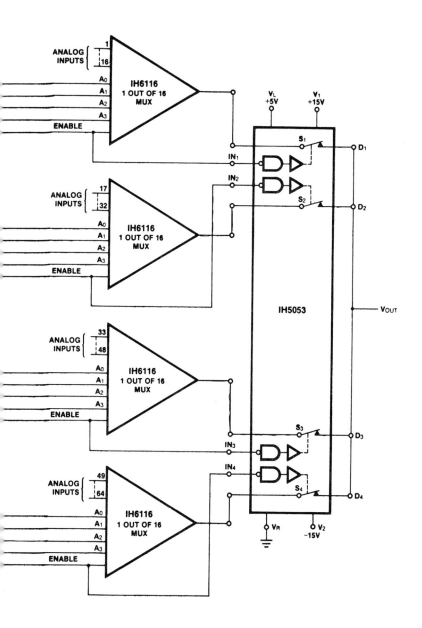

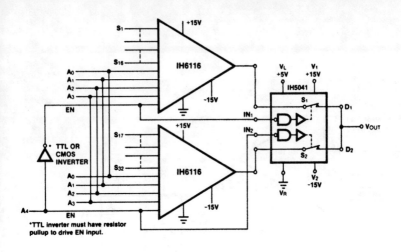

A_4	A_3	A_2	A_1	A_0	ON SWITCH
0	0	0	0	0	S1
0	0	0	0	1	S2
0	0	0	1	0	S3
0	0	0	1	1	S4
0	0	1	0	0	S5
0	0	1	0	1	S6
0	0	1	1	0	S7
0	0	1	1	1	S8
0	1	0	0	0	S9
0	1	0	0	1	S10
0	1	0	1	0	S11
0	1	0	1	1	S12
0	1	1	0	0	S13
0	1	1	0	1	S14
0	1	1	1	0	S15
0	1	1	1	1	S16
1	0	0	0	0	S17
1	0	0	0	1	S18
1	0	0	1	0	S19
1	0	0	1	1	S20
1	0	1	0	0	S21
1	0	1	0	1	S22
1	0	1	1	0	S23
1	0	1	1	1	S24
1	1	0	0	0	S25
1	1	0	0	1	S26
1	1	0	1	0	S27
1	1	0	1	1	S28
1	1	1	0	0	S29
1	1	1	0	1	S30
1	1	1	1	0	S31
1	1	1	1	1	S32

1-out-of-32 multiplexer using two IH6116s and one IH5041. The IH6116s are CMOS 16-channel analog multiplexers (courtesy Intersil, Inc.).

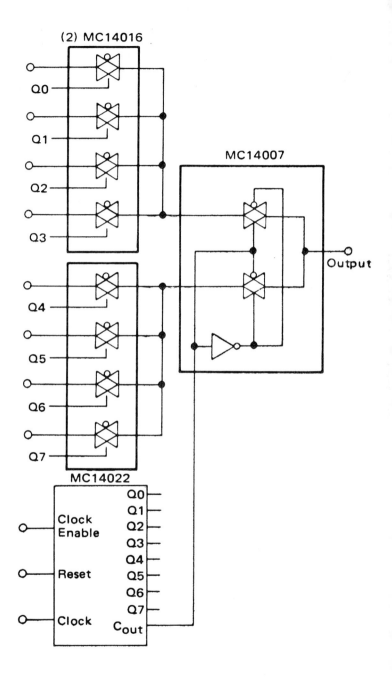

2-level 8-channel multiplexer (courtesy Motorola Semiconductor Products Inc.).

Transmitter & Receiver Radio Circuits

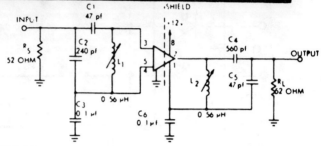

30 MHz RF amplifier using an ECG703A IC. The input signal is coupled through tank L1–C1–C2 to the ECG703A. Capacitors C1 and C2 form an impedance matching network. Network L2–C4–C5 coupled the output signal to the load. Use a shield between the input and output to prevent spurious oscillations (courtesy GTE Sylvania Incorporated).

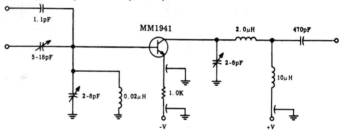

250 MHz RF to 50 MHz IF mixer using an MM1941 transistor. The external local oscillator injection frequency is 300 MHz. The output of the mixer is tuned to the difference between the two inputs (courtesy Motorola Semiconductor Products Inc.).

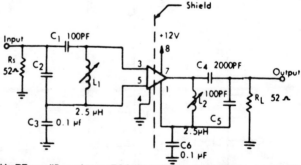

10 MHz RF amplifier using an ECG703A IC. The input signal is coupled to the amplifier through tank L1–C1–C2, tuned to 10 MHz. The capacitive divided C1–C2 provides impedance matching. Network L2–C4–C5 connects the amplifier to the load. A shield should be used between the input and output to prevent oscillations (courtesy GTE Sylvania Incorporated).

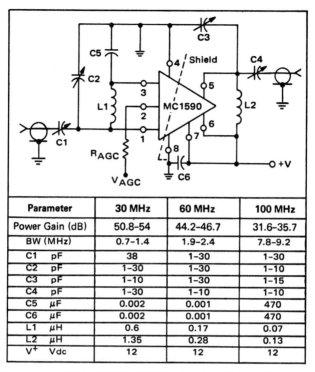

Parameter	30 MHz	60 MHz	100 MHz
Power Gain (dB)	50.8–54	44.2–46.7	31.6–35.7
BW (MHz)	0.7–1.4	1.9–2.4	7.8–9.2
C1 pF	38	1–30	1–30
C2 pF	1–30	1–30	1–10
C3 pF	1–10	1–30	1–15
C4 pF	1–30	1–10	1–10
C5 μF	0.002	0.001	470
C6 μF	0.002	0.001	470
L1 μH	0.6	0.17	0.07
L2 μH	1.35	0.28	0.13
V^+ Vdc	12	12	12

RF-IF amplifier for 30 MHz/60 MHz/100 MHz (courtesy Motorola Semiconductor Products Inc.).

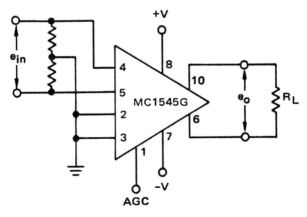

Wide-band differential amplifier with AGC using an MC1545G (courtesy Motorola Semiconductor Products Inc.).

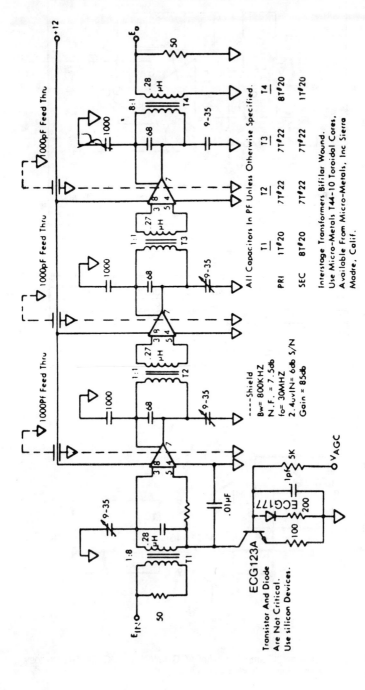

30 MHz IF strip for microwave or radar receivers using three ECG703A ICs (courtesy GTE Sylvania Incorporated).

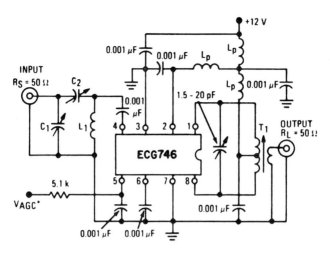

*Connect to ground for maximum power gain test.
All power-supply chokes (L_p), are self-resonate at
input frequency. $L_p \geq 20\,k\Omega$.

See Figure 10 for frequency response curve.

L_1 @ 45 MHz = 7 1/4 Turns on a 1/4" coil form.
@ 58 MHz = 6 Turns on a 1/4" coil form
T_1 Primary Winding = 18 Turns on a 1/4" coil form, center-tapped
Secondary Winding = 2 Turns centered over Primary Winding @ 45 MHz
= 1 Turn @ 58 MHz
Slug = Arnold TH Material 1/2" Long

	45 MHz		58 MHz	
L_1	0.4 µH	Q ≥ 100	0.3 µH	Q ≥ 100
T_1	1.3 - 3.4 µH	Q ≥ 100 @ 2 µH	1.2 - 3.8 µH	Q ≥ 100 @ 2 µH
C_1	50 - 160 pF		8 - 60 pF	
C_2	8 - 60 pF		3 - 35 pF	

RF amplifier for 45 MHz or 58 MHz. See table for component selection (courtesy GTE Sylvania Incorporated).

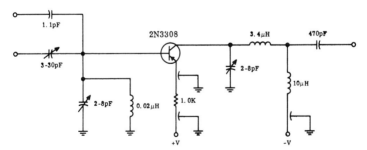

250 MHz RF to 50 MHz IF mixer using a 2N3308 transistor. The external local oscillator injection frequency is 300 MHz. The output of the 2N3308 is tuned to the difference between the two signals (courtesy Motorola Semiconductor Products Inc.).

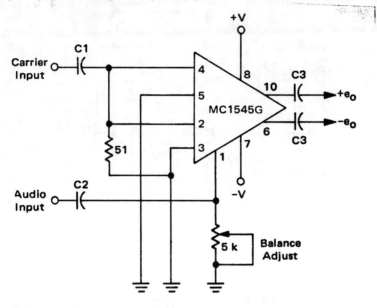

Balanced modulator using an MC1545G wide-band amplifier (courtesy Motorola Semiconductor Products Inc.).

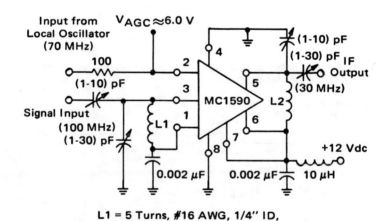

100 MHz mixer using an MC1590 RF-IF amplifier (courtesy Motorola Semiconductor Products Inc.).

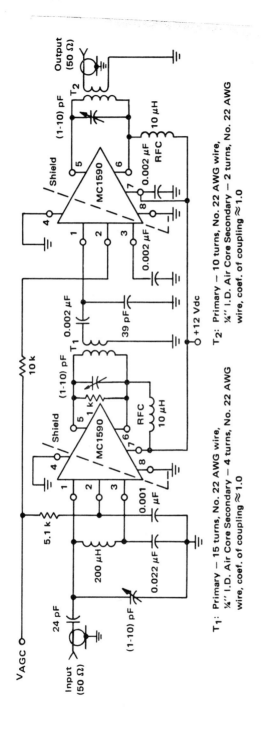

Two-stage 60 MHz IF amplifier with power gain of 80 dB and bandwidth of 1.5 MHz (courtesy Motorola Semiconductor Products Inc.).

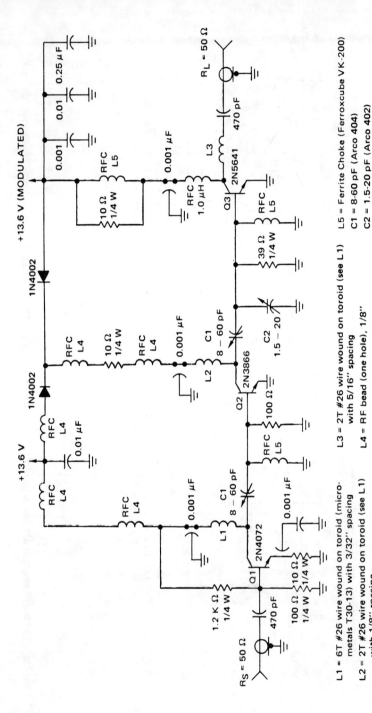

2.5-watt AM transmitter designed for 118 MHz to 136 MHz operation (courtesy Motorola Semiconductor Products Inc.).

L1 = 6T #26 wire wound on toroid (micro-metals T30-13) with 3/32" spacing
L2 = 2T #26 wire wound on toroid (see L1) with 1/8" spacing
L3 = 2T #26 wire wound on toroid (see L1) with 5/16" spacing
L4 = RF bead (one hole), 1/8"
L5 = Ferrite Choke (Ferroxcube VK-200)
C1 = 8-60 pF (Arco 404)
C2 = 1.5-20 pF (Arco 402)

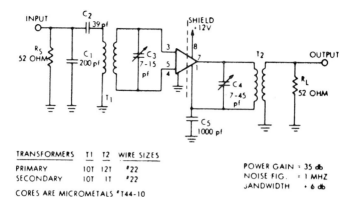

30 MHz RF amplifier with transformer coupling using the ECG703A (courtesy GTE Sylvania Incorporated).

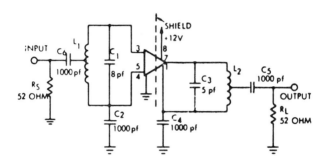

L_1 = 3T #20 AIR SPACED. TAPPED AT 1.5T
L_2 = 5T #16 AIR SPACED. TAPPED AT 1T
COIL DIA. 1/4 INCH

POWER GAIN = 14 db
NOISE FIG. = 7.5 db
BANDWIDTH = 10 MHZ

200 MHz RF amplifier using the ECG703A IC (courtesy of GTE Sylvania Incorporated).

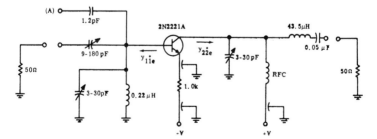

30 MHz RF to 5 MHz IF mixer. Injection point for the external local oscillator is at point A. The injection frequency is 35 MHz. The output of the 2N221A mixer is tuned to the difference of the two signals (courtesy Motorola Semiconductor Products Inc.).

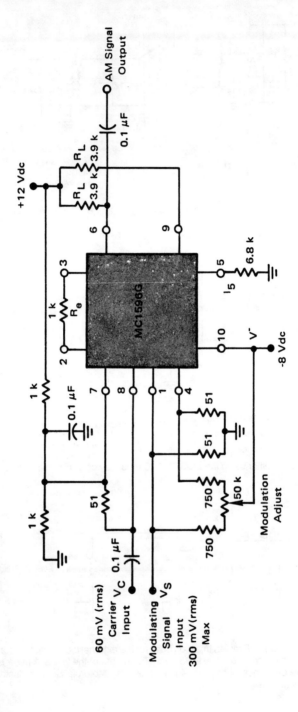

Amplitude modulator using an MC1596G. Modulation levels from zero to greater than 100% can be obtained with this circuit (courtesy Motorola Semiconductor Products Inc.).

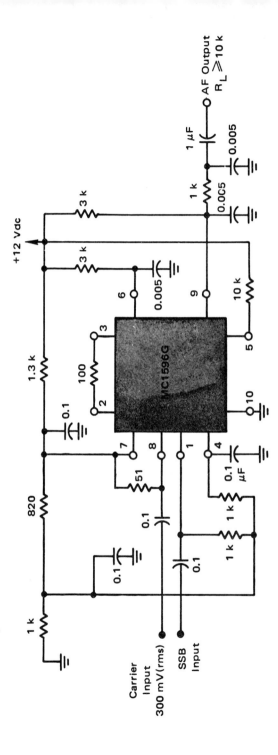

Product detector using an MC1596G. This circuit requires a single +12-volt supply. The circuit performs well with carrier level of 100 to 500 mV RMS. No transformers or tuned circuits are required for excellent performance from very low frequencies up to 100 MHz. The audio output at pin 6 can be used to drive AGC (courtesy Motorola Semiconductor Products Inc.).

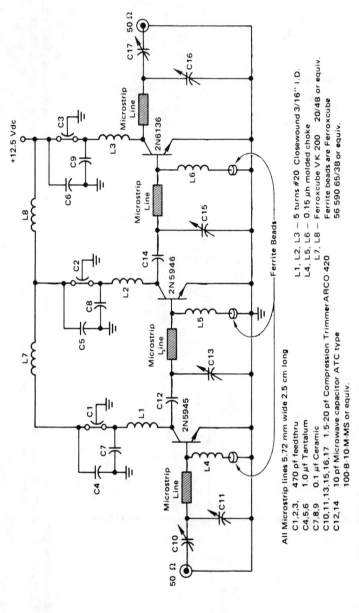

25-watt UHF amplifier for 450 to 512 MHz (courtesy Motorola Semiconductor Products Inc.).

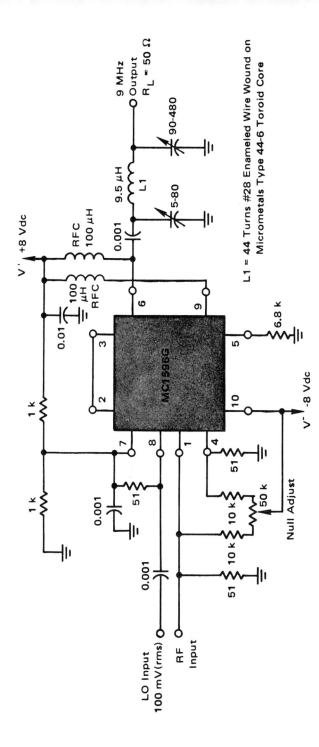

Double balanced mixer with broadband inputs and 9 MHz tuned output. The 3 dB bandwidth of the 9 MHz output tank is 450 kHz. Since the input is broadband the circuit can operate from HF to VHF. The local oscillator frequency should be +9 MHz (courtesy Motorola Semiconductor Products Inc.).

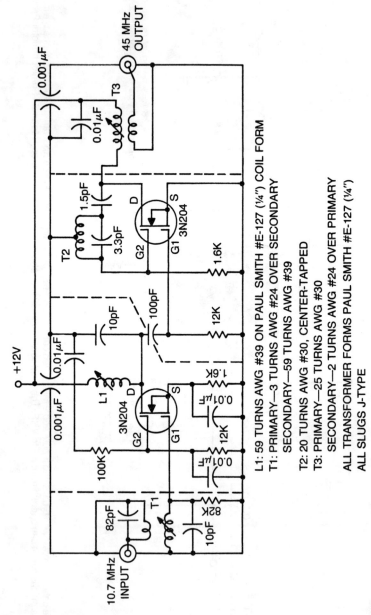

10.7 MHz to 45 MHz converter using two 3N204 dual-gate MOSFETs (courtesy Texas Instruments Incorporated).

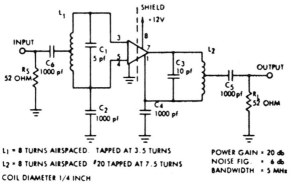

100 MHz RF amplifier using the ECG703A IC. Reverse AGC may be obtained at pin 5 (courtesy GTE Sylvania Incorporated).

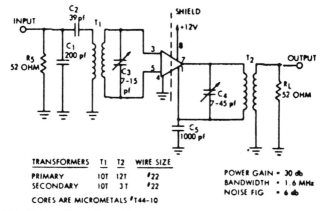

30 MHz RF amplifier with limiting using the EGC703A IC (courtesy of GTE Sylvania Incorporated).

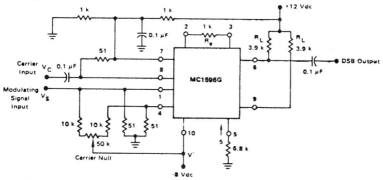

Balanced modulator circuit using an MC1596G. For a maximum modulating signal input of 300 mV RMS, the suppression of spurious sideband is typically 55 dB at a carrier frequency of 500 kHz (courtesy Motorola Semiconductor Products Inc.).

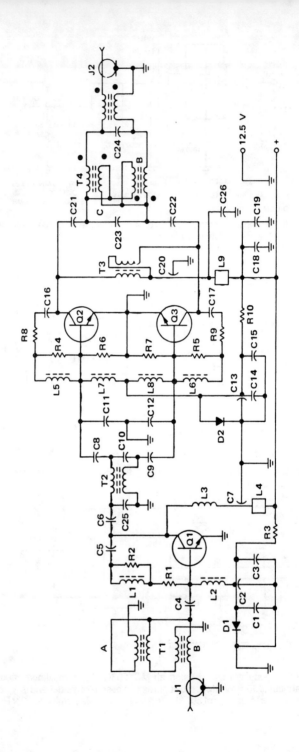

530

C1, C14, C18 – 0.1 µF ceramic.
C2, C7, C13, C20 – 0.001 µF feed through.
C3 – 100 µF/3V.
C4, C6 – 0.033 µF mylar
C5 – 0.0047 µF mylar.
C8, C9 – 0.015 and 0.033 µF mylars in parallel.
C10 – 470 pF mica.
C11, C12 – 560 pF mica.
C15 – 1000 µF/3 V
C16, C17 – 0.015 µF mylar
C19 – 10 pF 15 V
C21, C22 – two 0.068 µF mylars in parallel.
C23 – 330 pF mica
C24 – 39 pF mica
C25 – 680 pF mica
C26 – .01 µF ceramic

R1, R6, R7 – 10 Ω, 1/2 W carbon.
R2 – 51 Ω, 1/2 W carbon
R3 – 240 Ω, 1 wire W
R4, R5 – 18 Ω, 1 W carbon
R8, R9 – 27 Ω, 2 W carbon
R10 – 33 Ω, 6 W wire W

L1 – 0.22 µh molded choke
L2, L7, L8 – 10 µh molded choke
L5, L6 – 0.15 µh
L3 – 25 t, #26 wire, wound on a 100 Ω, 2 W resistor. (1.0 µh)
L4, L9 – 3 ferrite beads each.

T1 – 2 twisted pairs of #26 wire, 8 twists per inch. A = 4 turns, B = 8 turns. Core--Stackpole 57-9322-11, Indiana General F627-8Q1 or equivalent

T2 – 2 twisted pairs of #24 wire, 8 twists per inch, 6 turns. (Core as above.)

T3 – 2 twisted pairs of #20 wire, 6 twists per inch, 4 turns. (Core as above.)

T4 – A and B = 2 twisted pairs of #24 wire, 8 twists per inch. 5 turns each. C = 1 twisted pair of #24 wire, 8 turns. Core--Stackpole 57-9074-11, Indiana General F624-19Q1 or equivalent.

Q1 – 2N6367

Q2, Q3 – 2N6368

D1 – 1N4001
D2 – 1N4997

J1, J2 – BNC connectors

80-watt PEP broadband linear amplifier for 12.5-volt operation (courtesy Motorola Semiconductor Products Inc.).

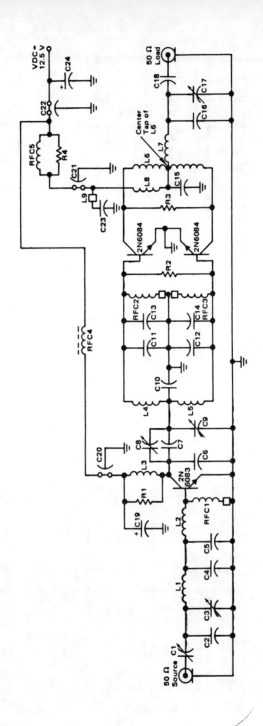

—Amplifier Constructed on 0.062",
Single Sided, G10 Circuit Board—

R1- 100 Ohm, 2.0 W, Carbon Resistor
R2,R3-10 Ohm, 1.0 W, Carbon Resistor
R4- 10 Ohm, 2.0 W, Carbon Resistor
L1- 1 T, #16 AWG Wire, 0.25" I.D. (18 nH)
L3- 3 T, #16 AWG Wire, Wound on R1 (60 nH)
L4,L5-1.1" Long, #14 AWG Wire, Formed Around 0.6" Dia. Cylinder (12 nH)
L8- 3T, #14 AWG Wire, 0.25" I.D. (50 nH)
L6- Cut from .031" Single Sided G10 Circuit Board (8 nH to Center Tap)

L2- Cut from 0.031" Single Sided G10 Circuit Board (5 nH)

L7- #12 AWG Wire, Approximately 1.1" Long, 0.7" (10 nH)
L9- Ferrite Bead, Ferroxcube 5659065/3B
RFC1,2,3- 0.15 µH Molded Choke with Ferroxcube 5659065/3B Ferrite Bead on Ground Lead
RFC4- Ferroxcube VK200 19/4B Ferrite Choke
RFC5- 10T, #14 AWG Wire, Wound on R4
C1,C3,C9,C17- #462 ARCO Trim Caps (5-80 pf)
C2 - 25 pF
C8 #403 ARCO trim Cap (4-40 pF)
C4, C5, C6, C11, C12, C13, C14 — 100 pF } Underwood Electric Co. Type J-101
C7, C10, C16 - 40 pF
C15- 250 pF
C18- 30 pF
C19-1.0 µF-Tantalum
C20, C21, C22- 680 pF, Allen Bradley Type FA5C
C23- 0.1 µF, 75 V, Ceramic Disc
C24- 5.0 µF Tantalum

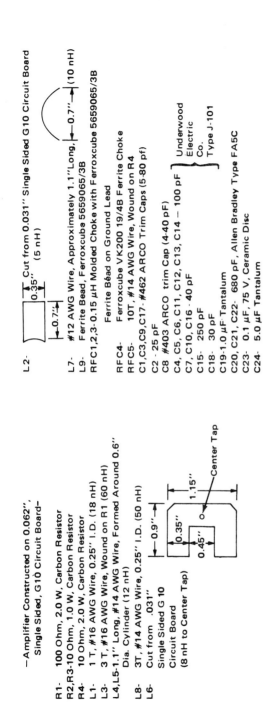

Two-stage 80-watt RF power amplifier for 144 to 175 MHz FM operation. About 5.5 watts of drive are required for the rated output (courtesy Motorola Semiconductor Products Inc.).

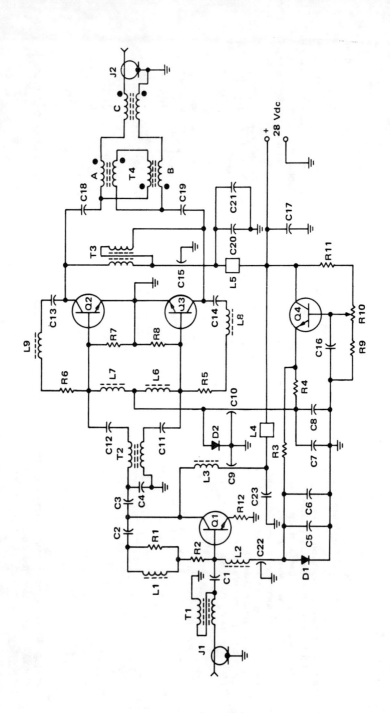

C1 – 0.033 µF mylar
C2, C3 – 0.01 µF mylar
C4 – 620 pF dipped mica
C5, C7, C16 – 0.1 µF ceramic
C6 – 100 µF/15 V electrolytic
C8 – 500 µF/6 V electrolytic
C9, C10, C15, C22 – 1000 pF feed through
C11, C12 – 0.01 µF
C13, C14 – 0.015 µF mylar
C17 – 10 µF/35 V electrolytic
C18, C19, C21 – Two 0.068 µF mylars in parallel
C20 – 0.1 µF disc ceramic
C23 – 0.1 µF disc ceramic
R1 – 220 Ω, 1/4 W carbon
R2 – 47 Ω, 1/2 W carbon
R3 – 820 Ω, 1 W wire W
R4 – 35 Ω, 5 W wire W
R5, R6 – Two 150 Ω, 1/2 W carbon in parallel
R7, R8 – 10 Ω, 1/2 W carbon
R9, R11 – 1 k, 1/2 W carbon
R10 – 1 k, 1/2 W potentiometer
R12 – 0.85 Ω (6 5.1 Ω or 4 3.3 Ω 1/4 W resistors in parallel, divided equally between both emitter leads)

T1 – 4:1 Transformer, 6 turns, 2 twisted pairs of #26 AWG enameled wire (8 twists per inch)
T2 – 1:1 Balun, 6 turns, 2 twisted pairs of #24 AWG enameled wire (6 twists per inch)
T3 – Collector choke, 4 turns, 2 twisted pairs of #22 AWG enameled wire (6 twists per inch)
T4 – 1:4 Transformer Balun, A&B – 5 turns, 2 twisted pairs of #24, C – 8 turns, 1 twisted pair of #24 AWG enameled wire (All windings 6 twists per inch). (T4 – Indiana General F624-19Q1, – All others are Indiana General F627-8Q1 ferrite toroids or equivalent.)

PARTS LIS'

L1 – .33 µH, molded choke
L2, L6, L7 – 10 µH, molded choke
L3 – 1.8 µH (Ohmite Z-144)
L4, L5 – 3 ferrite beads each
L8, L9 – .22 µH, molded choke

Q1 – 2N6370
Q2, Q3 – 2N5942
Q4 – 2N5190
D1 – 1N4001
D2 – 1N4997

J1, J2 – BNC connectors

160-watt PEP broadband linear amplifier for 28-volt operation (courtesy Motorola Semiconductor Products Inc.).

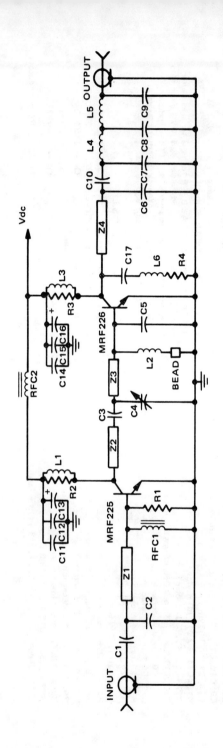

Amplifier/Filter Test Data for f = 225 MHz, Vdc = 12.5 Volts

P_{out} = 12 W at Filter Output

P_{in} for P_{out} = 12 W at Filter Output	125 mW
Power Gain	19.6 dB
L. P. Filter insertion loss	0.35 dB
2nd Harmonic attenuation	60 dB
Bandwidth	4 MHz (for ±0.1 dB)
	10 MHz (for ±0.5 dB)
Efficiency	58 percent
Input VSWR	1.5:1
DC Current:	
Total	1.66 A
Output Stage	1.50 A
Input Stage	0.16 A

Stability: Stable for input drive levels from zero to more than 30% overdrive for Vdc values from 8.0 to 15.5 V.

Ruggedness: No transistor damage for open and short circuit load conditions for all load phases angles and high line of 15.5 V. Total amplifier current peaks and nulls at approximately 2.3 and 0.5 A respectively as the load phase angle is cycled.

C1 — 10 pF
C2, C3 — 15 pF
C5 — 68 pF
C6 — 18 pF
C7, C9 — 12 pF
C8, C10 — 20 pF
C17 — 82 pF

} Dipped Silvered Mica, El-Menco Case DM10

C11, 14 — .01 µF Ceramic Disc
C12, 15 — 220 pF Ceramic Disc
C13, 16 — 5 µF, 25 V, Aluminum Electrolytic

C4 — 3-35 pF Trimmer, ARCO NO. 403
L1 — 5 T NO. 20 AWG Wire, Wound on R2 (50 nH)
L2 — 1.5 T NO. 20 AWG Wire, 0.25-In. I.D. (30 nH) with Ferroxcube 5659065/3B Ferrite Bead
L3 — 2 T NO. 20 AWG Wire, Wound on R3 (35 nH)
L4, 5 — 2 T NO. 18 AWG Wire, 0.25-In. I.D. (44 nH)
L6 — 3.5 T NO. 18 AWG Wire, 0.25-In. I.D. (90 nH)

R1 — 100 Ohm, 1/4 W, ±10% Carbon Resistor
R2 — 820 Ohm, 1/2 W, ±10% Carbon Resistor
R3 — 330 Ohm, 1 W, ±10% Carbon Resistor
R4 — 22 Ohm, 1/4 W, ±10% Carbon Resistor
RFC1, 2 — Ferroxcube VK200 19/4B Choke

Z1 — Microstrip Line, 2200 × 62 Mils
Z2 — Microstrip Line, 1200 × 62 Mils
Z3 — Microstrip Line, 1000 × 62 Mils
Z4 — Microstrip Line, 1600 × 62 Mils

Board — G10 Epoxy-Glass, ϵ_r = 5, t = 62 Mils
1 oz. Copper

13-watt microstrip UHF amplifier for 220 to 225 MHz (courtesy Motorola Semiconductor Products Inc.).

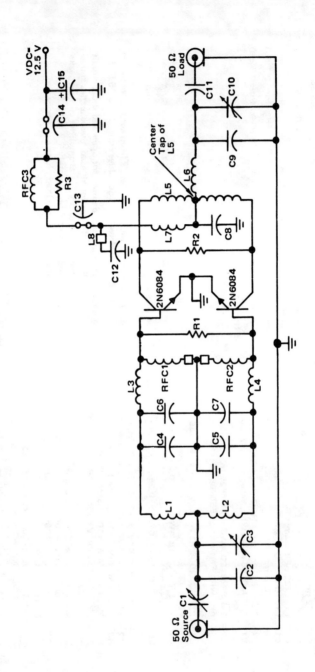

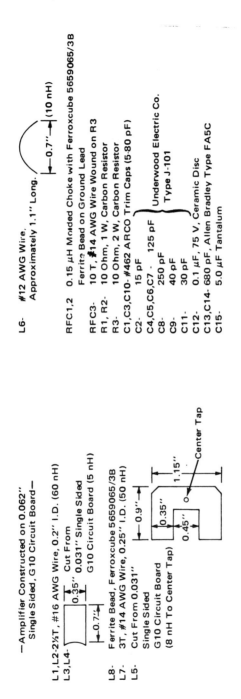

—Amplifier Constructed on 0.062"
Single Sided, G10 Circuit Board—

L1, L2- 2½T, #16 AWG Wire, 0.2" I.D. (60 nH)
L3, L4- 0.35" Cut From 0.031" Single Sided G10 Circuit Board (5 nH)
L8- Ferrite Bead, Ferroxcube 5659065/3B
L7- 3T, #14 AWG Wire, 0.25" I.D. (50 nH)
L5- Cut From 0.031" Single Sided G10 Circuit Board (8 nH To Center Tap)
L6- #12 AWG Wire. Approximately 1.1" Long. (10 nH)

RFC1,2 0.15 µH Moded Choke with Ferroxcube 5659065/3B Ferrite Bead on Ground Lead
RFC3-   10 T, #14 AWG Wire Wound on R3
R1, R2- 10 Ohm, 1 W, Carbon Resistor
R3-   10 Ohm, 2 W, Carbon Resistor
C1, C3, C10- #462 ARCO Trim Caps (5-80 pF)
C2-   15 pF
C4, C5, C6, C7 - 125 pF ⎫
C8-   250 pF   ⎬ Underwood Electric Co. Type J-101
C9-   40 pF   ⎪
C11-   30 pF   ⎭
C12-   0.1 µF, 75 V, Ceramic Disc
C13, C14- 680 pF, Allen Bradley Type FA5C
C15-   5.0 µF Tantalum

Single-stage 80-watt RF power amplifier for 144 to 175 MHz FM operation. About 25 watts of drive are required for the rated output (courtesy Motorola Semiconductor Products Inc.).

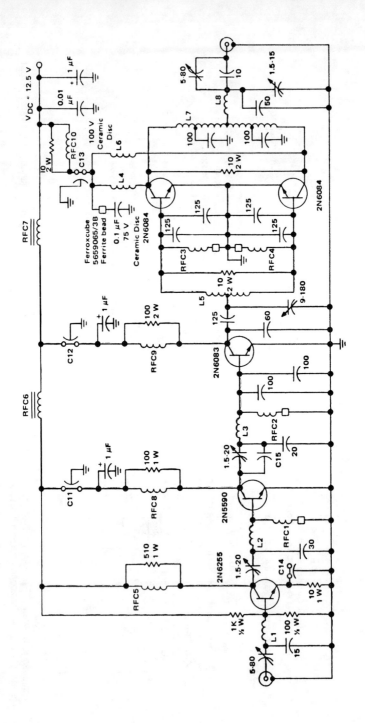

NOTES: 1. All resistors in Ohms
2. All capacitors in pf unless noted otherwise
3. All fixed value capacitors from 10 to 125 pF are Underwood Type J-101.
4. All trimmer capacitors are ARCO compression mica or equivalent.
5. Constructed on 0.062", single sided, G10, circuit board

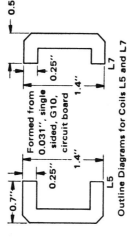

Outline Diagrams for Coils L5 and L7

RFC1,2,3,4 — 0.15 µH, molded choke with Ferroxcube 5659065/3B ferrite bead on ground lead
RFC5 — 0.15µH molded choke
RFC6,7 — Ferroxcube VK-200 19/4B ferrite choke
RFC8 — 4T #16 awg wire, wound on 100Ω 1 W resistor (75 nH)
RFC9 — 2T #15 awg wire, wound on 100Ω 2 W resistor (45 nH)
RFC10 — 10T #14 awg wire wound on 10Ω 2 W resistor
L1,2,3 — 1T #18 awg, ½" dia, ¾" L (25 nH)
L4,6 — 2T #15 awg wire, ¼" dia, ½" L (30 nH)
L5,7 — See outline diagram.
L8 — #12 awg wire approximately 1" Long (9 nH)
C11,12,13 — 680 pF, Allen Bradley Type FA5C
C14 — 470 pF, Allen Bradley Type SS5D
C15 — 5 pF, Dipped Silvered Mica

80-watt 175 MHz FM transmitter for 12.5-volt operation. For the rated output of 80 watts 180 mW of RF drive is required (courtesy Motorola Semiconductor Products Inc.).

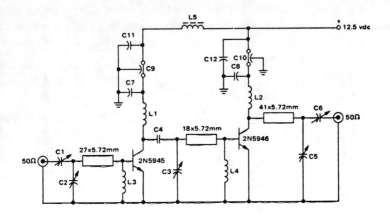

C1,2,3,5,6 – 1.5-20 pF, Arco 402 or equiv.
C4 – 10 pf dipped mica
C7, 8 – 0.1 µF ceramic
C9, 10 – 470 pf Feed thru
C11, 12 – 1 µf Tantalum
L1, 2 – 5 turns #20 AWG Closewound 3/16" I.D.
L3, 4 – 3.9 µhy molded choke w/ferrite core
L5 – Ferroxcube VK200 20/4B or equiv.
Board is 1/16" thick epoxy-glass
"G-10" Dielectric with 1oz copper on both sides

10-watt UHF microstrip amplifier for 450 to 470 MHz (courtesy Motorola Semiconductor Products Inc.).

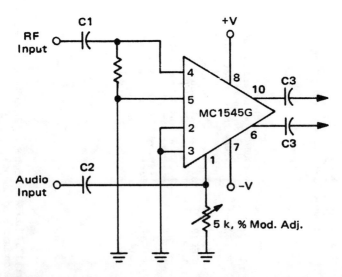

Amplitude modulator using an AM1545G wide-band amplifier (courtesy Motorola Semiconductor Products Inc.).

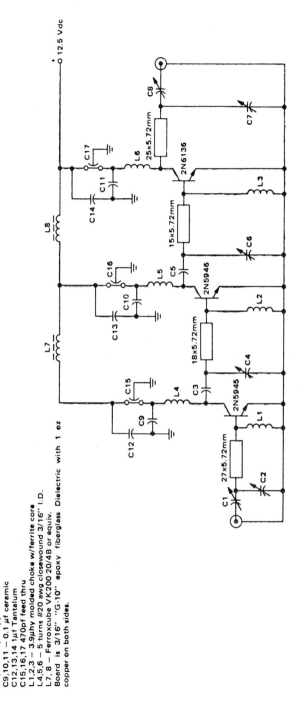

C1,2,4,6,7,8 — 1.5-20pf Arco 402 or equiv.
C3, 5 — 10pf dipped mica
C9,10,11 — 0.1 µf ceramic
C12,13,14 1µf Tantalum
C15,16,17 470pf feed thru
L1,2,3 — 3.9µhy molded choke w/ferrite core
L4,5,6 — 5 turns #20 awg closewound 3/16" I.D.
L7, 8 — Ferroxcube VK200 20/4B or equiv.
Board is 3/16" "G-10" epoxy fiberglass Dielectric with 1 oz copper on both sides.

25-watt UHF microstrip amplifier for 450 to 470 MHz (courtesy Motorola Semiconductor Products Inc.).

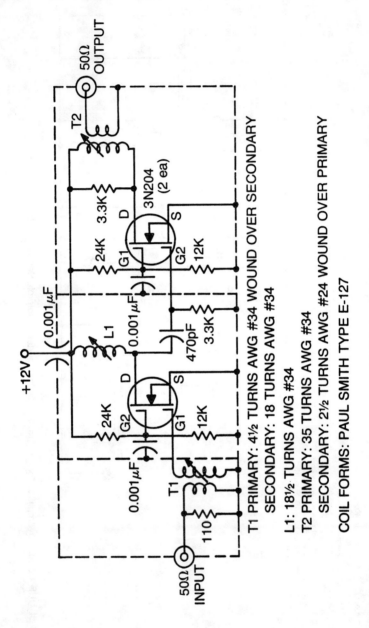

45 MHz RF post amplifier using two 3N204 dual-gate MOSFETs (courtesy Texas Instruments Incorporated).

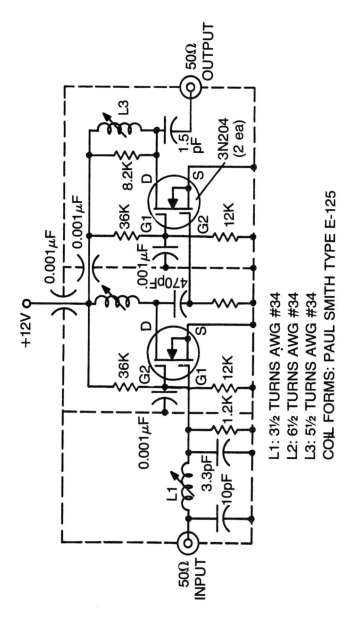

200 MHz RF post amplifier using two 3N204 dual-gate MOSFETs (courtesy Texas Instruments Incorporated).

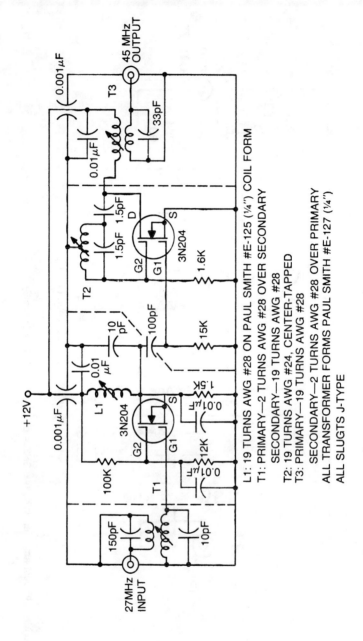

27 MHz to 45 MHz converter using two 3N204 MOSFETs (courtesy Texas Instruments Incorporated).

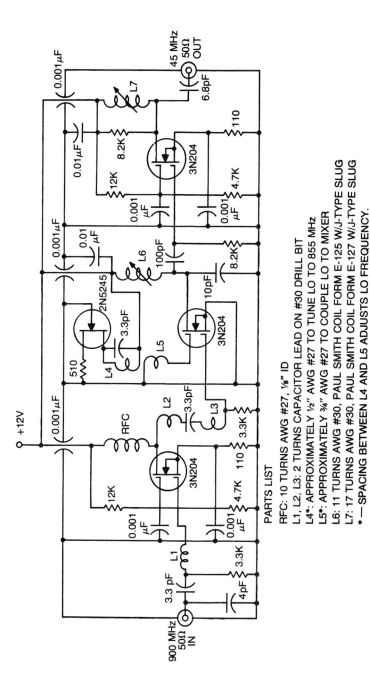

PARTS LIST

RFC: 10 TURNS AWG #27, 1/8" ID
L1, L2, L3: 2 TURNS CAPACITOR LEAD ON #30 DRILL BIT
L4*: APPROXIMATELY 1/2" AWG #27 TO TUNE LO TO 855 MHz
L5*: APPROXIMATELY 3/4" AWG #27 TO COUPLE LO TO MIXER
L6: 11 TURNS AWG #30, PAUL SMITH COIL FORM E-125 W/J-TYPE SLUG
L7: 17 TURNS AWG #30, PAUL SMITH COIL FORM E-127 W/J-TYPE SLUG
*—SPACING BETWEEN L4 AND L5 ADJUSTS LO FREQUENCY.

900 MHz to 45 MHz converter using dual-gate 3N204 MOSFETs and a 2N5245 (courtesy Texas Instruments Incorporated).

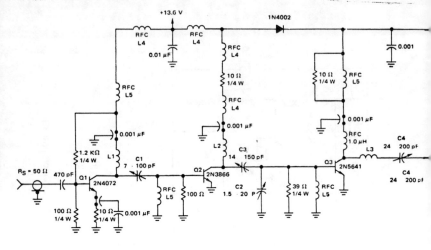

C1 = 7–100 pF (Arco 423)
C2 = 1.5 –20 pF (Arco 402)
C3 = 14–150 pF (Arco 424)
C4 = 24–200 pF (Arco 425)
C5 = 25–280 pF (Arco 464)

L1 = 6T #26 wire wound on toroid (Micrometals T30-13) with 3/32" spacing.
L2 = 2T #26 wire wound on toroid (see L1) with 1/8" spacing
L3 = 1T #26 wire wound on toroid (see L1) with 5/16" spacing

L1 = 6T #26 wire wound on toroid (Micrometals T30-13) with 3/32" spacing
L2 = 2T #26 wire wound on toroid (see L1) with 1/8" spacing

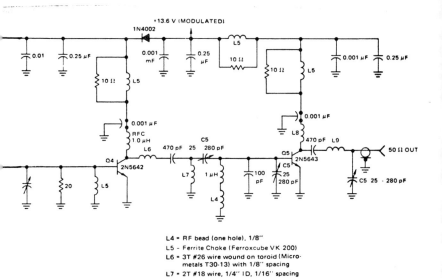

13-watt AM transmitter for aircraft use. Designed to operate between 118 MHz and 136 MHz (courtesy Semiconductor Products Inc.).

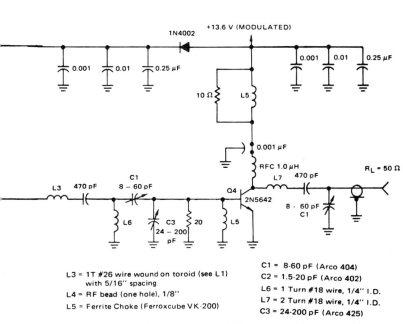

7-watt AM transmitter designed for 118 MHz to 136 MHZ operation (courtesy Motorola Semiconductor Products Inc.).

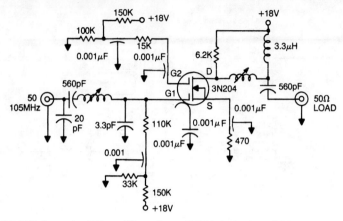

105 MHz low-noise RF amplifier using a 3N204 dual-gate MOSFET (courtesy Texas Instruments Incorporated).

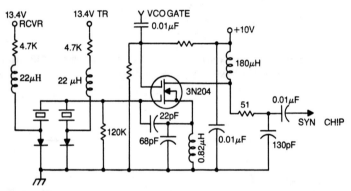

Frequency synthesizer mixer for CB operation using a dual-gate 3N204 dual-gate MOSFET (courtesy Texas Instruments Incorporated).

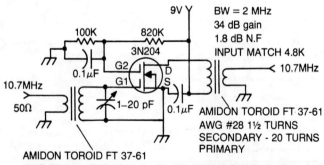

10.7 MHz IF amplifier using a 3N204 dual-gate MOSFET (courtesy Texas Instruments Incorporated).

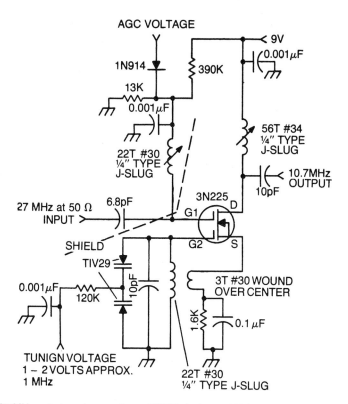

27 MHz autodyne tuner using a 3N225 dual-gate MOSFET (courtesy Texas Instruments Incorporated).

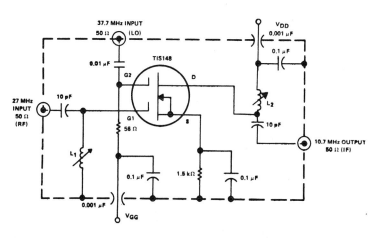

27 MHz mixer for CB operation using a TIS148 dual-gate MOSFET (courtesy Texas Instruments Incorporated).

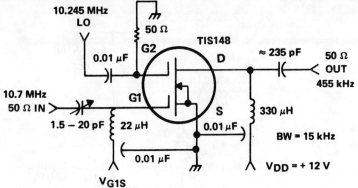

10.7 MHz to 455 kHz mixer for CB operation using a TIS148 dual-gate MOSFET (courtesy Texas Instruments Incorporated).

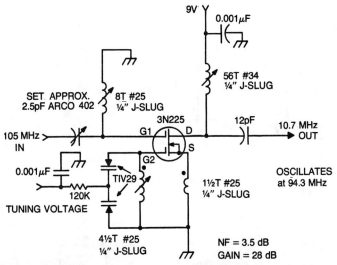

105 MHz to 10.7 MHz converter using a 3N225 dual-gate MOSFET (courtesy Texas Instruments Incorporated).

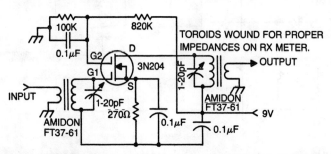

10.7 MHz IF amplifier using a 3N204 dual-gate MOSFET (courtesy Texas Instruments Incorporated).

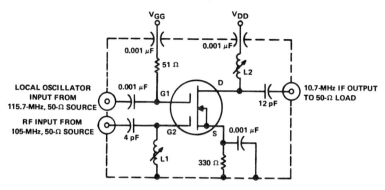

CIRCUIT COMPONENT INFORMATION

ALL 0.001 μF CAPACITORS ARE LEADLESS-DISC OR FEED-THRU TYPE
L1: 5 1/2 TURNS #30 WIRE
L2: 56 TURNS #34 WIRE, 35 TURNS IN INNER LAYER SEPARATED BY 2 LAYER 1 MIL MYLAR TAPE FROM 21 TURNS IN OUTER LAYER
PAUL SMITH CO. E-127 COIL FORMS, TYPE J SLUGS

105 MHz to 10.7 MHz mixer using a TIS152 dual-gate MOSFET (courtesy Texas Instruments Incorporated).

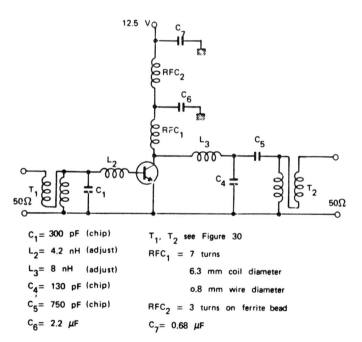

C_1 = 300 pF (chip)
L_2 = 4.2 nH (adjust)
L_3 = 8 nH (adjust)
C_4 = 130 pF (chip)
C_5 = 750 pF (chip)
C_6 = 2.2 μF

T_1, T_2 see Figure 30
RFC_1 = 7 turns
    6.3 mm coil diameter
    0.8 mm wire diameter
RFC_2 = 3 turns on ferrite bead
C_7 = 0.68 μF

118 to 136 MHz broadband RF amplifier using a 2N6083 bipolar transistor. This circuit is ideal for mobile operation since the recommended supply voltage is 12.5 volts. It provides 30 watts output for a 4-watt input at 125 MHz (courtesy Motorola Semiconductor Products Inc.).

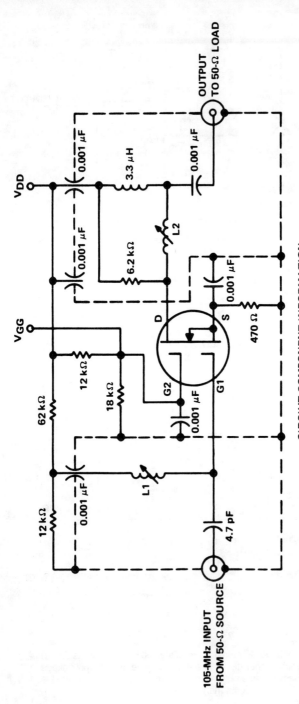

105 MHz gate-2-controlled RF amplifier using a TIS152 dual-gate MOSFET. This circuit has been optimized for third-order intermodulation distortion (courtesy Texas Instruments Incorporated).

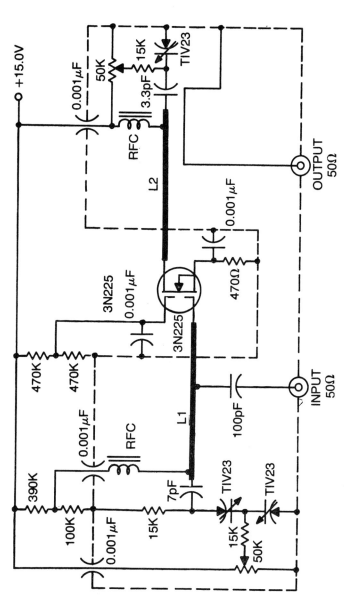

900 MHz RF amplifier using a 3N225 dual-gate MOSFET. Coils L1 and L2 consist of the leads of the 1 pF and 3 pF capacitors (courtesy Texas Instruments Incorporated).

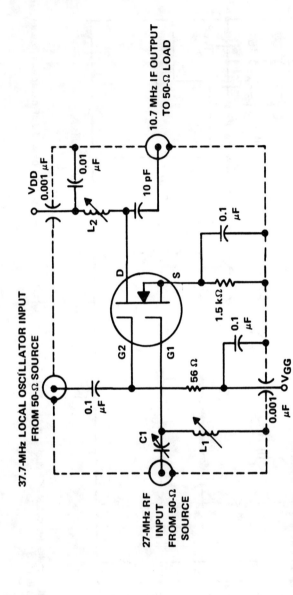

27 MHz to 10.7 MHz mixer for CB operation using a TIS148 dual-gate MOSFET (courtesy Texas Instruments Incorporated).

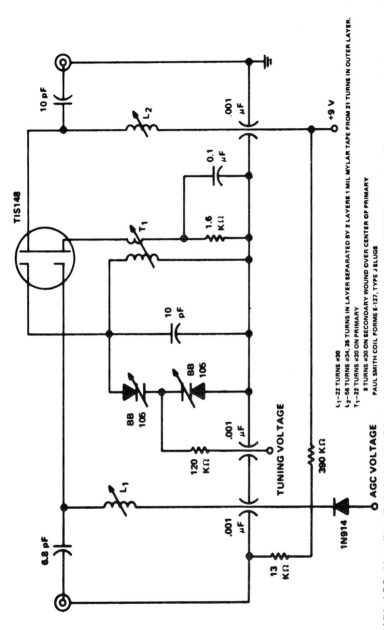

27 MHz AGC-able self-oscillating mixer for CB operation using a TIS148 dual gate MOSFET (courtesy Texas Instruments Incorporated).

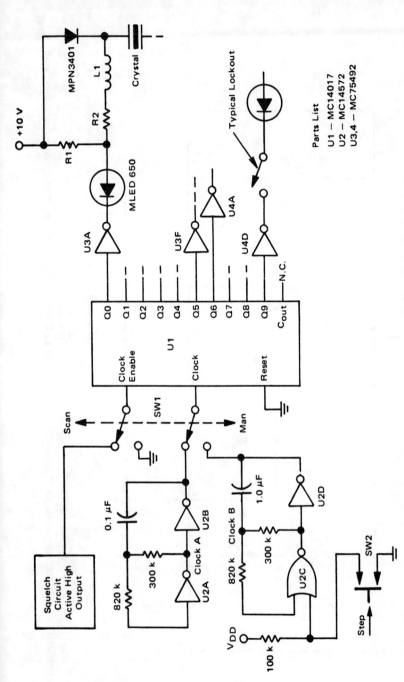

10-channel scanning logic for RF scanner/monitor receiver (courtesy Motorola Semiconductor Products Inc.).

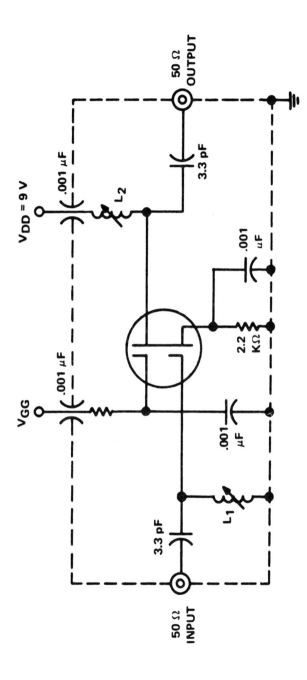

105 MHz gate-2-controlled RF amplifier using a TIS152 dual-gate MOSFET (courtesy Texas Instruments Incorporated).

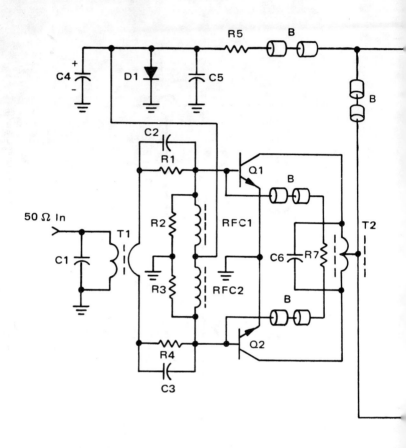

R1, R4 — 10 Ohms, 1/4 W
R2, R3 — 30 Ohms, 1/4 W
R5, R6 — 82 Ohms, 3 W (Nom.)
R7 — 47 Ohms, 1/4 W
R8, R11 — 6.8 Ohms, 1/4 W
R9, R10 — 15 Ohms, 1/4 W
R12 — 130 Ohms, 1/4 W

C1 — 39 pF Dipped Mica
C2, C3 — 680 pF Ceramic Disc
C4, C10 — 220 µF, 4 V, Tantalum
C5, C7, C11, C13 — 0.1 µF Ceramic Disc
C6 — 56 pF Dipped Mica
C8, C9 — 1200 pF Ceramic Disc
C12, C14 — 10 µF, 25 V Tantalum

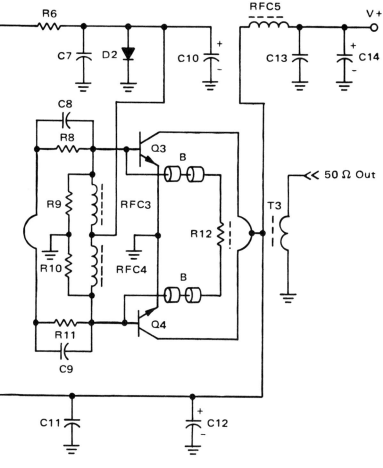

RFC5 — Ferroxcube VK200 19/4B
RFC1, 2, 3, 4 — 10 µH Molded Choke

B — Ferrite Beads (Fair-Rite Prod. Corp. #2643000101 or Ferroxcube #56 590 65/3B)

D1, D2 — 1N4001

Q1, Q2 — MRF476

Q3, Q4 — MRF475

T1, T2 — 4:1 Impedance Transformer
T3 — 1:4 Impedance Transformer

20-watt 25 dB 1.6 to 30 MHz SSB linear amplifier. Supply voltage is 13.8 volts (courtesy Motorola Semiconductor Products Inc.).

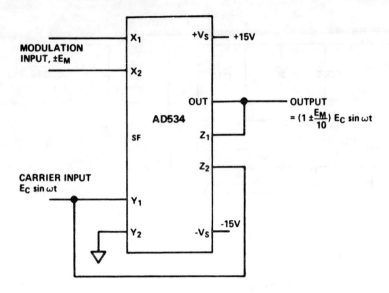

THE SF PIN OR A Z-ATTENUATOR CAN BE USED TO PROVIDE OVERALL SIGNAL AMPLIFICATION. OPERATION FROM A SINGLE SUPPLY IS POSSIBLE; BIAS Y_2 TO $V_S/2$.

Linear amplitude modulator circuit using an AD534 multiplier/divider chip (courtesy Analog Devices, Inc.).

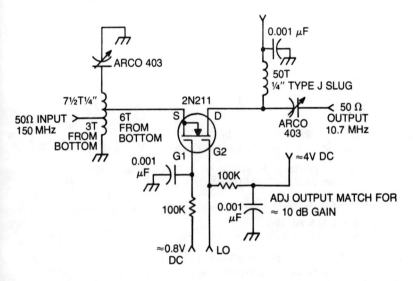

105 MHz to 10.7 MHz mixer using a 3N211 dual-gate MOSFET (courtesy Texas Instruments Incorporated).

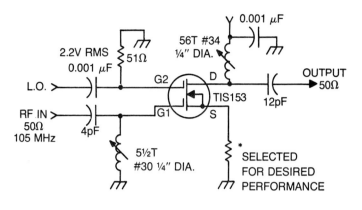

105 MHz to 10.7 MHz mixer for FM operation (courtesy Texas Instruments Incorporated).

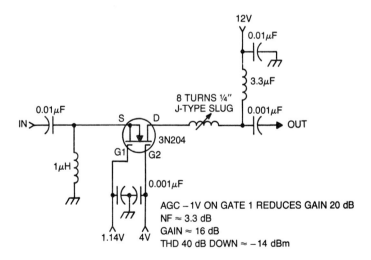

105 MHz RF amplifier using a 3N204 dual-gate MOSFET (courtesy Texas Instruments Incorporated).

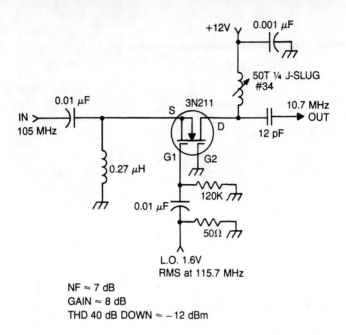

105 MHz to 10.7 MHz mixer (courtesy Texas Instruments Incorporated).

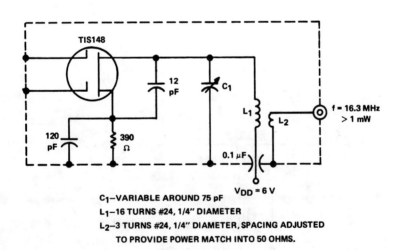

16.3 MHz oscillator for low-side injection in CB operation using a TIS148 dual-gate MOSFET (courtesy Texas Instruments Incorporated).

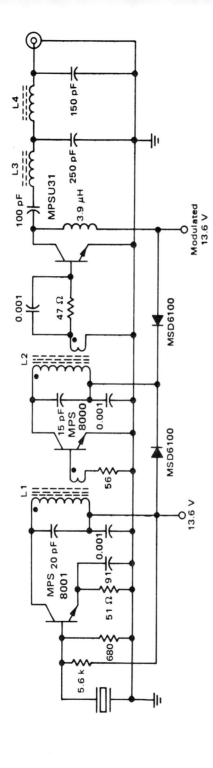

27 MHz transmitter for CB operation. Power output is 3.5 watts minimum. Typical efficiency is 70%. Second harmonic suppression is typically 38 dB down, while third harmonic suppression is 55 dB down. All coils are on ¼-inch forms with AWG #22 wire. Slugs are ¼- by ⅜-inch J-types. Secondary windings are overwound on the bottom of the primary. L1 primary is 12 turns close wound, while the secondary is 2 turns. L2 primary is 18 turns close wound, while its secondary is 2 turns. L3 is 7 turns close wound. L4 is 5 turns close wound (courtesy Motorola Semiconductor Products Inc.).

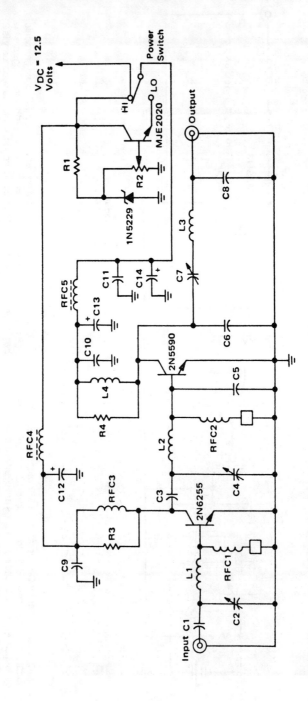

10-watt marine band transmitters. See page—for parts list.

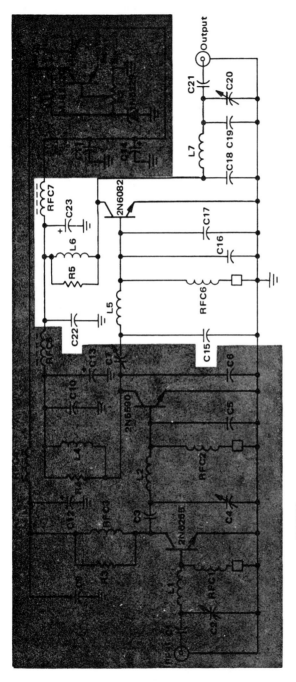

NOTES:
1. Shaded components are common to both 25 Watt and 10 Watt systems.
2. Component designations C8 and L3 are not used in 25 Watt system.
3. Amplifiers constructed on 0.62″, single sided, G10, printed circuit board.

25-watt marine band transmitter. See page 258 for parts list.

C1, C3	—	10 pF
C5, C6, C16, C17, C18	—	56 pF } Dipped Silvered
C8	—	36 pF } Mica
C15, C21	—	22 pF
C19	—	33 pF
C9, C10, C22	—	0.001 μF Ceramic Disc
C11	—	0.01 μF Ceramic Disc
C12, C13, C23	—	1.0 μF, 35 V, Tantalum
C14	—	10 μF, 25 V, Aluminum Electrolytic
C2, C4, C7, C20	—	8-60 pF Compression Mica Trimmer ARCO #404 or Equivalent
L1, L2	—	1-1/2 T, #16 AWG Wire, 0.25'' I.D. (30 nH)
L3	—	1-1/2 T, #16 AWG Wire, 0.30'' I.D. (35 nH)
L4, L6	—	3 T, #16 AWG Wire, Wound on 100 Ohm Resistor (45 nH)
L5	—	#16 AWG Wire, 0.8'' Long, "U" Shaped (12 nH)
L7	—	#16 AWG Wire, 1.1'' Long, Formed Around 0.6'' Dia. Cyl. (15 nH)
RFC1, 2, 6	—	0.15 μH Molded Choke with Ferroxcube 5659065/3B Ferrite Bead on Ground Lead
RFC3	—	7 T, #20 AWG Wire, Wound on R3 (100 nH)
RFC4, 5, 7	—	Ferroxcube VK200 19/4B Ferrite Choke
R1	—	91 Ohm, 2 W, ±5% Carbon Resistor
R2	—	100 Ohm, 0.25 W, Potentiometer, CTS Type R101B or Equivalent
R3	—	560 Ohm, 1 W, ±10% Carbon Resistor
R4, R5	—	100 Ohm, 1 W, ±10% Carbon Resistor

Quantity	25 Watt Transmitter			10 Watt Transmitter			Units
	V_{DC} = 11.0 V	V_{DC} = 12.5 V	V_{DC} = 15.5 V	V_{DC} = 11.0 V	V_{DC} = 12.5 V	V_{DC} = 15.5 V	
P_{LO}	0.72	0.80	0.92	0.70	0.80	0.90	Watts
I_T	0.87	0.95	1.05	1.00	1.08	1.20	Adc
I_I	98	100	110	185	200	205	mAdc
I_{R1}	75	91	125	75	91	125	mAdc
I_C	697	759	815	740	789	870	mAdc
V_{LP}	2.75	2.80	2.90	1.8	1.9	2.0	Volts
P_D MJE2020	5.8	7.4	10.3	6.8	8.3	11.7	Watts
P_DR1	0.51	0.75	1.41	0.51	0.75	1.41	Watts
*P_D 1N5229	138	206	352	138	206	352	mW
*P_D R2	185	185	185	185	185	185	mW

*For zero base current conditions (high power operation)

NOTES:
1. With V_{DC} = 12.5 Volts, Amplifier Input Power Set for Rated High Power Output at 158 MHz. Low Power Output (P_{LO}) then set for 0.8 Watts by Adjusting R2.
2. A Nominal Zener Voltage of 4.3 Volts has been used for all calculations.
3. P_D notes the DC power being dissipated by the indicated component.

10-watt and 25-watt marine band transmitters (courtesy Motorola Semiconductor Products Inc.).

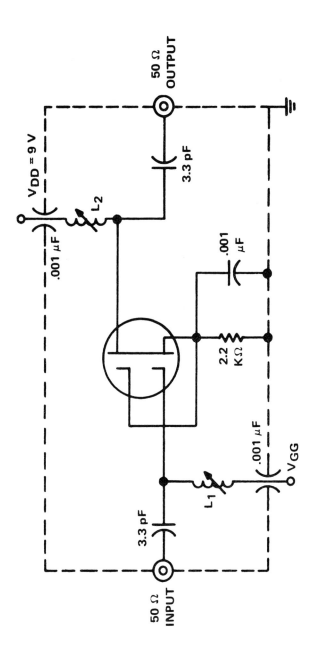

105 MHz gate-1-controlled RF amplifier using a TIS152 dual-gate MOSFET (courtesy Texas Instruments Incorporated).

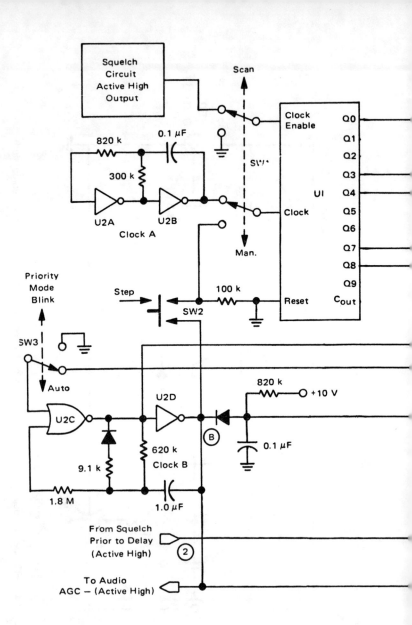

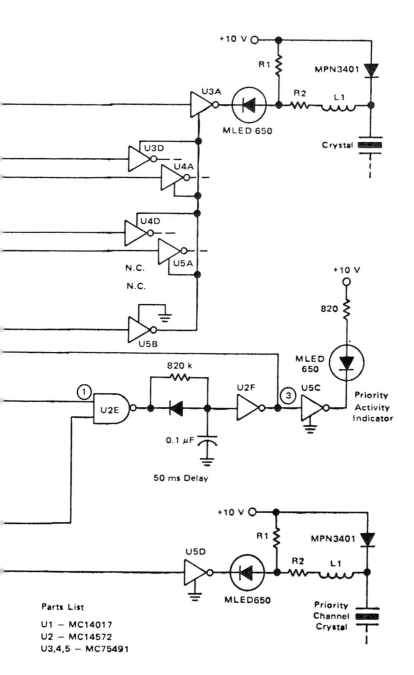

9-channel plus fixed-priority-channel logic for RF scanner/monitor receiver (courtesy Motorola Semiconductor Products Inc.).

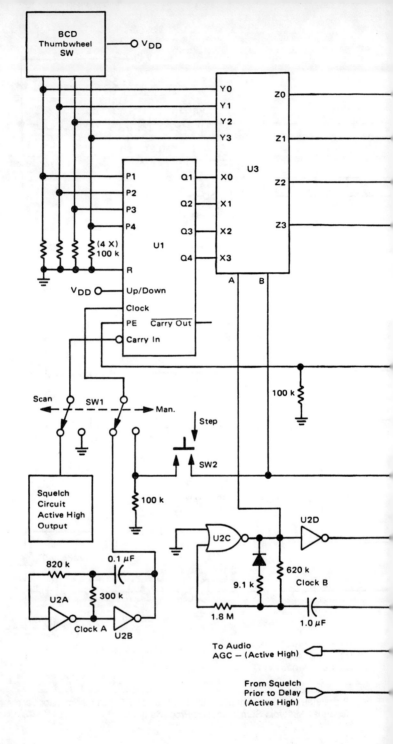

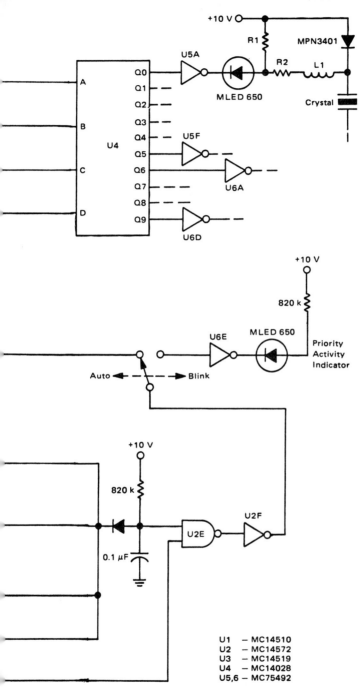

10-channel with thumbwheel-programmable-priority logic for scanner/monitor receiver (courtesy Motorola Semiconductor Products Inc.).

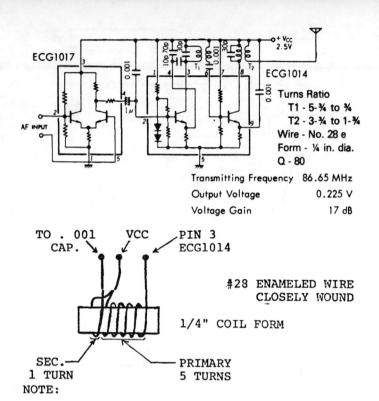

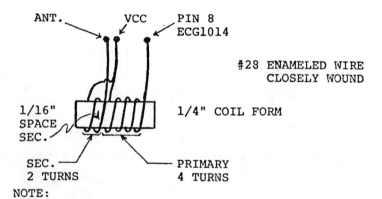

NOTE:

1. PUSH PRIMARY AGAINST SECONDARY.

FM wireless microphone using an ECG1014 and ECG1017. Circuit works well up to 250 to 300 feet. For maximum stability cement coil forms to chassis or board. Recommended supply voltage is 4.5 volts using three cells (courtesy GTE Sylvania Incorporated).

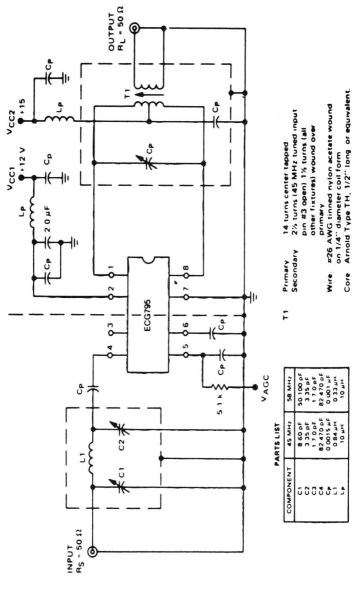

45 MHz/58 MHz RF amplifier with tuned input. See listing for component values. (courtesy GTE Sylvania Incorporated).

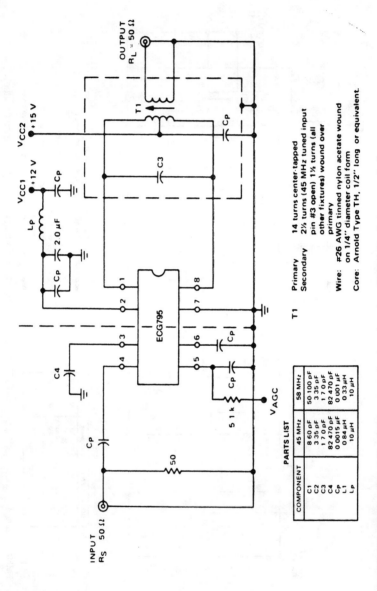

45 MHz/58 MHz RF amplifier. See listing for component values (courtesy GTE Sylvania Incorporated).

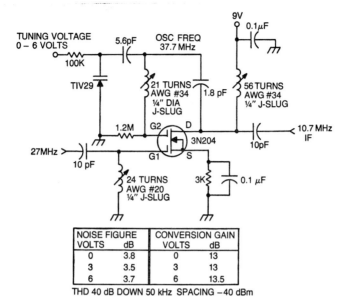

27 MHz autodyne tuner using a 3N204 dual-gate MOSFET (courtesy Texas Instruments Incorporated).

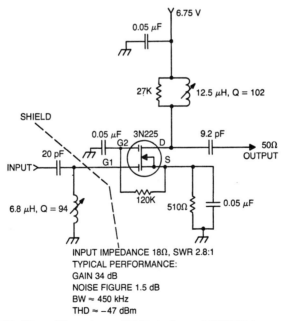

12.5 MHz IF amplifier using a 3N225 dual-gate MOSFET (courtesy Texas Instruments Incorporated).

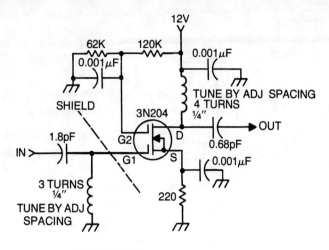

GAIN 19.5 dB
NF 1.4 dB
BANDWIDTH 3 dB DOWN 27 MHz
UNITY GAIN 130 MHz

450/432 MHz RF amplifier using a 3N204 dual-gate MOSFET (courtesy Texas Instruments Incorporated).

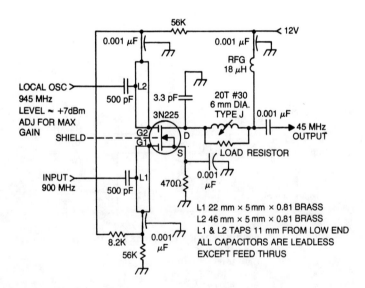

900 MHz to 45 MHz mixer using a 3N225 dual-gate MOSFET (courtesy Texas Instruments Incorporated).

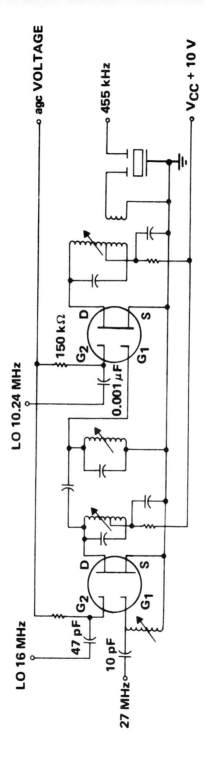

First and second mixer stages for a 27 MHz CB receiver. Note that these two stages replace three in conventional receivers not employing dual-gate MOSFETs (courtesy Texas Instruments Incorporated).

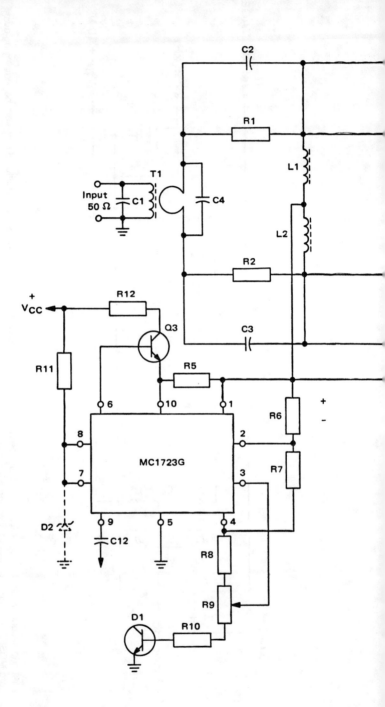

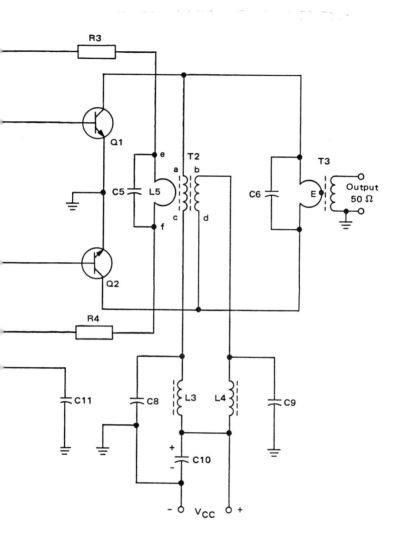

100-/140-/180-watt linear amplifier for 1.6 to 30 MHz 13.6-volt mobile operation (courtesy Motorola Semiconductor Products Inc.).

PARTS LIST FOR 100-/140-/180-WATT LINEAR AMPLIFIER

	100 W AMPLIFIER	140 W AMPLIFIER	180 W AMPLIFIER
C1	51 pF	51 pF	82 pF
C2, C3	5600 pF	5600 pF	6800 pF
C4	—	390 pF	1000 pF
C5	680 pF	680 pF	680 pF
C6	1620 pF (2 × 470 pF chips + 680 pF dipped mica in parallel)	1760 pF (2 × 470 pF chips + 820 pF dipped mica in parallel)	1940 pF (2 × 470 pF chips + 1000 pF dipped mica in parallel)
C8, C9	0.68 µF	0.68 µF	0.68 µF
C10	100 µF/20 V electrolytic	100 µF/20 V electrolytic	100 µF/20 V electrolytic
C11	500 µF/3 V electrolytic	500 µF/3 V electrolytic	500 µF/3 V electrolytic
C12	1000 pF disc ceramic	1000 pF disc ceramic	1000 pF disc ceramic
R1, R2	2 × 3.9 Ω/½ W in parallel	2 × 3.6 Ω/½ W in parallel	2 × 3.3 Ω/½ W in parallel
R3, R4	2 × 4.7 Ω/½ W in parallel	2 × 5.6 Ω/½ W in parallel	2 × 3.9 Ω/½ W in parallel
R5	1.0 Ω/½ W	0.5 Ω/½ W	0.5 Ω/½ W
R6	1.0 kΩ/½ W	1.0 kΩ/½ W	1.0 kΩ/½ W
R7	18 kΩ/½ W	18 kΩ/½ W	18 kΩ/½ W
R8	8.2 kΩ/½ W	8.2 kΩ/½ W	8.2 kΩ/½ W
R9	1.0 kΩ trimpot	1.0 kΩ trimpot	1.0 kΩ trimpot
R10	150 Ω/½ W	150 Ω/½ W	150 Ω/½ W
R11	1.0 kΩ/½ W	1.0 kΩ/½ W	1.0 kΩ/½ W
R12	20 Ω/5 W	20 Ω/5 W	20 Ω/5 W

	100 W AMPLIFIER	140 W AMPLIFIER	180 W AMPLIFIER
L1, L2 L3, L4 L5 T1	Ferroxcube VK200 19/4B ferrite choke Two Fair-Rite Products ferrite beads 2673021801 or equivalent on AWG #16 wire each. 1 separate turn through toroid of T2. 9:1 (3:1 turns ratio) 9:1 (3:1 turns ratio) 16:1 (4:1 turns ratio) Ferrite core: Stackpole 57-1845-24B, Fair-Rite Products 2873000201 or two Fair-Rite Products 0.375″ OD x 0.200″ ID x 0.400″, Material 77 beads for type A (Figure 3) transformer. See text.		
T2	6 turns of AWG #18 enameled, bifilar wire Ferrite core: Stackpole 57-9322, Indiana General F627-8 Q1 or equivalent. 16:1 (4:1 turns ratio) 16:1 (4:1 turns ratio) 25:1 (5:1 turns ratio)		
T3	Ferrite core: 2 Stackpole 57-3238 ferrite sleeves (7D material) or number of toroids with similar magnetic characteristics and 0.175″ sq. total cross sectional area. See text. All capacitors except C12, part of C5 and the electrolytics are ceramic chips. Values over 82 pF are Union Carbide type 1225 or Varadyne size 14. Others are type 1813 or size 18 respectively.		
Q1, Q2	HEPS3037, MRF453, MRF460, MRF455	MRF454, MRF458, 2N5989 or equivalent	MRF421
Q3 D1 D2		2N5190 or equivalent Not Used	
	c. Dotted line in performance data.	b. Dashed line in performance data.	a. Solid line in performance data.

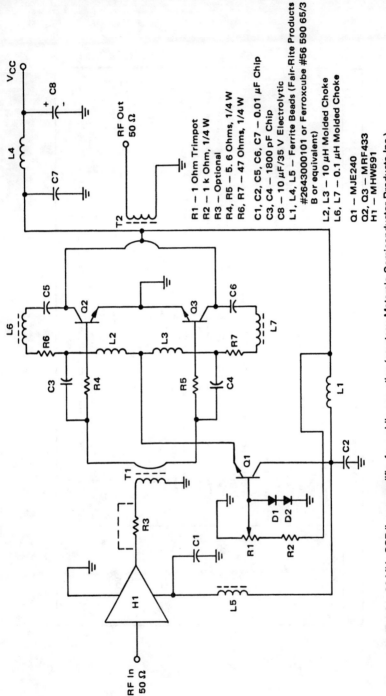

20-watt 55 dB 1.6 to 30 MHz SSB linear amplifier for mobile operation (courtesy Motorola Semiconductor Products Inc.).

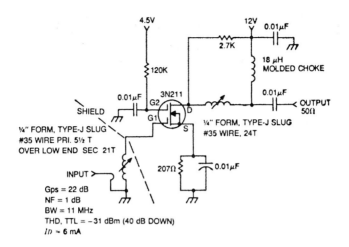

36 MHz RF amplifier using a 3N211 dual-gate MOSFET (courtesy Texas Instruments Incorporated).

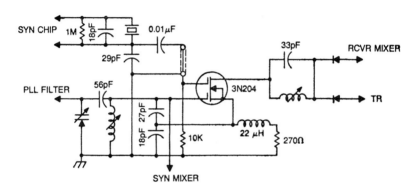

VCO and mixer for CB operation using a single 3N204 dual-gate MOSFET (courtesy Texas Instruments Incorporated).

Miscellaneous Circuits

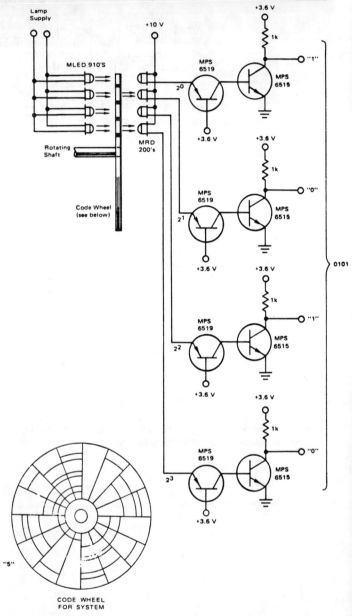

Electro-optical shaft encoder (courtesy Motorola Semiconductor Products Inc.).

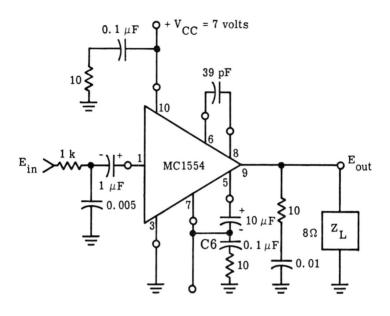

1-watt noninverting power amplifier for split power supply operation using an MC1554. As shown voltage gain is nine (courtesy Motorola Semiconductor Products Inc.).

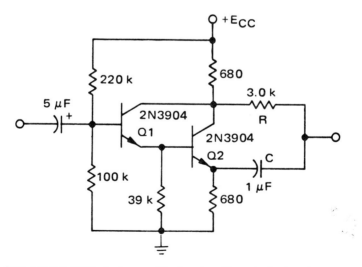

Push-pull discrete Darlington amplifier (courtesy Motorola Semiconductor Products Inc.).

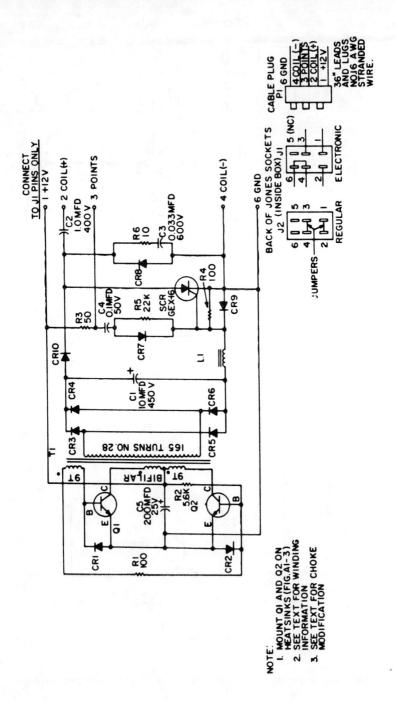

Parts List

C1 – 10-mfd, 450-volt electrolytic capacitor
C2 – 1.0-mfd, 400-volt non-polarized capacitor
C3 – 0.033-mfd, 600-volt capacitor
C4 – 0.1-mfd, 50-volt capacitor
C5 – 200-mfd, 25-volt electrolytic capacitor
CR1 – CR10 – GE-504A rectifier diodes
L1 – 250-millihenry choke; modified 0.5 millihenry choke with 0.030-inch shim (Triad C36X, or equivalent). See text.
Q1, Q2 – GE-X18 transistor

R1 – 180-ohm, 2-watt resistor
R2 – 5600-ohms, 1-watt resistor
R3 – 50-ohms, 5-watt wire-wound resistor
R4 – 100-ohms, 1/2-watt resistor
R5 – 20K-ohm, ¾-watt resistor
R6 – 10-ohm, ¾-watt resistor
SCR – GE-X16 silicon controlled rectifier
T1 – Pulse transformer * See text for winding information.
J1, J2 – Jones S-306-AB, 6-pin sockets, or equivalent
P1 – Jones, P-306-CCT, plug, or equivalent
Heatsinks (2) – See Figure A1-3
Box – 4" x 5" x 6"; Bud CU-3007A, or equivalent

*Available from General Electric Co., 3800 N. Milwaukee Avenue, Chicago, Ill. 60611. Specify ETRS-5450 for the two E cores and ETRS-5451 for the bobbin.

Note: Printed-wiring boards with completely wound and mounted T1 transformers may be purchased from Felmox Electronics, N. Division Street Road, Auburn, N.Y. 13021.

Capacitor discharge SCR ignition system. The circuit draws less than 1 ampere at high RPM (courtesy General Electric Company).

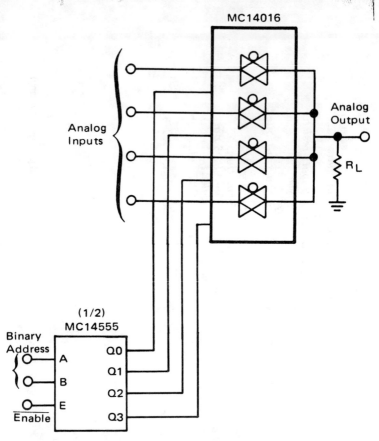

4-channel analog data selector (courtesy Motorola Semiconductor Products Inc.).

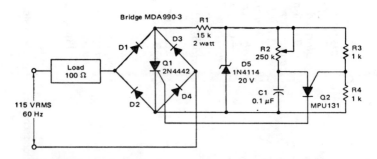

SCR phase control circuit with a PUT. Relaxation oscillator Q2 provides conduction control of Q1 from 1 ms to 7.8 ms (courtesy Motorola Semiconductor Products Inc.).

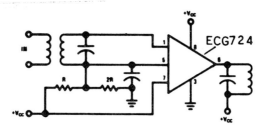

Balance differential amplifier with a controlled constant-current-source drive and AGC capability. Operation can be from DC to 120 MHz. Supply voltage is typically 9 to 12 volts (courtesy GTE Sylvania Incorporated).

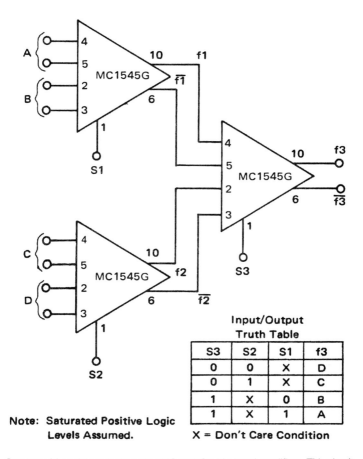

Input/Output Truth Table

S3	S2	S1	f3
0	0	X	D
0	1	X	C
1	X	0	B
1	X	1	A

Note: Saturated Positive Logic Levels Assumed. X = Don't Care Condition

One-out-of-four data selector using MC1545G wide-band amplifiers. This circuit can be modified to be a one-out-of-N data selector (courtesy Motorola Semiconductor Products Inc.).

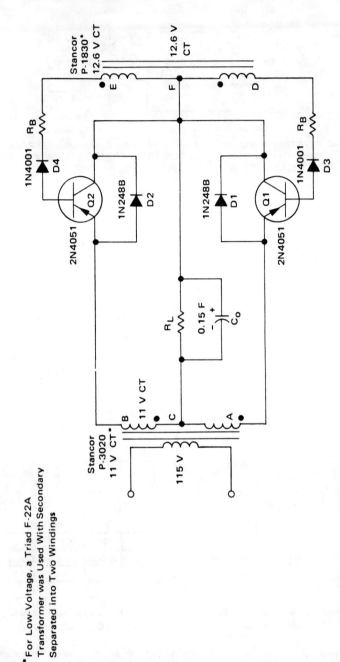

Full-wave synchronous rectification circuit (courtesy Motorola Semiconductor Products Inc.).

*For Low-Voltage, a Triad F-22A Transformer was Used With Secondary Separated into Two Windings

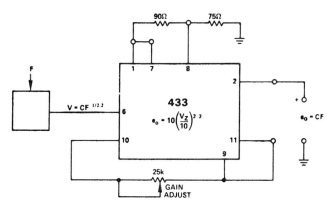

Transducer linearization circuit using the 433 multiplier/divider chip (courtesy Analog Devices, Inc.).

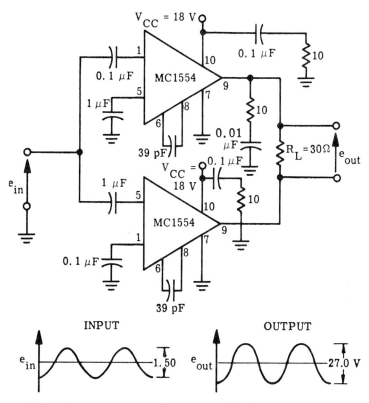

3-watt differential output power amplifier using two MC1554s (courtesy Motorola Semiconductor Products Inc.).

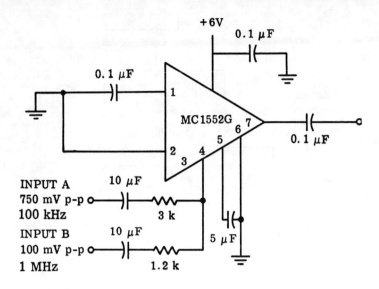

Summing/scaling amplifier using an MC1552G. Scaling considerations are accomplished by adjustment of the input resistors (courtesy Motorola Semiconductor Products Inc.).

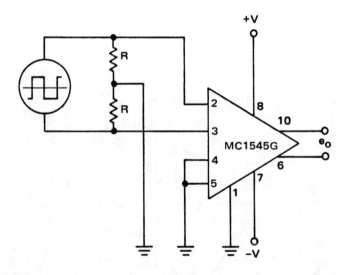

Pulse amplifier using an MC1545G for applications in radar IFs, pulse width modulation and pulse amplitude modulation systems (courtesy Motorola Semiconductor Products Inc.).

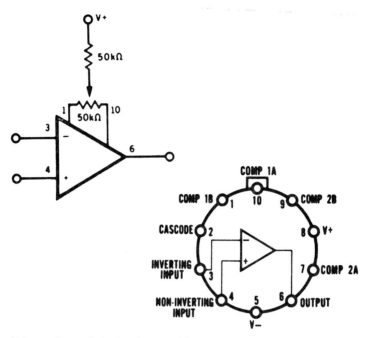

Voltage offset null circuit using an ECG915 operational amplifier. See basing diagram for supply connections. Supply voltage is ±15 volts (courtesy GTE Sylvania Incorporated).

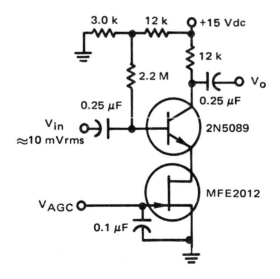

General-purpose amplifier with FET AGC circuit (courtesy Motorola Semiconductor Products Inc.).

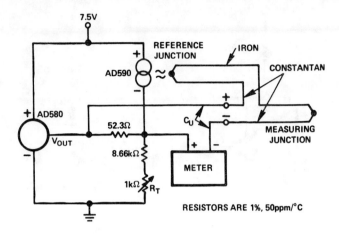

Cold junction compensation circuit for type J thermocouple (courtesy Analog Devices, Inc.).

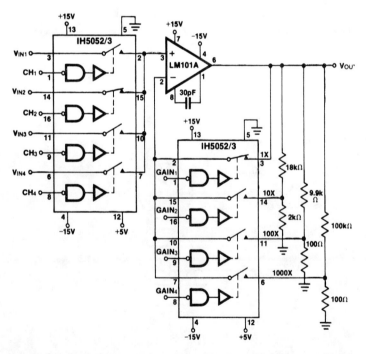

Programmable gain noninverting amplifier with selectable inputs. The IH5052/53 chips are 16-pin DIP CMOS analog gates (courtesy Intersil, Inc.).

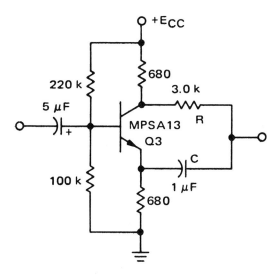

Push-pull monolithic Darlington amplifier (courtesy Motorola Semiconductor Products Inc.).

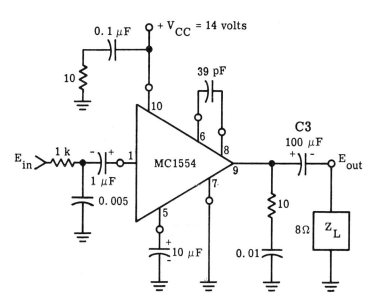

NOTE: Voltage Gain = 9 For Connection Shown

1-watt noninverting power amplifier using an MC1554 connected to a single supply. As shown voltage gain is nine (courtesy Motorola Semiconductor Products Inc.).

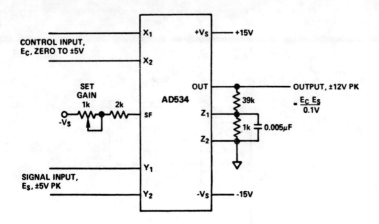

NOTES:
1) GAIN IS X10 PER VOLT OF E_C, ZERO TO X50
2) WIDEBAND (10Hz – 30kHz) OUTPUT NOISE IS 3mV RMS, TYP CORRESPONDING TO A F.S. S/N RATIO OF 70dB
3) NOISE REFERRED TO SIGNAL INPUT, WITH E_C = ±5V, IS 60µV RMS, TYP
4) BANDWIDTH IS DC TO 20kHz, -3dB, INDEPENDENT OF GAIN

Voltage-controlled amplifier using an AD534 multiplier/divider chip (courtesy Analog Devices, Inc.).

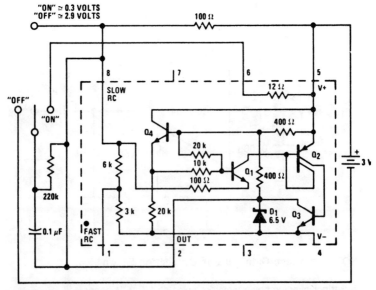

Latch circuit using an LM3909 chip. Circuitry inside dashed lines is the LM3909. The circuit switches to and holds its condition whenever the switch changes sides (courtesy National Semiconductor Corporation).

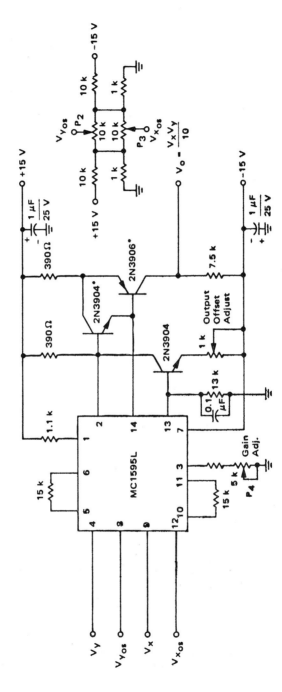

Discrete level-shifting circuit (courtesy Motorola Semiconductor Products Inc.).

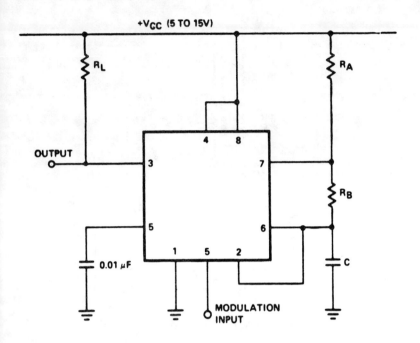

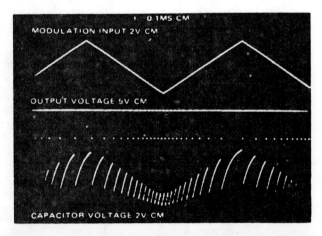

Pulse position modulator using an ECG955M timer/oscillator chip. The timer is connected as an astable multivibrator. With a modulating signal applied to pin 5 the pulse position will vary with the modulating signal (courtesy GTE Sylvania Incorporated).

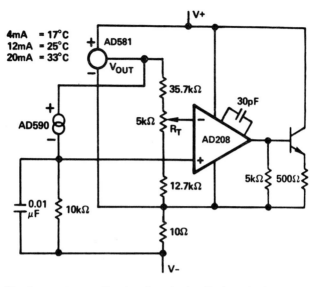

4-to-20 mA current transmitter (courtesy Analog Devices, Inc.).

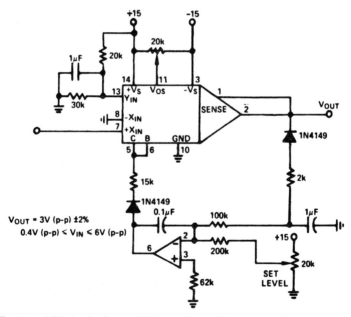

Precision AGC circuit using an AD531 multiplier/divider and an AD741 op amp. The circuit works by rectifying a signal and comparing it with the voltage at the set level pot. The comparator output is then applied as a control signal to the AD531. The circuit can regulate V_{OUT} to 3 volts peak to peak for inputs from 0.4 volt peak to peak to 6.0 volts peak to peak from 30 hertz to 400 kilohertz (courtesy Analog Devices, Inc.).

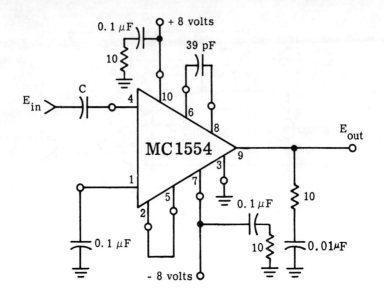

1-watt inverting power amplifier using an MC1554. As shown the voltage gain is 35. An external heat sink is required for dissipation greater than 350 mW (courtesy Motorola Semiconductor Products Inc.).

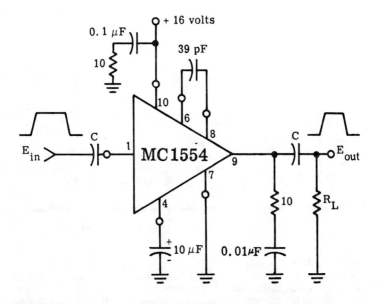

Pulse power amplifier using an MC1554. As shown voltage gain is 18. Peak power of 3 watts is possible (courtesy Motorola Semiconductor Products Inc.).

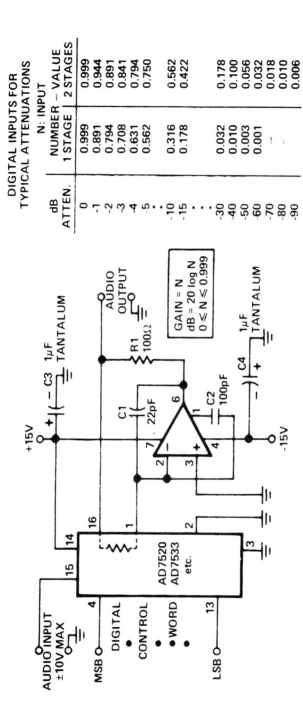

Inexpensive high-performance gain control circuit (courtesy Analog Devices, Inc.).

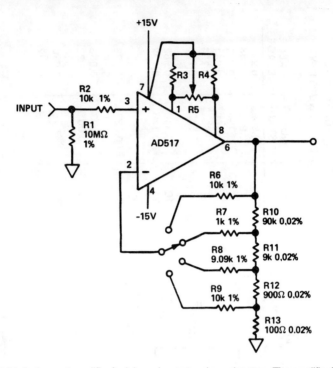

Stable instrument amplifier for lab equipment and panel meters. The amplifier is noninverting and offers selectable gains from 1 to 1000 in decade steps. Input impedance is 10M (courtesy Analog Devices, Inc.).

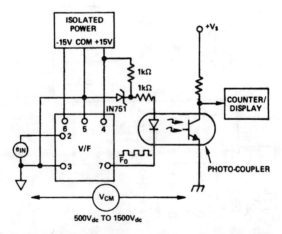

Precision high CMV (common mode voltage) analog isolator. This circuit provides up to 1500 volts DC CMV isolation. The V/F converter shown is a 452 with a full scale output of 100 kHz for 10 volts input (courtesy Analog Devices, Inc.).

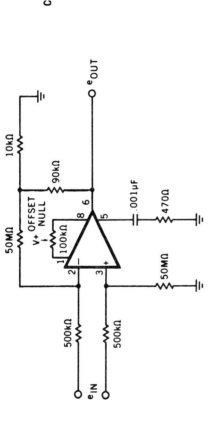

CHARACTERISTICS:
$A_{VCL} = 1000 = 60$ dB
DC Gain Error = 0.05%
Bandwidth = 1 kHz for −0.05% error
Diff. Input Res. = 1 MΩ
Typical amplifying capability
$e_{IN} = 10$ μV on $V_{CMI} = 1.0$ V
Caution: Minimize Stray Capacitance

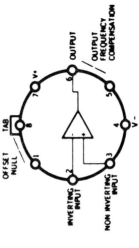

NOTE: Pin 4 connected to case

Precision amplifier with closed loop voltage gain of 1000 using an ECG925. Typical supply voltage is ±15 volts, but it can be powered with supplies from ±3 volts to ±22 volts (courtesy GTE Sylvania Incorporated).

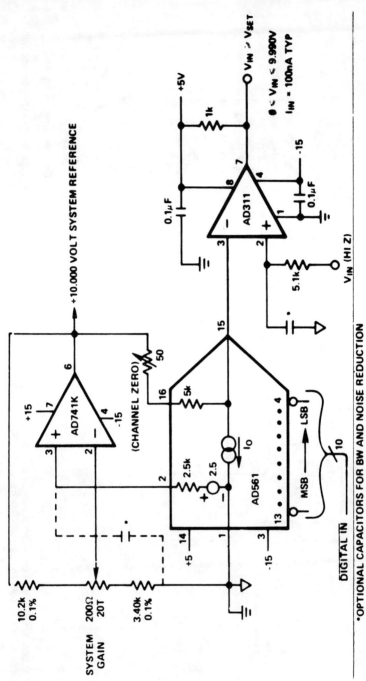

Digitally programmed set point comparator (courtesy Analog Devices, Inc.).

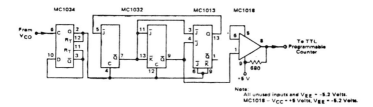

Fixed divide-by-ten prescaler using an MC1034, MC1032, MC1013 and MC1018 (courtesy Motorola Semiconductor Products Inc.).

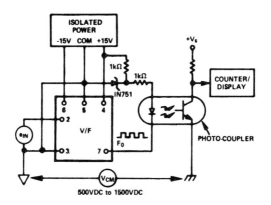

Precision high CMV (common mode voltage) analog isolator. This circuit provides up to 1500 volts DC CMV isolation (courtesy Analog Devices, Inc.).

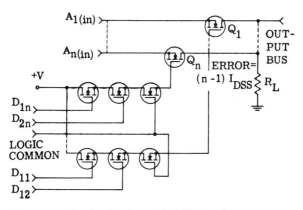

Commutator network using N-channel MOSFETs. All semiconductors are 2N4351s. Each switch has a three-input AND gate in series with the gate drive (courtesy Motorola Semiconductor Products Inc.).

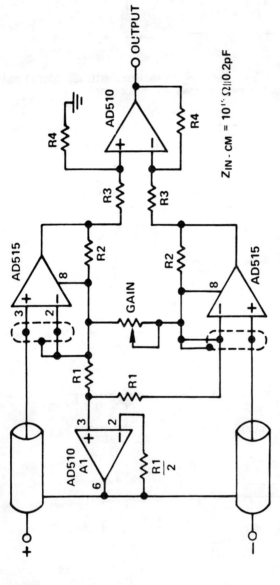

Very-high impedance instrumentation amplifier using two AD510s and two AD515s (courtesy Analog Devices, Inc.).

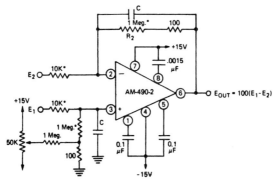

Differential amplifier using a Datel AM-490-2 8-pin TO-99 chip. For a 0.01% match between 1M resistors and 10K resistors, common mode rejection ratio is approximately 120 dB. Capacitors C are used only to reduce bandwidth and hence output noise (courtesy Datel Systems, Inc.).

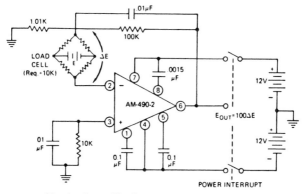

Battery-powered load-cell amplifier for discontinuous service. The AM-490-2 is an 8-pin TO-99 differential-input chopper-stabilized op amp. For power interrupt applications the amplifier has a warm-up period of 200 ms (courtesy Datel Systems, Inc.).

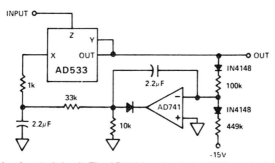

Automatic level control circuit. The AD533 is set up in the divide mode. Its output is rectified and compared with the power supply derived −15 volt reference. The net current is integrated by the AD741 and fed into the denominator input of the AD533, maintaining the level of the output at the AC average value programmed by the feedback circuitry, 7 volts in this case (courtesy Analog Devices, Inc.).

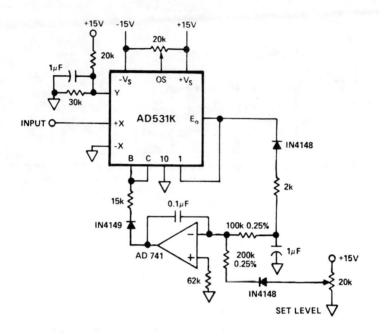

Automatic gain control circuit. The AD531 programmable multiplier/divider maintains a 3-volt peak-to-peak output for inputs from 0.1 volt to over 12 volts with 2% regulation for the range from 0.4 volt peak to peak to 6 volts peak to peak. The input frequency can range from 30 hertz to 400 hertz (courtesy Analog Devices, Inc.).

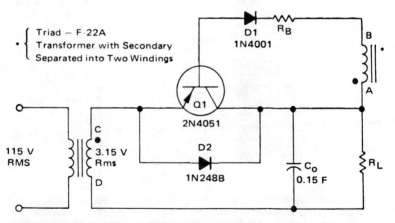

Half-wave synchronous rectification circuit (courtesy Motorola Semiconductor Products Inc.).

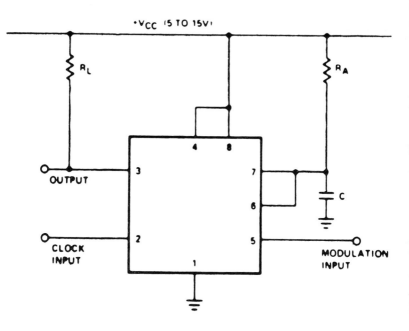
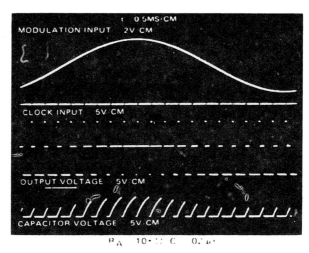

Pulse width modulator using an ECG955M timer/oscillator chip. The flip-flop is connected in a monostable mode. The circuit is triggered with a continuous pulse train and the threshold voltage is modulated by the signal applied to control voltage pin 5. This has the effect of modulating the pulse width as control voltage varies (courtesy GTE Sylvania Incorporated).

TV Circuits

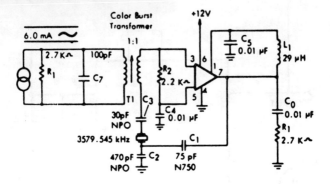

3.58 MHz injection-locked oscillator for color TV, using one ECG703A IC. The burst transformer induces the chroma reference directly into the feedback loop of the oscillator, thereby locking the oscillator in phase with the transmitted burst (courtesy GTE Sylvania Incorporated).

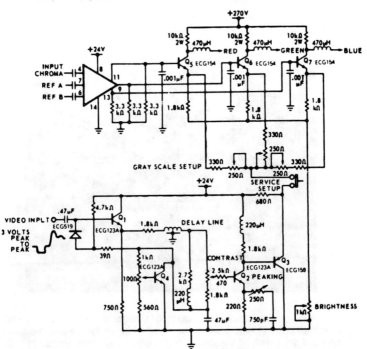

Complete RGB video output stage for color TV using the ECG713 chip (courtesy GTE Sylvania Incorporated).

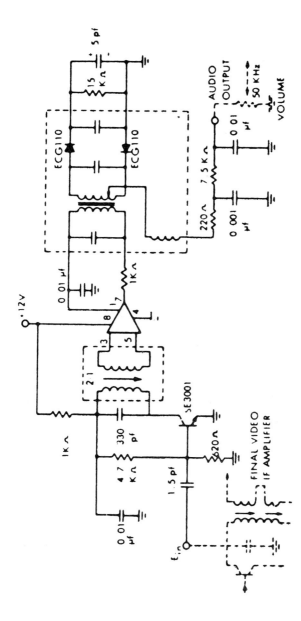

4.5 MHz color TV sound IF amplifier using ECG703A. In a color TV the 4.5 MHz IF is not generated in the video detector as in B&W sets. A separate transistor sound detector, driven from the final video IF amplifier, is used to develop the 4.5 MHz IF. The driver transformer at the input of the chip is tuned to 4.5 MHz. The output drives a conventional ratio detector, which provides good AM rejection well below full limiting (courtesy GTE Sylvania Incorporated).

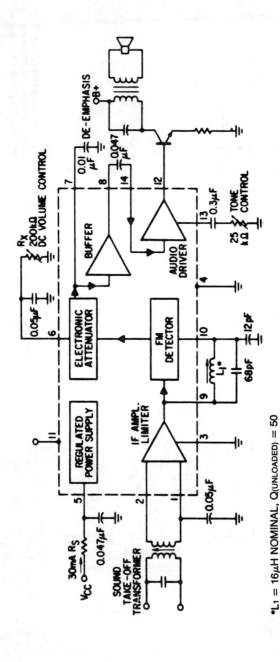

*L_1 = 16 μH NOMINAL, $Q_{(UNLOADED)}$ = 50

TV FM sound system using the ECG712. The supply terminal can be connected to any supply voltage with a suitable dropping resistor provided that the dissipation rating, 400 mW, is not exceeded since the ECG712 has internal zener regulation. Besides TV, the circuit can be used for FM mobile communications. It also employs DC volume control so no shielding is necessary (courtesy GTE Sylvania Incorporated).

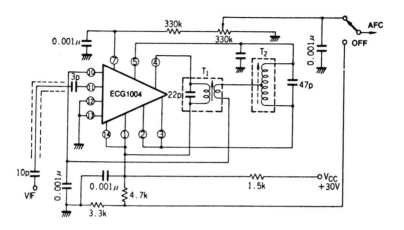

TV AFT/AFC circuit for 58.75 MHz (courtesy GTE Sylvania Incorporated).

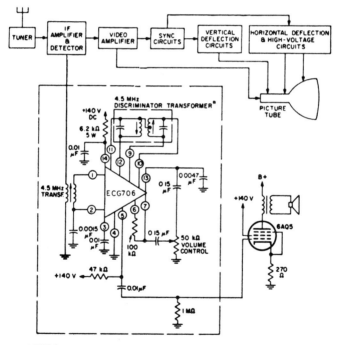

* TRW Electronics, Des Plaines, Illinois. Part No. E023874, or equivalent.

4.5 MHz sound section for TV using one ECG706 IC. The chip contains a wide-band IF amplifier-limiter section, an FM detector stage, a zener regulated power supply section, and an audio amplifier section specifically designed to drive a 6AQ5 (courtesy GTE Sylvania Incorporated).

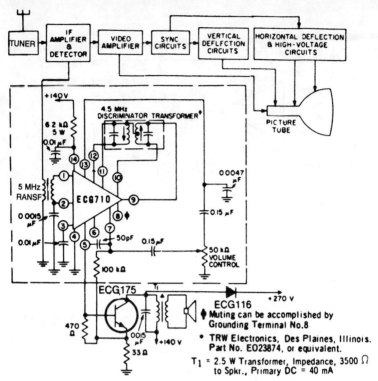

TV FM sound IF with transistor power output stage. The ECG710 contains a multisection wide-band IF amplifier, a zener-regulated power supply section, and an AF amplifier section (courtesy GTE Sylvania Incorporated).

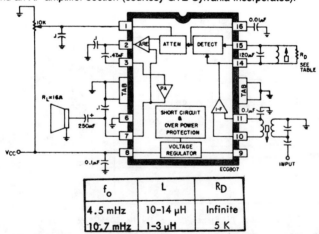

f_o	L	R_D
4.5 mHz	10–14 μH	Infinite
10.7 mHz	1–3 μH	5 K

TV sound channel with 1-watt output using an ECG807. The circuit features DC volume control with 70 dB attenuation typical, limiter gain of 70 dB, limiting threshold at 200 μV, automatic thermal shutdown, and audio output short-circuit protection. The circuit will drive 8-, 16- or 32-ohm speaker loads and provide a true 1-watt output with an 8-ohm load (courtesy GTE Sylvania Incorporated).

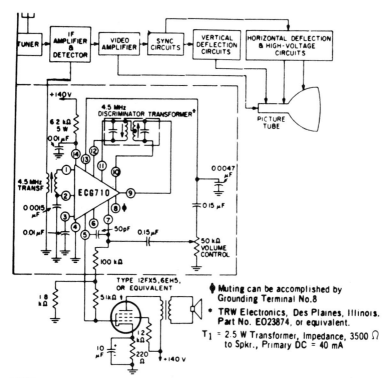

TV FM sound IF with tube-type audio output stage. The ECG710 contains a multistage wide-band amplifier section, a zener-regulated power supply section, and an AF amplifier section (courtesy GTE Sylvania Incorporated).

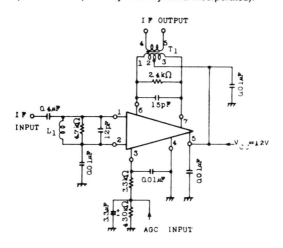

TV video IF amplifier using an ECG1128 7-pin module. Typical power gain is 48 dB. Maximum output voltage is 180 mV (courtesy GTE Sylvania Incorporated).

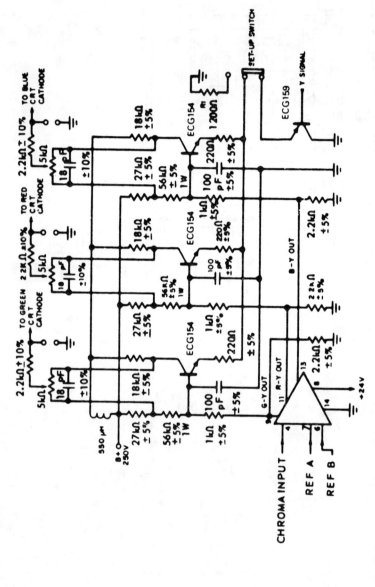

Complete RGB video output stage for color TV using one ECG713 chip (courtesy GTE Sylvania Incorporated).

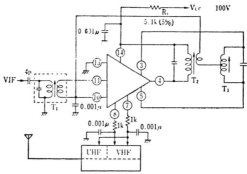

TV AFT system using an ECG1096 14-pin DIP with tab. This type of AFT is for a double-ended system, where both positive and negative correction voltages are required by the VHF tuner. For the UHF tuner one of the correction voltages can be used. Transformers T1, T2 and T3 are selected and tuned for the specific video IF, for example, 45.75 MHz (courtesy GTE Sylvania Incorporated).

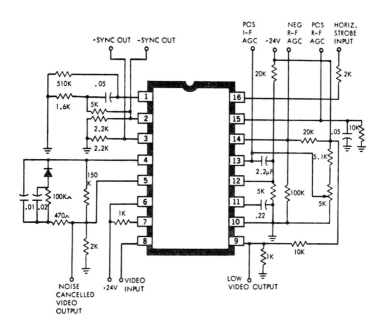

Video signal processor for either B&W or color TV. The chip contains a keyed AGC amplifier that provides forward AGC voltage for video IF amplifiers and tuners equipped with NPN bipolar transistors. It provides composite sync outputs with positive and negative going signals. There's also reverse AGC voltage for tuners equipped with MOSFETs, PNP bipolar transistors, or tubes in the RF stage. The video output is a low-impedance emitter follower. In addition there's a preset noise detector for gating the chip's sync separator and AGC detector (courtesy GTE Sylvania Incorporated).

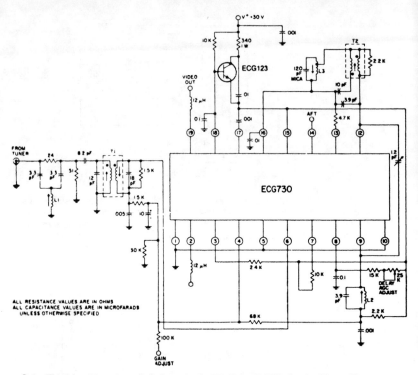

Color TV video IF system. Gain is typically 75 dB at 45 MHz for the IF amplifier and 12 dB for the video amplifier. The chip contains a zener reference diode for convenient and economical power supply regulation (courtesy GTE Sylvania Incorporated).

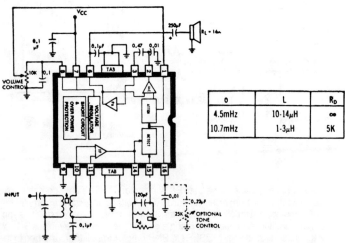

O	L	R_D
4.5 mHz	10-14 μH	∞
10.7 mHz	1-3 μH	5K

TV sound channel with 2-watt output using an ECG742. Supply voltage is from 18 to 27 volts. The chip is a 16-pin DIP. The circuit can be connected to an 8-ohm load and still provide the 2-watt output (courtesy GTE Sylvania Incorporated).

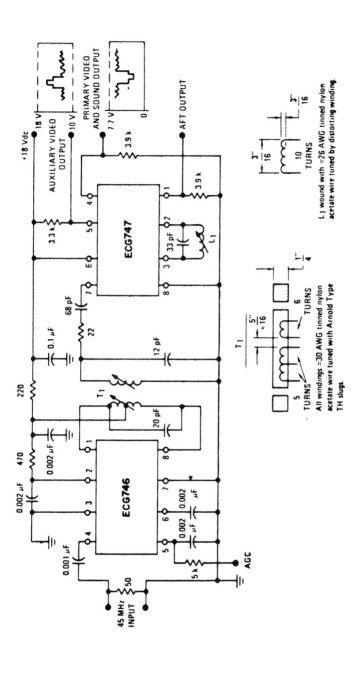

Typical ECG746 Video IF Amplifier and ECG747 Low-Level Video Detector Circuit

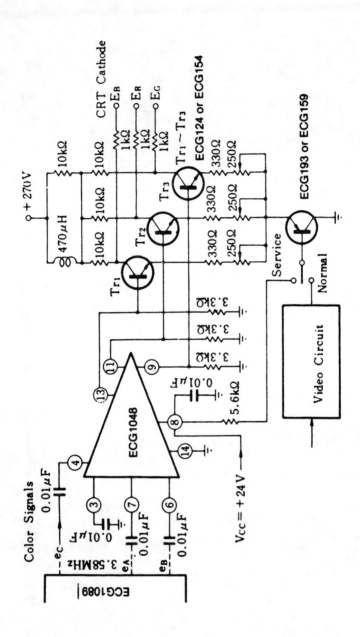

Color TV chroma demodulator. The ECG1048 is a 14-pin DIP with an end metal tab. The ECG1089 partially shown is a color processor chip (courtesy GTE Sylvania Incorporated).

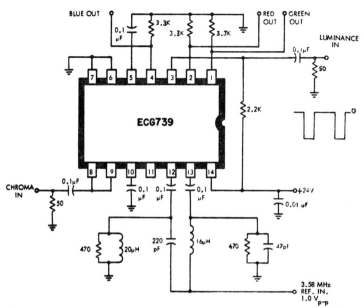

Color TV chroma demodulator with RGB output matrix. By applying a positive going blanking pulse to pin 6, blanking of the picture during line and frame flyback is achieved (courtesy GTE Sylvania Incorporated).

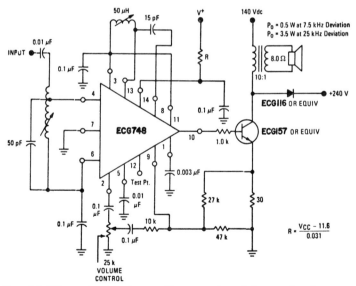

Complete TV sound system that provides approximately 2 watts of audio power. Sound detection is accomplished through a coincidence discriminator in the chip that requires only one RLC phase shift network. Supply for the chip should be approximately 16 volts. Select the value of resistor R to obtain 11.6 volts at pin 14 of the ECG748 (courtesy GTE Sylvania Incorporated).

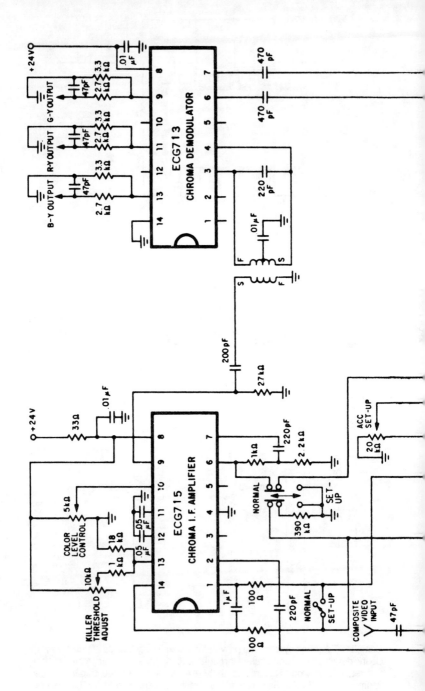

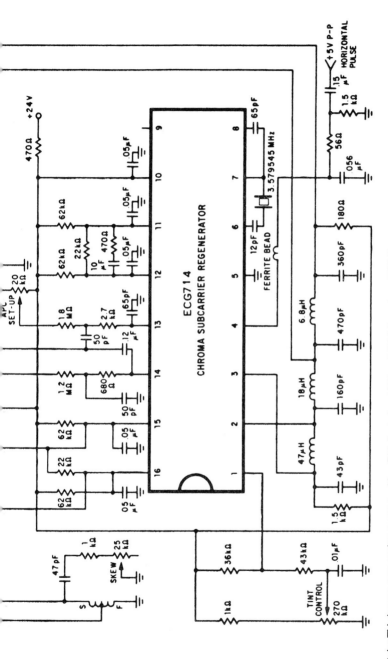

Color TV chroma processing system with a PLL featuring APC and ACC (courtesy GTE Sylvania Incorporated).

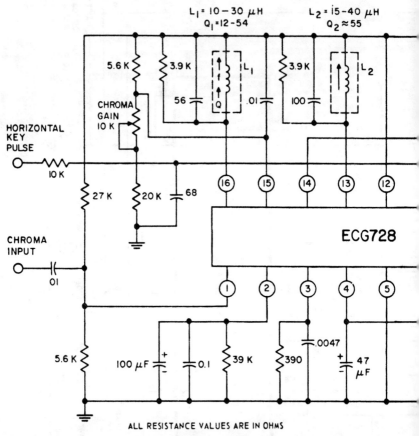

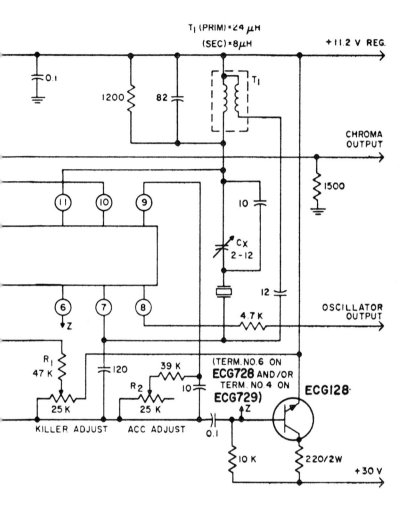

Color TV chroma signal processor. The ECG728 provides subcarrier regeneration and total chroma signal processing prior to demodulation. The coils of the chroma amplifier and bandpass amplifier are stagger-tuned to provide a combined typical bandpass of 3.08 to 4.08 MHz. A burst separator amplifier injects the burst signal into the 3.58 MHz oscillator. The ACC detector and killer detector sense the burst level or absence of burst, respectively, by monitoring the oscillator's response to the burst injection level. The thresholds for the ACC and killer are independently adjusted by resistors R2 and R1 at terminals 9 and 4. The chroma input is at pin 14 and the oscillator output is at pin 8. Pin 6 is a zener diode for use as a regulated voltage reference at 11.9 volts (courtesy GTE Sylvania Incorporated).

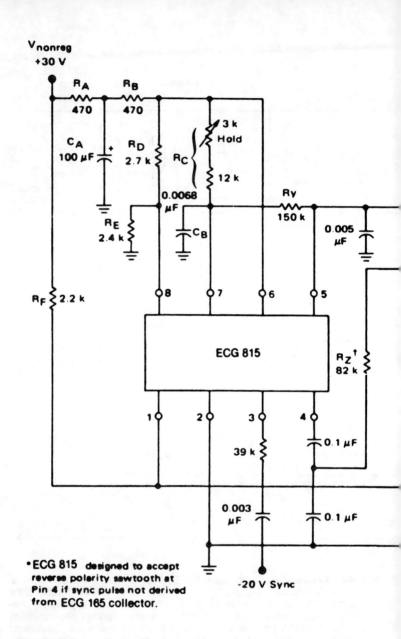

- ECG 815 designed to accept reverse polarity sawtooth at Pin 4 if sync pulse not derived from ECG 165 collector.

This circuit has an oscillator pull-in range of ±300 Hz, a noise bandwidth of 320 Hz, and a damping factor of 0.8.

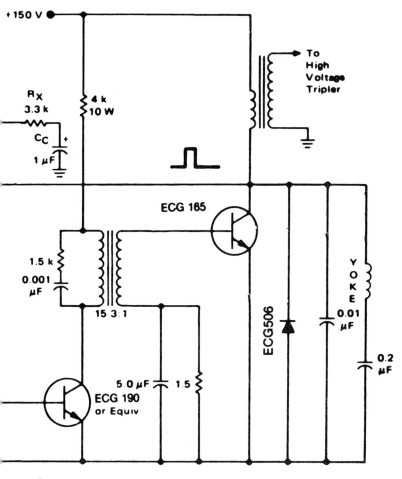

†R_Z = 6.8 k per 100 V of flyback amplitude.

TV horizontal processor with phase detector, oscillator, and predriver. Suited for all types of TV receivers, this circuit features internal shunt regulator, preset hold control capability, ±300-hertz pull-in, linear balanced phase detector, variable output duty cycle for driving tube or transistor, low thermal frequency drift, adjustable DC loop gain and positive flyback inputs (courtesy GTE Sylvania Incorporated).

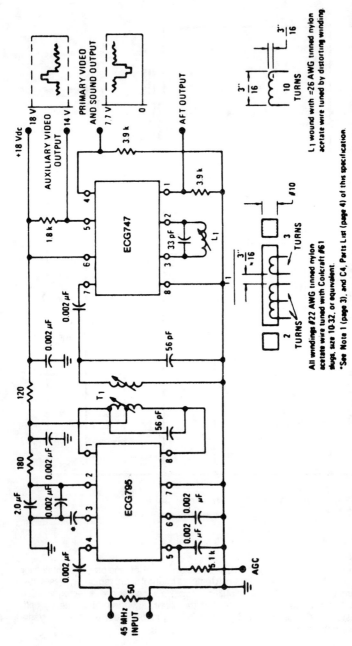

TV video IF amplifier and detector using two 8-pin DIPs. Power gain is as follows: 60 dB at 45 MHz (pin 3 open), 61 dB at 45 MHz (pin 3 bypassed), 56 dB at 58 MHz (pin 3 open), 59 dB at 58 MHz (pin 3 bypassed). AGC range is 80 dB from DC to 45 MHz. C4 should be 0.002 μF at 45 MHz (courtesy GTE Sylvania Incorporated).

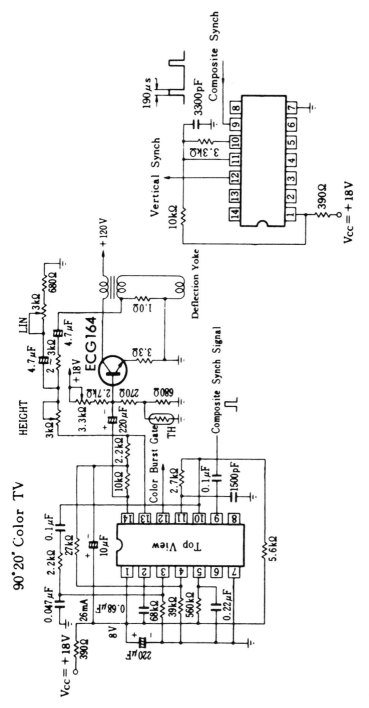

Complete TV vertical circuit with AFC for 90-degree 20-inch receivers (courtesy GTE Sylvania Incorporated).

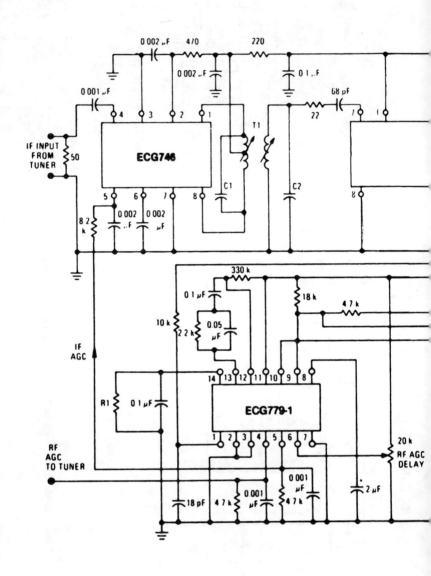

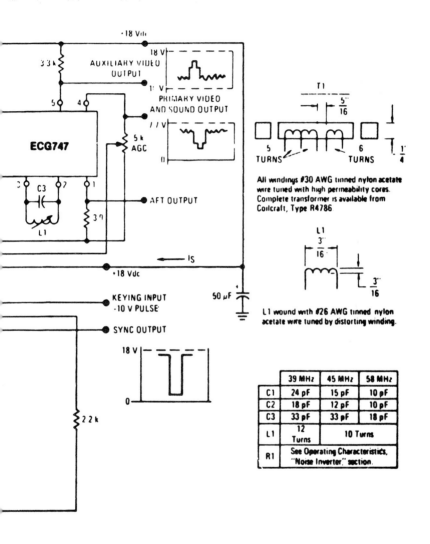

TV video IF amplifier, video detector, and signal processor circuits. Used in color and B&W TVs (courtesy GTE Sylvania Incorporated).

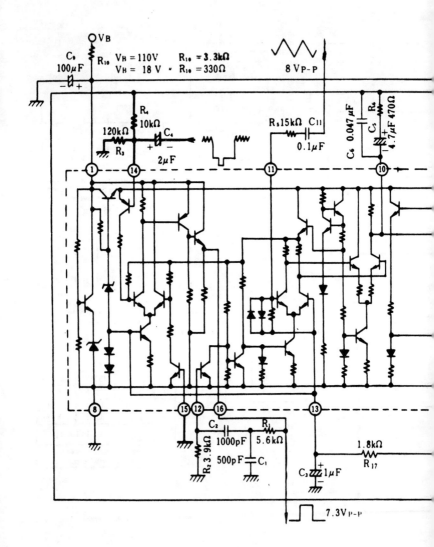

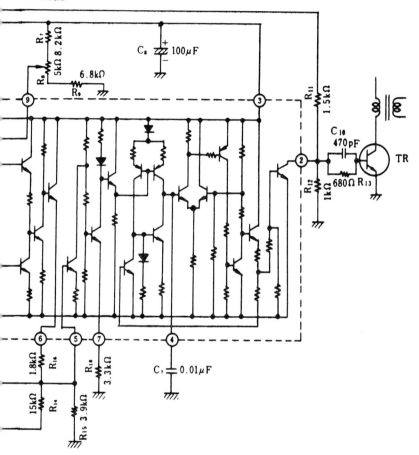

TV horizontal AFC and oscillator with sync separator for positive sync, using an ECG1086 16-pin DIP. The single transistor shown is an ECG199, a horizontal driver stage (courtesy GTE Sylvania Incorporated).

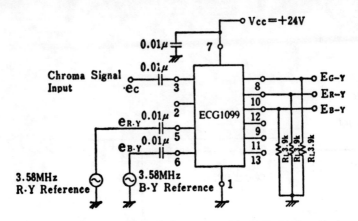

Color TV chroma demodulator with direct output. The ECG1099 is a 14-pin DIP with tab. R-Y, B-Y, and G-Y output voltages are 12 volts, 3.8 volts, and 1 volt peak to peak, respectively (courtesy GTE Sylvanic Incorporated).

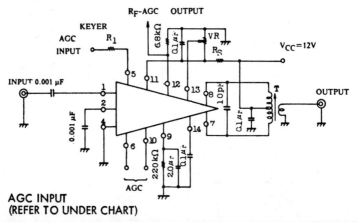

AGC INPUT
(REFER TO UNDER CHART)

VIDEO POLARITY	PIN 6 VOLTAGE	PIN 10 VOLTAGE	R_1
NEGATIVE	5.5 V ⎍ 2.0 V - - - - - - 0 V _____	1.0 - 4.0 V ADJ. 1.0-4.0 V NOM. 2.0 V	0
POSITIVE	1.0 - 8.0 V ADJ. 1.0-8.0V NOM. 4.5 V	4.5 V - - - - - 0 V ⎍	3.9 k Ohms

Linear TV IF amplifier using an ECG1080 chip, which provides 46 dB power gain. Transformer T is 0.32 mm enamel wire with a center-tapped primary of eight turns close wound. The secondary is one turn of the same wire. The coil form diameter is 5.5 mm (courtesy GTE Sylvania Incorporated).

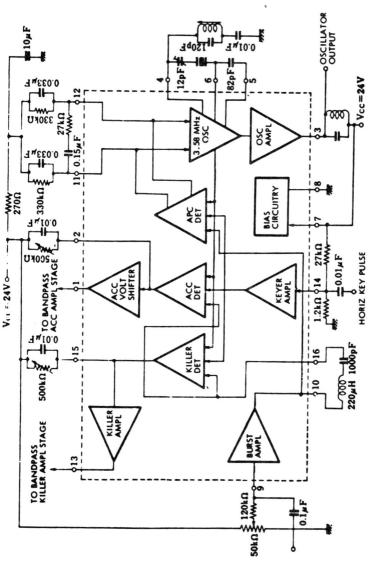

Color TV subcarrier generator using an ECG1105 16-pin DIP. The ECG1105 consists of a keyed APC, an ACC, a killer detector amplifier, a burst amplifier and a subcarrier amplifier (courtesy GTE Sylvania Incorporated).

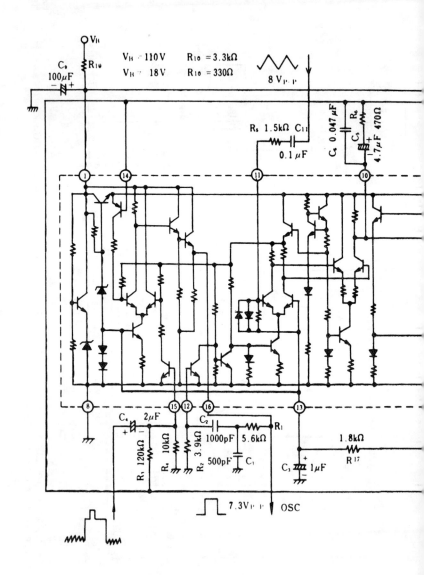

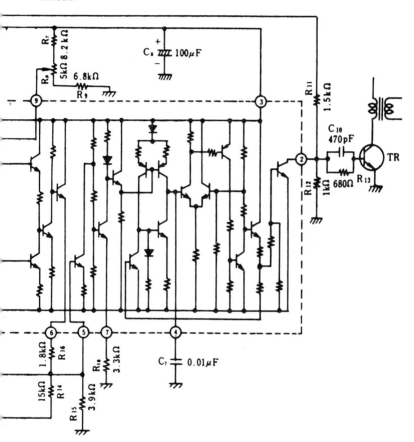

TV horizontal AFC and oscillator with sync separator for negative sync, using an ECG1086 16-pin DIP. The single transistor stage is the horizontal driver and can be an ECG199 bipolar transistor (courtesy GTE Sylvania Incorporated).

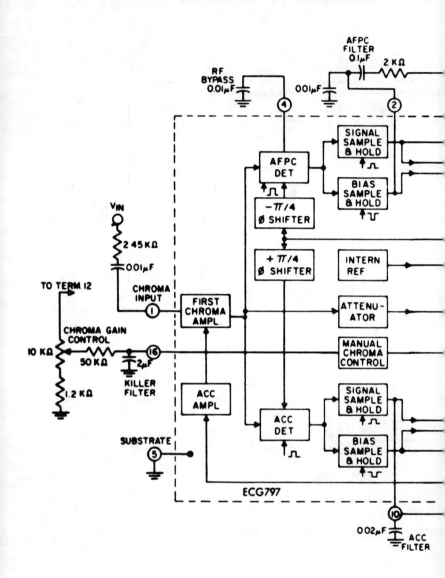

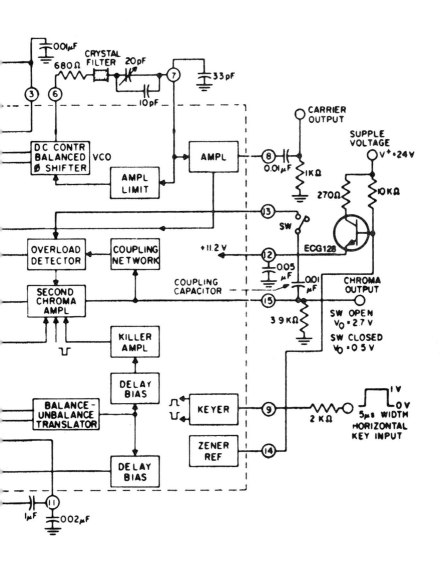

Color TV chroma processor using an ECG797 16-pin QIP with AFPC, ACC, and killer (courtesy GTE Sylvania Incorporated).

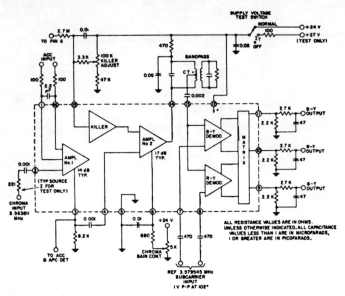

Color TV chroma amplifier/demodulator using an ECG791 16-pin DIP. The circuit features short-circuit protection, ACC, and color killer. Typical sensitivity of chroma input is 10 mV RMS with a 50 mV sensitivity for amplifier 2 at pin 4 (courtesy GTE Sylvania Incorporated).

2. V_{OUT}, $V_{OD(Max)}$, V_{IN}, CARRIER ATTENUATION, CARRIER OUTPUT VOLTAGE, BW_{IF}, BW_{VIDEO}, $V_O(AFT)$, V_1

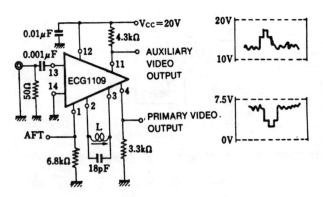

L: 0.4 mm$^\phi$ SILK WOUND COPPER WIRE 7 TURNS
COIL FORM OUTSIDE DIAMETER 5.5 mm$^\phi$

TV low-level video detector. Typical voltage output is 7.5 volts peak to peak. The 14-pin DIP contains a double balanced detector circuit (courtesy GTE Sylvania Incorporated).

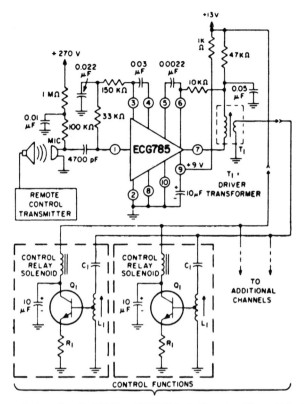

Remote control receiver for TV. Use ECG103As for the relay drivers. Select the coils and transformers for the desired frequency. The ECG785 has a voltage gain of 129 dB at 40 kHz with the three internal amplifiers in cascade (courtesy GTE Sylvania Incorporated).

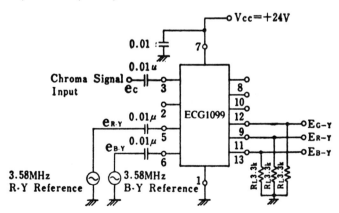

Color TV chroma demodulator with level-shifted output. The ECG1099 is a 14-pin DIP with tab (courtesy GTE Sylvania Incorporated).

TV video signal processor. The ECG1070 is a 16-pin DIP (courtesy GTE Sylvania Incorporated).

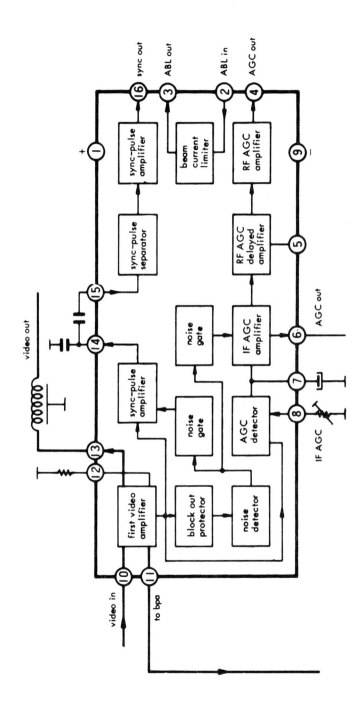

The ECG1070 16-pin DIP (courtesy GTE Sylvania Incorporated).

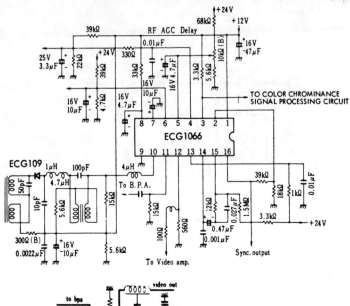

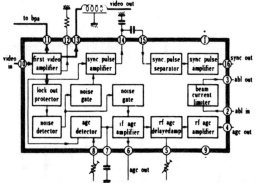

TV video signal processor circuit. The ECG1066 is a 16-pin DIP (courtesy GTE Sylvania Incorporated).

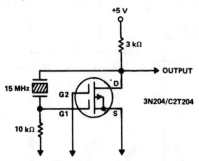

Color TV 3.58 MHz crystal oscillator using a dual-gate MOSFET (courtesy Texas Instruments Incorporated).

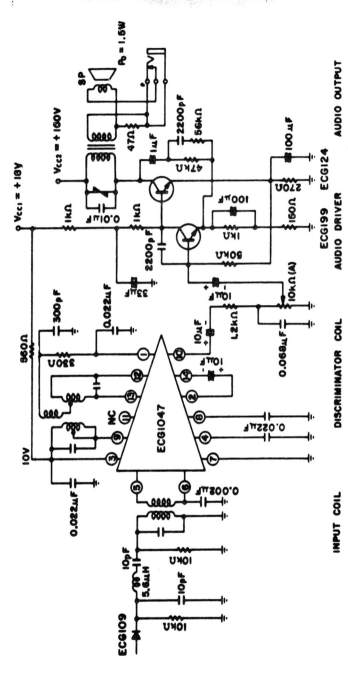

4.5 MHz TV sound channel with audio output stages. The ECG1047 has a three-stage differential amplifier for the IF amplifier section, a ratio detector and first and second audio amplifiers. The IF transformer, ratio detector coil and audio output transformer are standard and can be purchased at Radio Shack. This circuit provides 1.5 watts of audio power into an 8-ohm load (courtesy GTE Sylvania Incorporated).

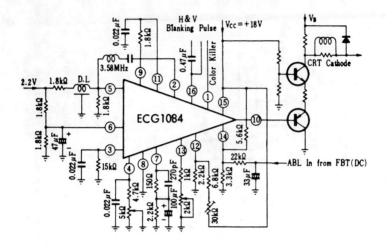

Color TV automatic resolution control (courtesy GTE Sylvania Incorporated).

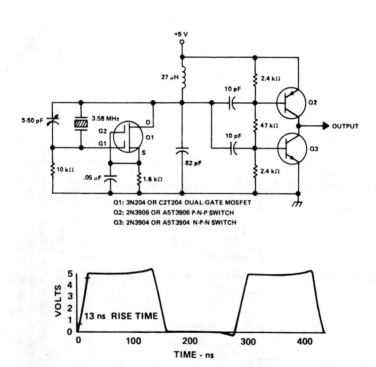

Color TV 3.58 MHz crystal oscillator using a dual-gate MOSFET (courtesy Texas Instruments Incorporated).

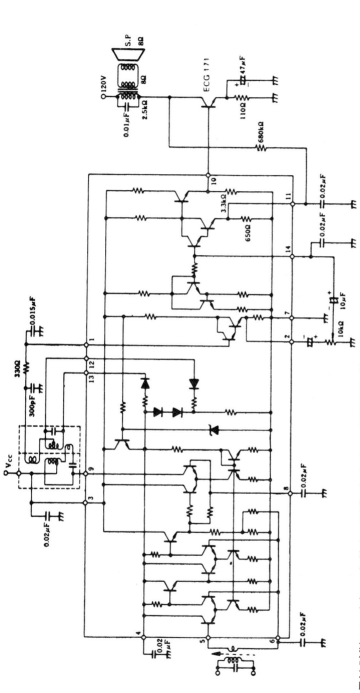

TV 4.5 MHz sound channel with 1-watt AF output using an ECG1051 chip and an ECG171 bipolar transistor. Typical supply voltage for the chip is 10 volts, while the output transistor can be operated from a 120-volt DC supply. Audio output from the chip is 3 volts RMS. The IF transformer, radio detector transformer and audio output transformer are all standard and can be purchased at Radio Shack (courtesy GTE Sylvania Incorporated).

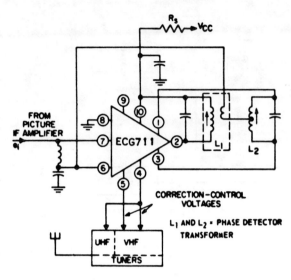

Color TV AFT circuit. The supply is +30 volts, resistor Rs is 1.5K, the bypass capacitor between pin 10 and ground is 0.001 µF, the capacitor across L1 is 91 pF, the capacitor across L2 is 68 pF, the series coil and capacitor at pins 6 and 7 are 3.3 µH and 0.001 µF and the input frequency at 45.75 MHz (courtesy GTE Sylvania Incorporated).

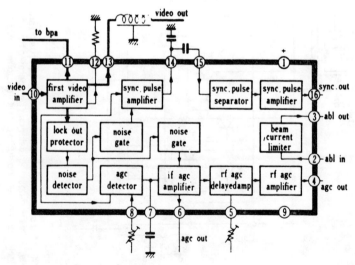

The ECG1064 chip (see page 651).

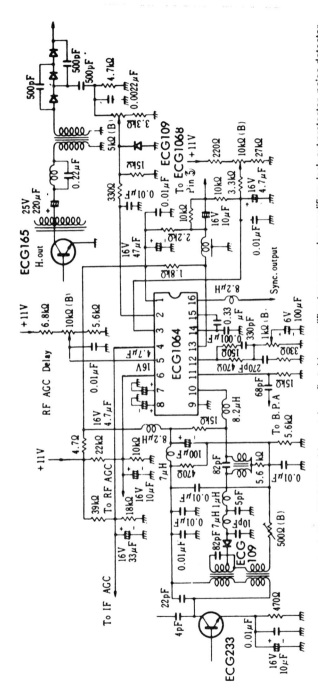

TV video signal processor circuit. The ECG1064 chip contains a first video amplifier, two sync pulse amplifiers, lock-out protector, noise detector, two noise gates, AGC detector, IF AGC amplifier, RF AGC delay clamp, RF AGC amplifier, beam current limiter and sync pulse separator (courtesy GTE Sylvania Incorporated).

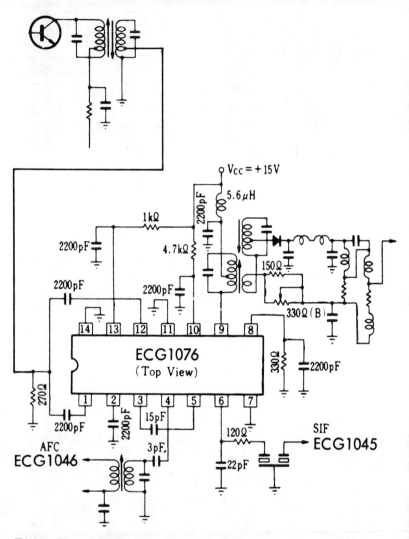

TV video IF output and detector with AFT and sound take-off output. All coils and transformers are standard and can be purchased at a local electronic parts jobber (Merit or Miller components). Sound output at 4.5 MHz is typically 300 mV RMS. AFT output is typically 210 mV RMS at a test frequency of 57 MHz. Typical voltage gain for the video IF is 36 dB (courtesy GTE Sylvania Incorporated).

**ALL RESISTORS IN OHMS
ALL CAPACITORS, UNLESS
OTHERWISE INDICATED
ARE IN μF**

Color TV chroma demodulator with tint control. The color demodulators consist of two sets of balanced detectors that receive their reference subcarrier from the internal demodulator drive amplifiers. The chroma signal input is applied to pin 14. The chroma signal differentially drives the demodulators. The demodulation components are matrixed and DC-shifted in voltage to give R−Y, G−Y and B−Y color difference components with close DC balance and proper amplitude ratios. When the zener reference element is not used the power supply should be maintained at +11.2 volts (courtesy GTE Sylvania Incorporated).

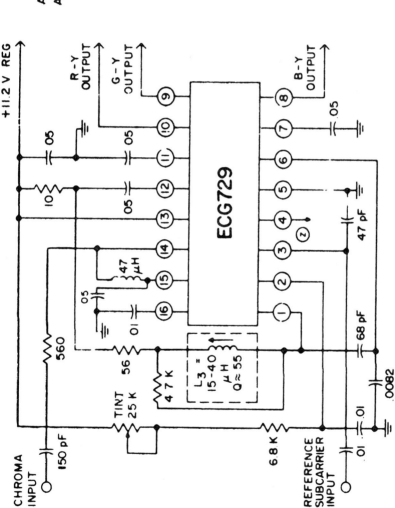

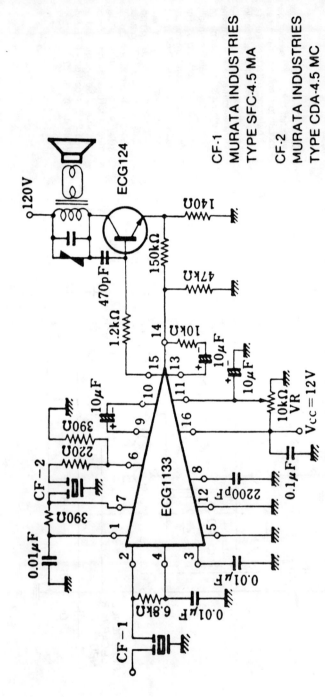

TV sound channel with 1-watt output. The ECG1133 contains three IF differential amplifiers, a phase detector, a DC operated volume control and an AF amplifier. The IF specified is 4.5 MHz (courtesy GTE Sylvania Incorporated).

CF-1
MURATA INDUSTRIES
TYPE SFC-4.5 MA

CF-2
MURATA INDUSTRIES
TYPE CDA-4.5 MC

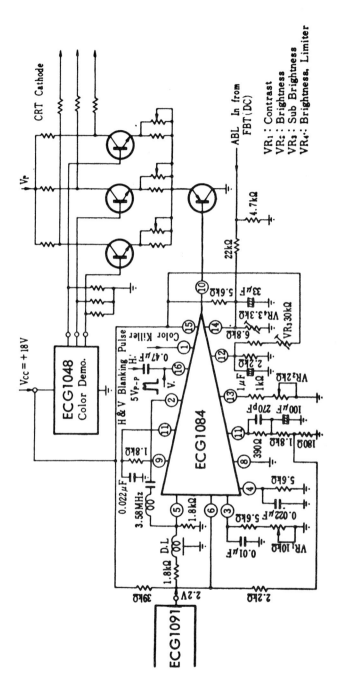

Color TV automatic resolution control. The ECG1091 partially shown is for reference only and is a video processor chip (courtesy GTE Sylvania Incorporated).

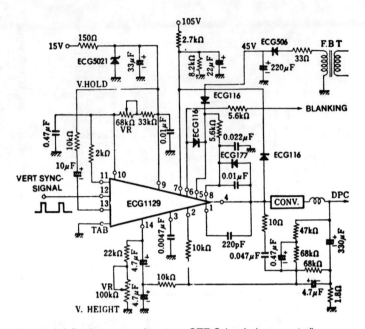

V vertical deflection system (courtesy GTE Sylvania Incorporated).

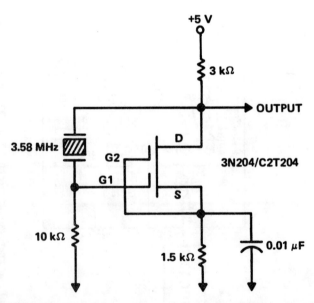

Color TV 3.58 MHz crystal oscillator using a dual-gate MOSFET (courtesy Texas Instruments Incorporated).

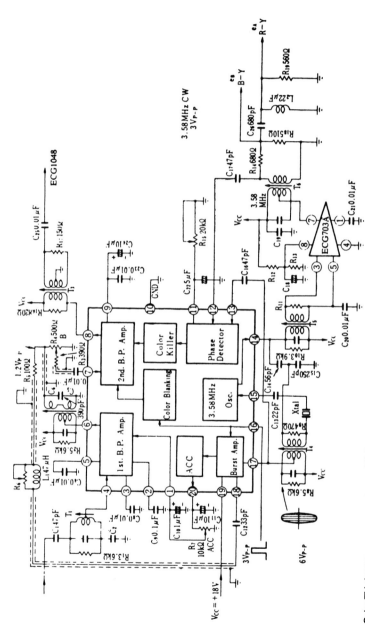

Color TV chroma processor using an ECG1089 20-pin DIP. The ECG1048 noted on the schematic is a 3.58 MHz amplifier (courtesy GTE Sylvania Incorporated). The ECG703A is a chroma demodulator.

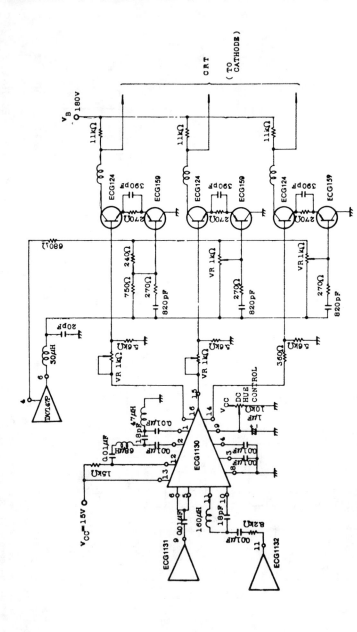

Color TV demodulator with color amplifiers. The ECG1131s shown are chroma signal amplifiers. The ECG1130 is a 16-pin DIP (courtesy GTE Sylvania Incorporated).

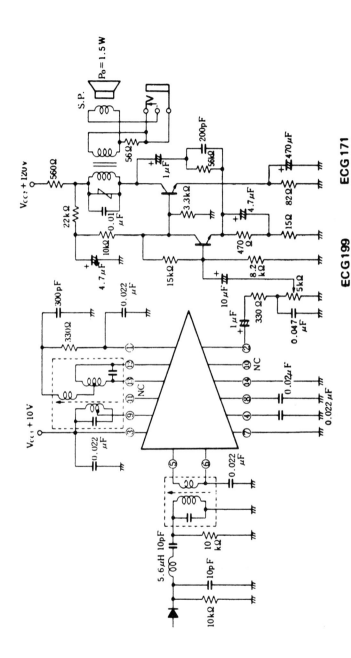

Complete FM/TV 4.5 MHz sound channel using an ECG1045 14-pin DIP. Audio power output is 1.5 watts. The ECG1045 contains a three-state high-gain directional IF amplifier and a ratio detector. IF transformers and the AF output transformer are standard items and can be purchased at Radio Shack. This circuit can be modified for 10.7 MHz operation by selecting the proper IF transformers (courtesy GTE Sylvania Incorporated).

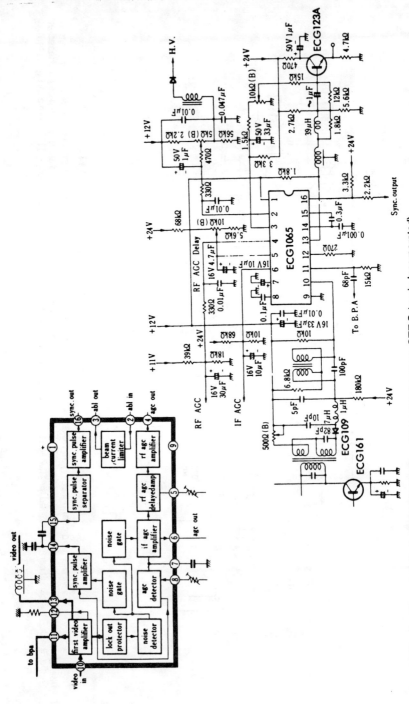

TV video signal processor circuit. The ECG1065 is a 16-pin DIP (courtesy GTE Sylvania Incorporated).

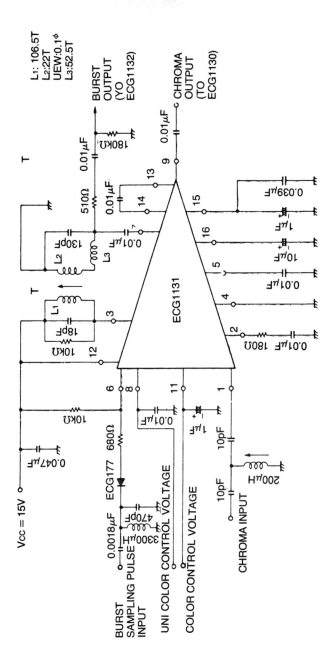

Color TV chroma signal amplifier. The ECG1131 is a 16-pin DIP and consists of a chroma amplifier, an ACC peak detector, a color killer, a DC chroma gain control, a DC uni-color control and a balanced sampling circuit for the burst signal. By connecting the control terminal of the uni-color and contrast terminal of the ECG1131 it is possible to control chroma gain and contrast simultaneously (courtesy GTE Sylvania Incorporated).

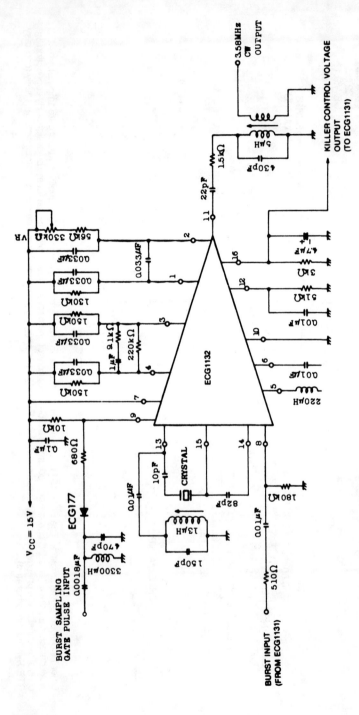

Color TV subcarrier generator. The crystal shown is for 3.58 MHz. The ECG1131 noted on the schematic is a chroma signal amplifier. The ECG1132 is a 16-pin DIP (courtesy GTE Sylvania Incorporated).

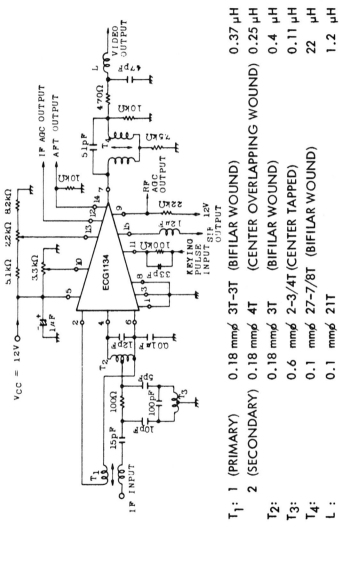

T₁: 1 (PRIMARY)   0.18 mmø 3T-3T (BIFILAR WOUND)   0.37 μH
    2 (SECONDARY) 0.18 mmø 4T (CENTER OVERLAPPING WOUND) 0.25 μH
T₂: 0.18 mmø 3T (BIFILAR WOUND)   0.4 μH
T₃: 0.6 mmø 2-3/4T (CENTER TAPPED)   0.11 μH
T₄: 0.1 mmø 27-7/8T (BIFILAR WOUND)   22 μH
L :   0.1 mmø 21T   1.2 μH

TV video processor. The ECG1134 contains a picture IF amplifier, a video detector, a video amplifier, keyed AGC with noise immunity, delayed AGC for the tuner, an AFT amplifier, a sound IF amplifier, a sound carrier detector, a 4.5 MHz sound carrier amplifier and a reverse AGC voltage for MOS FET tuner. Coil data is shown below schematic (courtesy GTE Sylvania Incorporated).

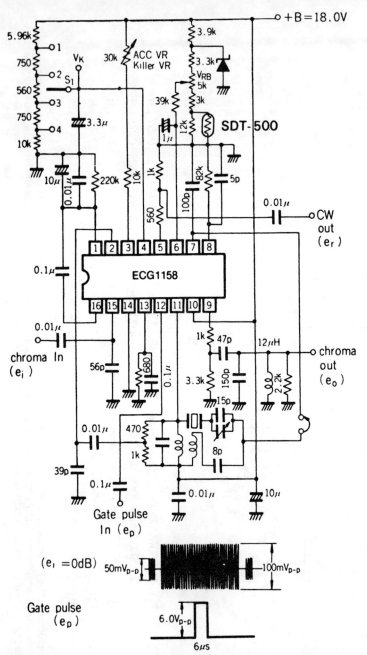

Color TV chroma processor. The ECG1158 includes the functions for chroma amplifier, oscillator, ACC, color killer and color control (courtesy GTE Sylvania Incorporated).

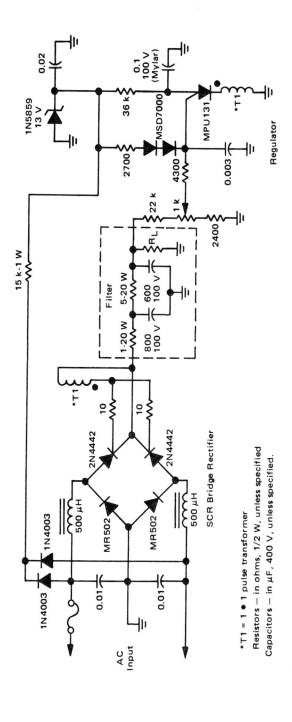

Full-wave SCR bridge power supply for color TV rated at 80 volts, 1.5 amperes. This circuit is designed for 19-inch color receivers (courtesy Motorola Semiconductor Products Inc.).

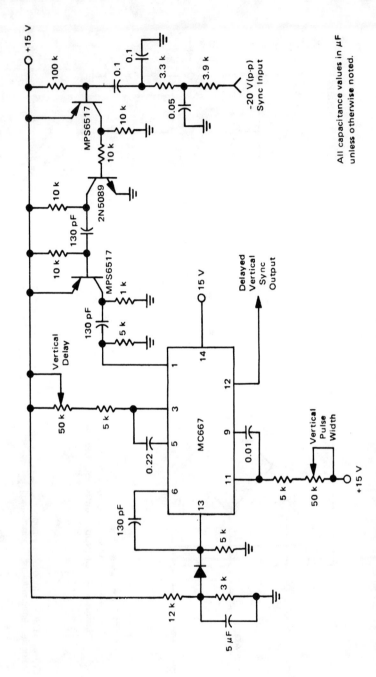

TV vertical sync delay circuit (courtesy Motorola Semiconductor Products Inc.).

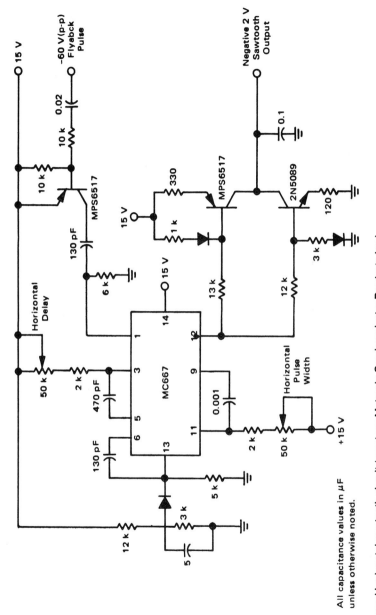

TV delayed horizontal sawtooth circuit (courtesy Motorola Semiconductor Products Inc.).

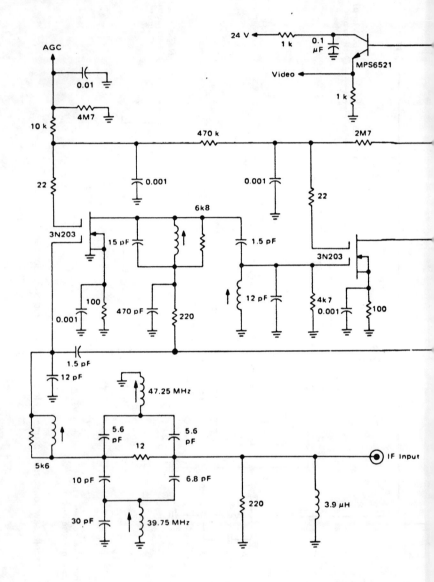

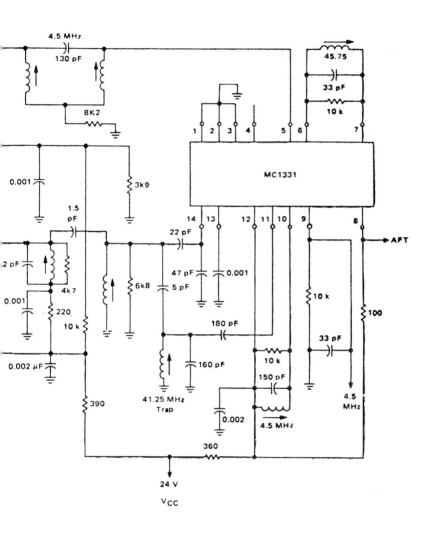

Low-noise TV IF system with two FET amplifiers (courtesy Motorola Semiconductor Products Inc.).

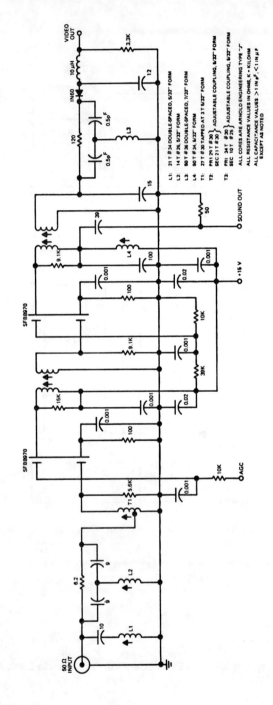

TV IF amplifier using dual-gate MOSFETs. The IF response is as follows: 39.75 MHz greater than 30 dB down, 41.25 MHz greater than 40 dB down, 47.25 MHz greater than 30 dB down, 42.17 nearly 8 dB down, 45.75 MHz equal to 6 dB down. Sensitivity is 700 μV for 1V of video carrier at 100% modulation. AGC range is 43 dB. AGC voltage is +4.0V to −6.0V. Sound output is 570 μV of 4.5 MHz signal (courtesy Texas Instruments Incorporated).

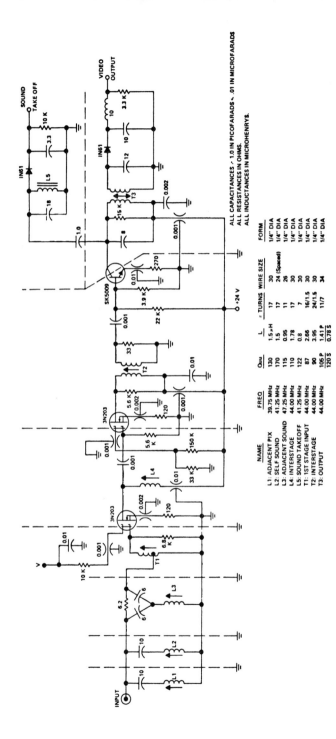

TV IF amplifier with video detector and sound takeoff using dual-gate MOSFETs (courtesy Texas Instruments Incorporated).

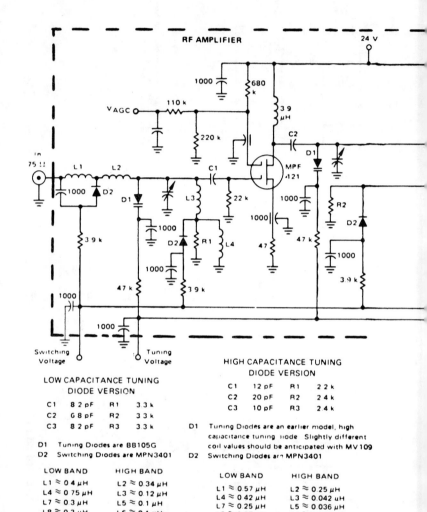

LOW CAPACITANCE TUNING DIODE VERSION

C1	8.2 pF	R1	3.3 k
C2	6.8 pF	R2	3.3 k
C3	8.2 pF	R3	3.3 k

D1 Tuning Diodes are BB105G
D2 Switching Diodes are MPN3401

LOW BAND	HIGH BAND
L1 ≈ 0.4 µH	L2 ≈ 0.34 µH
L4 ≈ 0.75 µH	L3 ≈ 0.12 µH
L7 ≈ 0.3 µH	L5 ≈ 0.1 µH
L8 ≈ 0.3 µH	L6 ≈ 0.1 µH
L9 ≈ 0.01 µH	L10 ≈ 0.05 µH
L11 ≈ 0.09 µH	

HIGH CAPACITANCE TUNING DIODE VERSION

C1	12 pF	R1	2.2 k
C2	20 pF	R2	2.4 k
C3	10 pF	R3	2.4 k

D1 Tuning Diodes are an earlier model, high capacitance tuning diode. Slightly different coil values should be anticipated with MV109
D2 Switching Diodes are MPN3401

LOW BAND	HIGH BAND
L1 ≈ 0.57 µH	L2 ≈ 0.25 µH
L4 ≈ 0.42 µH	L3 ≈ 0.042 µH
L7 ≈ 0.25 µH	L5 ≈ 0.036 µH
L8 ≈ 0.25 µH	L6 ≈ 0.036 µH
L9 ≈ 0.01 µH	L10 ≈ 0.02 µH
L11 ≈ 0.05 µH	

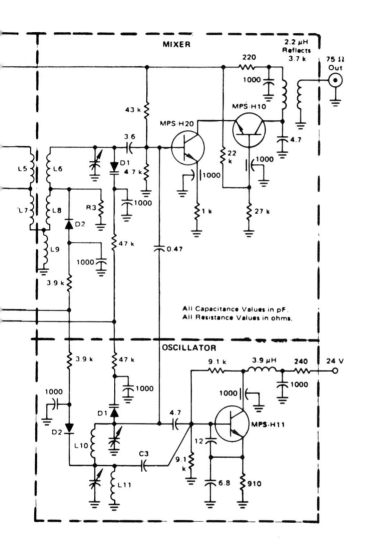

TV VHF varactor tuner with a dual-gate MOSFET RF amplifier (courtesy Motorola Semiconductor Products Inc.).

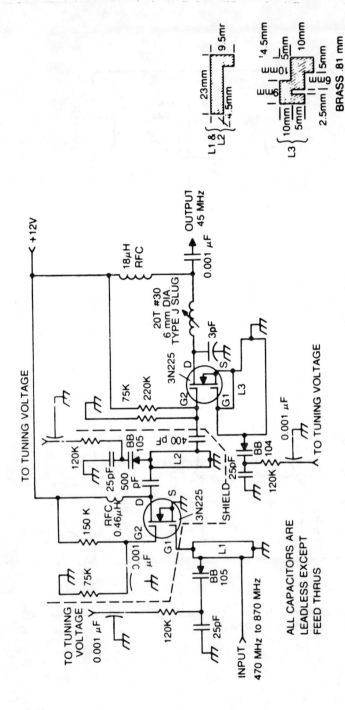

Varactor UHF TV tuner using a 3N225 dual-gate MOSFET (courtesy Texas Instruments Incorporated).

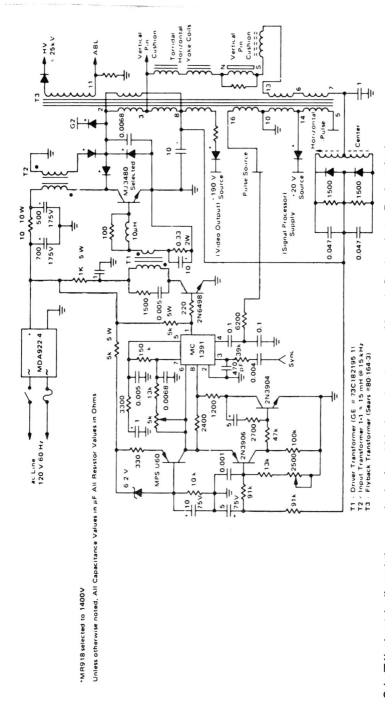

Color TV horizontal self-regulating scan circuitry for a 19-inch receiver (courtesy Motorola Semiconductor Products Inc.).

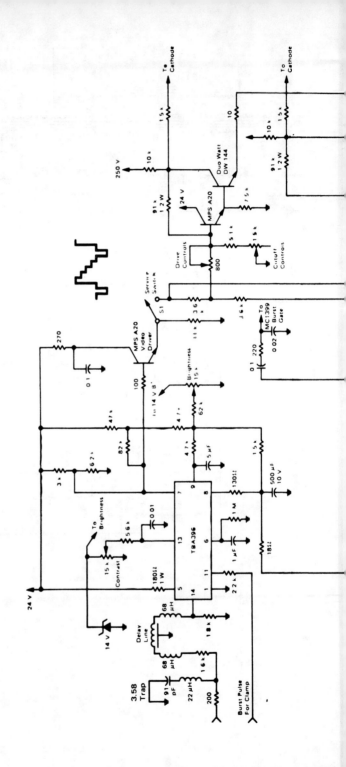

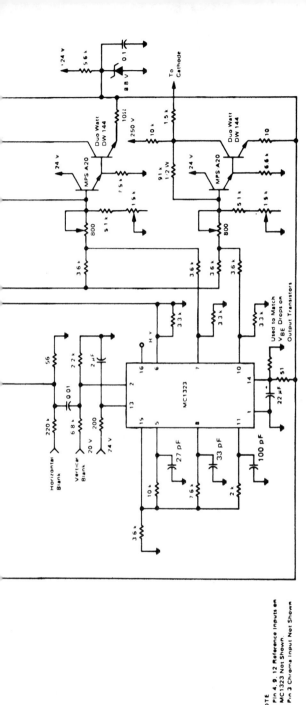

Color TV complete video amplifier system featuring DC contrast and brightness, DC restorer for black-level clamping, ganged contrast and color controls and ABL (courtesy Motorola Semiconductor Products Inc.).

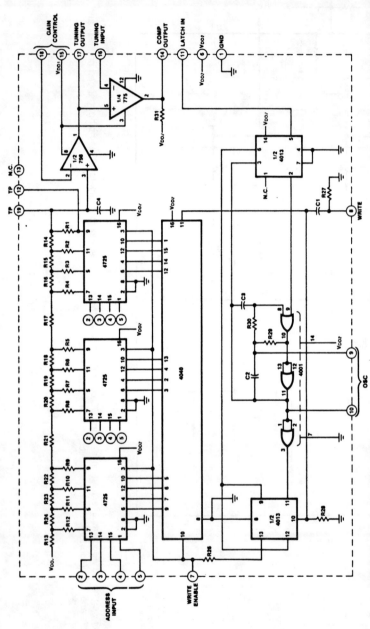

Electronic tuner control/memory for radio and TV. The SH1549 gives capability of storing and recalling up to 16 stations in addition to conventional varactor tuning. The SH1549 converts the analog tuning voltage into a 12-bit digital word and stores it in memory for future use (courtesy Fairchild Semiconductor).

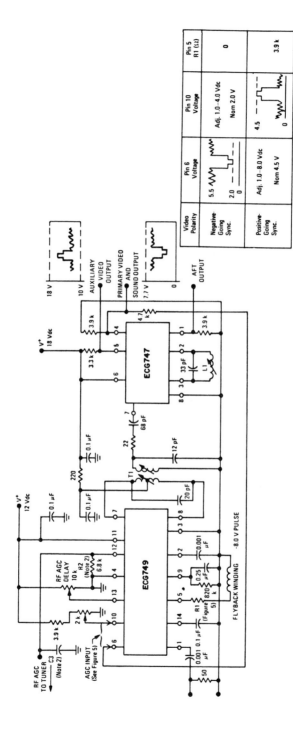

Complete TV video IF amplifier and detector system. The ECG749 contains two IF amplifiers, an AGC keyer, and an AGC amplifier. The circuit will work for both color and B&W. See table for AGC information (courtesy GTE Sylvania Incorporated).

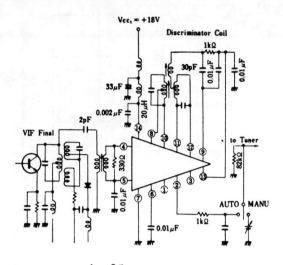
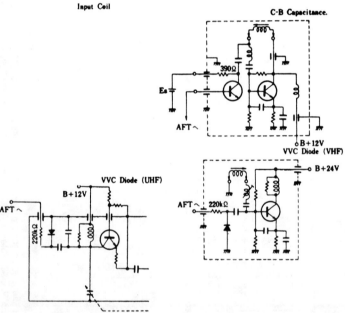

TV AFT/AFC circuit for three types of systems using an ECG1046 14-pin DIP. The ECG1046 contains two VIF differential amplifiers, a phase detector, and a DC amplifier, whose output is supplied to a varactor diode in a tuner. It has the additional feature of having an internal regulated voltage supply. Partial tuner circuits shown are only for example applications. Coils and transformers should be selected to match the IF and can be obtained from discarded TV receivers or purchased at a local electronics parts jobber (Merit or Miller components). Maximum supply voltage for the ECG1046 is 20 volts (courtesy GTE Sylvania Incorporated).

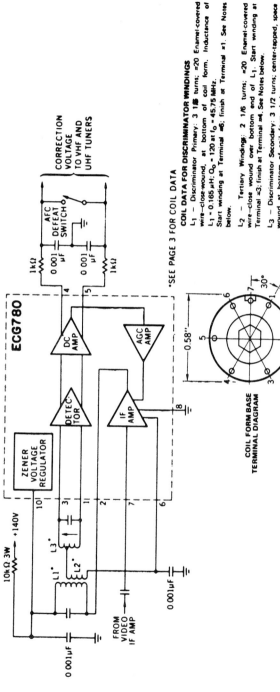

AFT circuit for TV with detailed coil winding data (courtesy GTE Sylvania Incorporated).

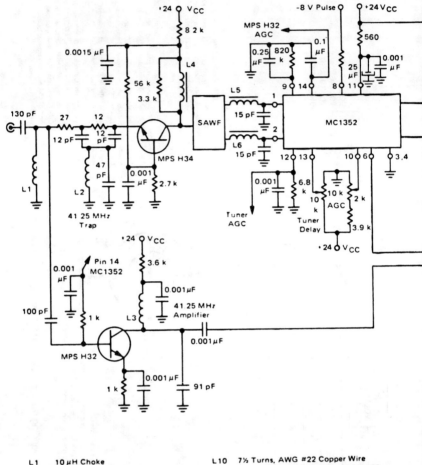

- L1 10 µH Choke
- L2 9½ Turns, AWG #18 Copper Wire
  PAUL SMITH CO Coil Form SK138
  3/8" 10 32 Core   Carbonyl J Material
- L3 4½ Turns, AWG #22 Copper Wire
  PAUL SMITH CO Coil Form SK478
  3/8" 10 32 Core   Carbonyl E Material
- L4 21 Turns, AWG #36 Copper Wire over
  3.3 kΩ Resistor, 1/2 W
- L5, L6 0.47 µH Choke
- L7, L8 8 Turns, Center Tapped, Air Core
  AWG #36 Single Cell Copper Wire
- L9 40 Turns, AWG #36 Copper Wire
  PAUL SMITH CO. Coil Form EF186
  3/8" 10-32 Core - Carbonyl E Material
- L10 7½ Turns, AWG #22 Copper Wire
  PAUL SMITH CO. Coil Form SK-478
  3/8" 10 32 Core - Carbonyl E Material
- L11 31/6 Turns, AWG #20 Copper Wire
  COILCRAFT T-Form (Start at Pin 3(MC1364)
  3/8" 10-32 Core - Carbonyl T Material
- L12 21/3 Turns, AWG #20 Copper Wire
  Over Bottom End of L11 (Finish at Pin 9 (MC
  3/8" 10-32 Core - Carbonyl T Material
- L13 31/6 Turns, AWG #20 Copper Wire
  Center Tapped.
  3/8" 10 32 Core - Carbonyl T Material

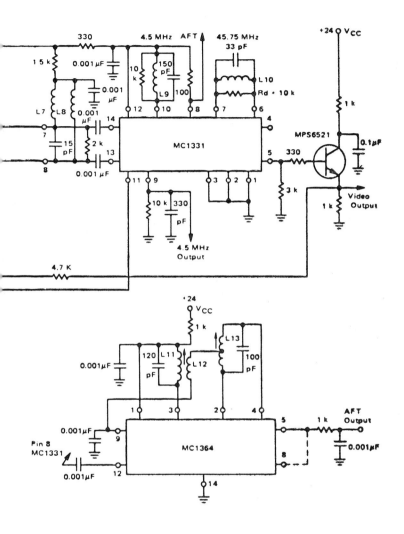

Complete TV video IF amplifier with sound and AFT outputs. The bandpass element used here (SAWF) has a differential output that directly drives the MC1352 (courtesy Motorola Semiconductor Products Inc.).

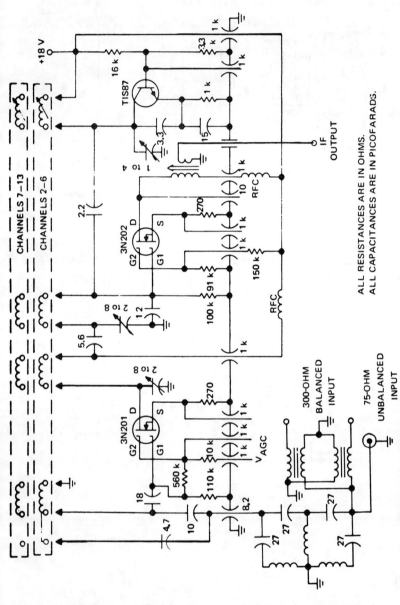

TV VHF tuner using dual-gate MOSFETs (courtesy Texas Instruments Incorporated).

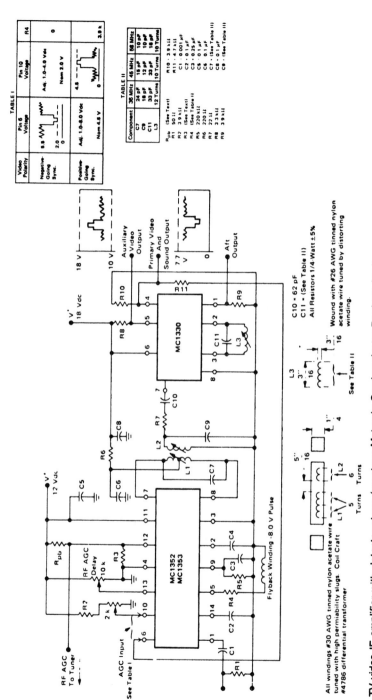

TV video IF amplifier with detector stage (courtesy Motorola Semiconductor Products Inc.).

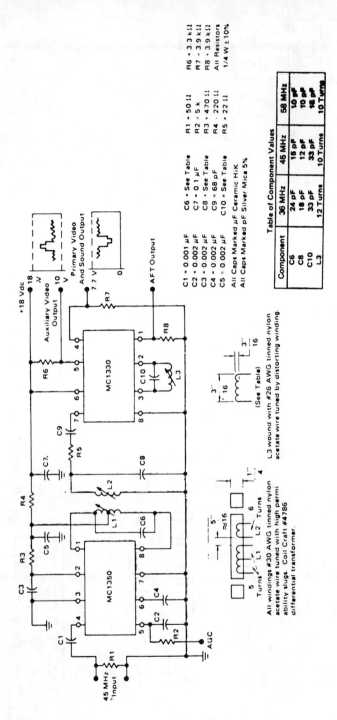

TV video IF amplifier with low-level detector (courtesy Motorola Semiconductor Products Inc.).

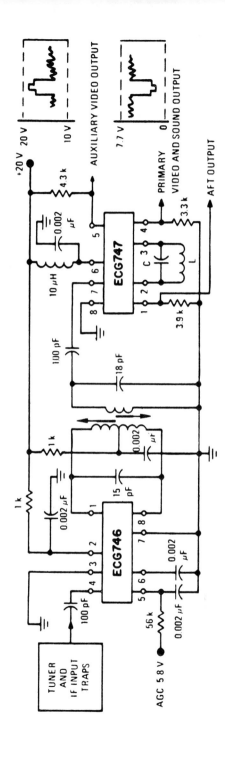

TV color IF amplifier and detector. This configuration will provide approximately 84 dB voltage gain. The nominal 3-volt peak-to-peak output can be varied between 0 and 7 volts with excellent linearity and freedom from spurious output products. Alignment is most easily accomplished with an AM generator set at a carrier frequency of 45.75 MHz, modulated with a video frequency sweep. The detector tank is first adjusted for maximum detected DC with a CW input. Next the video sweep modulation is applied in order to align the input and interstage circuits (courtesy GTE Sylvania Incorporated).

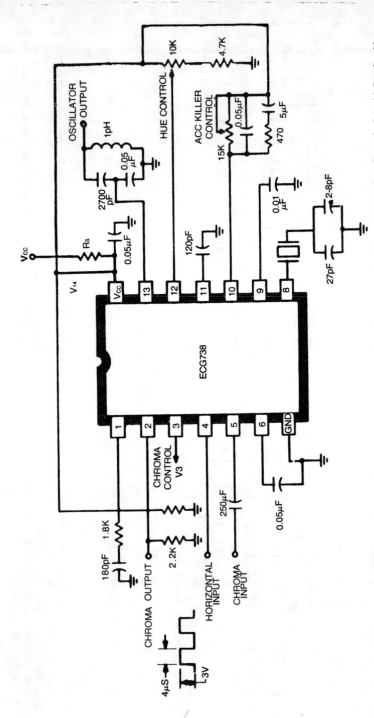

Color TV chroma processor with ACC, color killer, and an injection lock reference system. Typical power supply voltage is 9 to 11.5 volts. Hue control range is 100 degrees. The oscillator output signal is 180 mV RMS (courtesy GTE Sylvania Incorporated).

Index

Index

A

Absorbance circuit, light density	501
AC-controlled triac switch	424
AC flasher, sequential	273
AC power control	432
AC switch, high-low	413
Accumulator, 5-digit	103
Acoustic thermometer	503
Active filter	80
A/D converter	168
A/D converter, 3-digit I^{2}L	190, 191
4½-digit (2000-volt)	201
8-bit	203
10-bit	167
10-bit CMOS	199, 200, 204
SA	166
8-bit SA	202
12-bit SA	173, 179, 181
high-speed 10-bit	159
nearly 18-bit	193
A/D converter to 8-bit bus interface	255
A/D converter with sample-hold	178, 187
Adder, 4-bit look-ahead carry	29
8-bit parallel	28
16-bit look-ahead carry	27
32-bit look-ahead carry	27
A/D divider	205, 389, 395
Adjustable regulator, 5-volt	219
6-volt	219
12-volt	219
15-volt	219
24-volt	219
AF amplifier, 0.25-watt	281
0.5-watt	280, 285
0.7-watt	316
1-watt	283, 287, 290, 291, 313, 314
1.3-watt	314
1.5-watt/2-watt	295
2-watt	279, 287, 294, 299, 307, 308, 311
2.1-watt	300, 315
2.2-watt	289
2.5-watt	295
3.3-watt	300, 307
3.5-watt	310
4-watt	277, 285, 288, 319, 323
4.2-watt	317
4.4-watt	289
4.5-watt	296, 297, 314, 323
5-watt	303, 304
5.2-watt	301
5.5-watt	291
5.8-watt	320
6.5-watt	302, 306
7-watt AF	297
7-watt	296
8-watt	288
13-watt	316
15-watt	286, 309
20-watt	302
50-watt	305
75-watt	321
600 mW	315, 320
3-watt/4-watt	329
3-watt/5-watt	331
15-watt/20-watt	318
15-/20-/25-watt	337
20-/40-/50-/60-watt	322
35-/50-/60-/75-/100-watt	341
7-/10-/15-/20-/25-/	
35-watt	333, 335
15-/20-/25-/35-/50-/	
60-watt	325, 327
4-watt class A	282
2-watt cassette R/P	304
1-watt class B	282
2-watts-per-channel stereo	284, 317
cassette R/P	298, 312
equalizer	306
tape recorder	292
AFC and oscillator, TV horizontal	635, 639
AFC system, Smith	127
AF oscillator, variable	355
AF power amplifier, 13-watt	279
AF preamplifier	284
AF preamplifier, general-purpose	311
high-gain	309
stereo	276, 277, 278
AFT/AFC, TV	615, 680
AFT, color TV	650
AFT system, TV	619
AFT, TV	681
AF wattmeter	89
AGC	610
AGC circuit, precision	601
Alarm, automotive burglar	66
frequency-sensitive	74
light-operated SCR	72
light-relay-operated SCR	71
photo-activated	74
thermistor-operated temperature	504
water seepage	505
ALC	609
Alternating flasher	271
AM broadcast receiver	154
AM broadcast turner	127, 128, 131, 132, 133, 139, 146
AM/FM broadcast receiver	156
AM/FM IF amplifier	131, 133, 144, 147, 148
AM/FM IF amplifier, AM tuner with	146
AM/FM stereo tuner	145
AM/FM tuner	140
AM transmitter, 2.5 watt	522
AM transmitter, 7-watt 118 to 136 MHz	549
AM transmitter, 13-watt 118 to 136 MHz	549
AM transmitter, 118 MHz to 136 MHz	522
Amplifier, AM/FM IF	131, 133, 144, 147, 148
balance differential	591
battery-powered load-cell	609
cassette R/P AF	298, 312
clipping	117
color TV chroma	636
color TV chroma signal	661
complete color TV video	677
complete TV video IF	679, 683
compressor	121
differential	107, 108, 609
expander	121
FM IF	126, 128, 135, 136, 138, 141, 142, 143, 149, 154, 155
general-purpose	115, 275, 595
high-gain FM IF	130
high-gain inverting	106
high-impedance inverting	106
high-slew-rate power	105
IF	129
instrument	115
instrumentation	114, 608
isolation	55
inverting	109, 122
linear TV IF	636
line-operated servo motor	31
logarithmic	16, 17
low-drift low-noise	112
low-noise equalizing	286
Amplifier, NAB equalizer	301
Amplifier, noninverting	111, 116
precision	605
programmable gain noninverting	596
pulse	594
pluse power	602
push-pull Darlington	587
push-pull monolithic Darlington	597
record/playback	283
servo motor power	30
stable instrument	604
summing	120
summing/scaling	594
TV chroma IF	687
TV IF	670, 671
TV video IF	621, 630, 633, 652, 685, 686
video	60, 61
voltage-controlled	598
weighted averaging	108
wide-band differential	517
wide-band video	62
Amplifier/demodulator, color TV chroma	642
Amplitude modulator	524, 542

691

linear	559	
0.25-watt AF	281	
0.5-watt AF	280, 285	
0.7-watt AF	316	
1-watt AF	281, 283, 287, 291, 313, 314	
1.3-watt AF	314	
2-watt AF	279, 287, 294, 299, 307, 308, 311	
2.1-watt AF	300, 315	
2.2-watt AF	289	
2.5-watt AF	295	
3.3-watt AF	300	
3.5-watt AF	310	
4-watt AF	277, 285, 319, 323	
4.2-watt AF	317	
4.4-watt	289	
4.5-watt AF	296, 297, 314, 323	
5-watt AF	303, 304	
5.2-watt AF	301	
5.5-watt AF	291	
5.8-watt AF	320	
6.5-watt AF	302, 306	
7-watt AF	296, 297	
8-watt AF	288	
13-watt AF	316	
15-watt AF	286, 309	
20-watt AF	302	
50-watt AF	305	
75-watt AF	321	
1.6 to 30 MHz linear	581	
20W 1.6 to 30 MHz SSB linear	584	
20-watt 1.6 to 30 MHz SSB linear	561	
80-watt linear	531	
100-/140-/160-watt linear	581	
160-watt PEP linear	535	
10 MHz RF	516	
30 MHz RF	516, 523, 529	
36 MHz RF	585	
45/58 MHz RF	519, 575, 576	
100 MHz RF	529	
105 MHz RF	563	
105 MHz gate-1-controlled RF	559	
105 MHz gate-2-controlled RF	554, 569	
105 MHz low-noise RF	550	
118 to 136 MHz broadband RF	553	
144 MHz to 175 MHz FM RF	533	
200 MHz RF	523	
450/432 MHz RF	578	
900 MHz RF	555	
45 MHz RF post	544	
200 MHz RF post	545	
30 MHz video	63	
144 to 175 MHz 80-watt	539	
2-stage 60 MHz IF	521	
4.5 MHz color TV sound IF	613	
10.7 MHz IF	550, 552	
12.5 MHz IF	577	
30 MHz IF	518	
30/60/100 MHz RF-IF	517	
1-watt	290	
3.3-watt	307	
4-watt	288	
1-watt cassette R/P AF	304	
1-watt class B AF	282	
2-watts-per-channel stereo AF	284, 317	
600 mW AF	320	
1.5-watt/2-watt AF	295	
3-watt/5-watt AF	329, 331	
4-watt class A AF	282	
7-/10-/15-/20-/25-/35-watt AF	333, 335	
15-watt/20-watt AF	318	
15-/20-/25-watt AF	337	
15-/20-/25-/35-/50-/60-watt AF	325, 327	
20-/40-/50-/60-watt AF	322	

35-/50-/60-/75-/100-watt AF	341	
600 mW AF	315	
1-watt phonograph	280	
1-watt inverting power	602	
1-watt noninverting power	587, 597	
1-watt power	275	
13-watt power	279	
2-stage 80-watt RF power	533	
3W differential output power	593	
80-watt RF power	539	
10-watt 450 to 470 MHz UHF	542	
13-watt UHF	537	
25-watt UHF	526	
25-watt 450 to 470 MHz UHF	543	
220 to 225 MHz UHF	537	
450 MHz to 512 MHz UHF	526	
Analog data selector, 4-channel	590	
Analog isolator, precision high CMV	607	
Analog multiplexer, 2-channel	510	
Analog multiplexer, 8-channel	510	
Analog multiplier	377	
Analog selector, high CMV	604	
Analog switch, series-shunt	14	
Analog tachometer, precision	91	
Arc tangent function	373	
Audio dancing lights	67	
Audio wattmeter	89	
Autodyne tuner, 27 MHz	551, 577	
Automatic flasher light	270	
Automatic gain control	610	
Automatic level control	609	
Automatic liquid-level control	500	
Automatic night light	68	
Automatic resolution control, color TV	648, 655	
Automotive battery saver	68	
Automotive burglar alarm	66	
Automotive flasher, 12-volt 2-wire	271	
Automotive regulated battery charger	37	
Autoranging multimeter	102	
Average voltage control, half-wave	434	
Averaging amplifier, weighted	108	

B

Babysitter intercom, 200 mW	319	
Balance differential amplifier	591	
Balanced mixer, double	527	
Balanced modulator	520, 529	
Bandpass filter	80	
Bandpass filter insertion-loss tester	79	
Battery charger		
automotive regulated	37	
60-hertz nicad	38, 40	
12-volt	36, 37	
25 kHz nicad	36, 39	
Battery-operated flasher	267	
Battery operated fluorescent light	69	
Battery-powered load-cell amplifier	609	
Battery saver, automotive	68	
BCD-to-7 segment decoder display	481	
Binary D/V converter		
bipolar	162	
unipolar	163	
Bistable switch	14	
Blinker, 1.5-volt fast	272	
Booster, high-voltage	70	
Bridge amplifier	9	
high-impedance	9	
Bridge linearization function circuit	9	
Bridge transducer amplifier, prog gain	10	
Burglar alarm, automotive	66	
Buzz box continuity tester	88	
Bypass, current	210	

C

Calibrator, oscilloscope	88	
Capacitance meter	97	
Capacitor discharge SCR ignition	589	
Carry adder, 4-bit look-ahead	29	
Carry adder, 16-bit look-ahead	27	
Carry adder, 32-bit look-ahead	27	
Car turn signal reminder	274	
Cassette record/playback AF amplifier	298	
Cassette R/P AF amplifier	312	
Cassette R/P AF amplifier, 1-watt	304	
Celsius temp/freq converter	167	
Chopper, JFET	263	
Chopper, MOSFET	264	
Chopper, MOSFET analog	262	
Chopper, series	262	
Chopper, series-shunt	262, 263	
Chroma amplifier, color TV	636	
Chroma amplifier/demodulator, color TV	642	
Chroma demodulator, color TV	643, 653, 658, 622, 623	
Chroma IF amplifier, TV	687	
Chroma processing, color TV	625	
Chroma processor, color TV	641, 657, 664, 688	
Chroma signal amplifier, color TV	661	
Chroma signal processor, color TV	627	
Clamping circuit, diode-bridge	11	
Clipping amplifier	117	
Clock, PLL	473, 481	
Clock timebase with 1 MHz reference	460	
Clock, 24-hour	470	
Closed-loop frequency compensation	122	
CMOS A/D converter, 10-bit	199, 200, 204	
CMOS D/A converter	173	
CMOS D/A converter, 8-bit multiplying	197, 205	
CMOS D/A converter, low-power 10-bit	185	
CMOS gas discharge display, 4-digit	246	
CMOS gas discharge display, 12-digit	247	
CMOS keyboard data entry system	442	
CMOS static read/write memory	449	
CMOS to 4-digit display interface	248	
Code set, Morse	66	
Cold junction compensation	596	
Coil tester	88	
Color TV AFT	650	
Color TV automatic resolution control	648, 655	
Color TV chroma amplifier	636	
Color TV chroma amplifier, demodulator	642	
Color TV chroma demodulator	643, 653, 658, 622, 623	
Color TV chroma processor	641, 657, 664	
Color TV chroma signal amplifier	661	
Color TV chroma processing	625	
Color TV chroma processor	688	
Color TV chroma signal processor	627	
Color TV horizontal self-regulating scan	675	
Color TV sound IF, 4.5 MHz	613	
Color TV subcarrier generator	637, 662	
Color TV 3.58 MHz oscillator	646, 648, 656	
Color TV video amplifier, complete	677	

Color TV video IF system		620
Commutator network		607
Comparator, current		223
Comparator, digitally programmed set point		606
Compensation, closed-loop frequency		122
Compensation, cold junction		596
Compensation, feedforward frequency		109
Compensation, frequency		106
Compensation, single pole frequency		121
Compensation, two pole frequency		118
Components sorter		95
Composite op amp, differential-input		117
Composite op amp, inverting-only		119
Compressor amplifier		121
Continuity tester, buzz box		88
Continuous indicator		270
Control circuit, motor		403
Controller, tape recorder position		456
Converter, A/D		168
Converter, A/D with sample-hold		178, 187
Converter, buffered output D/A		201
Converter, cyclic		201
Converter, D/A		169
Converter, bipolar D/V		162
Converter, Celsius temp/freq		167
Converter, CMOS D/A		173
Converter, current-to-voltage		159
Converter, D/A		176, 206
Converter, D/F		158
Converter, differential-input V/F		165
Converter, 8-bit A/D		203
Converter, 8-bit CMOS multiplying D/A		205
Converter, 8-bit D/A		170, 172, 177, 183, 191
Converter, 8-bit multiplying D/A		203
Converter, 8-bit multiplying CMOS D/A		197, 200
Converter, 8-bit SA A/D		202
Converter, 4½-digit (2000-volt) A/D		201
Converter, 14-bit sign-magnitude		192
Converter, 4 to 20 mA		158, 167
Converter, F/V		164, 186, 187, 188, 189, 190, 192, 193, 197
Converter, gray-to-binary		200
Converter, high-crest-factor RMS-to-DC		165
Converter, high-speed 10-bit A/D		159
Converter, I/F		185, 186
Converter, Kelvin temp/freq		166
Converter, low-power 10-bit CMOS D/A		185
Converter, multiplying D/A		190
Converter, nearly 18-bit A/D		193
Converter, 900 MHz to 45 MHz		547
Converter, 1-bit SA A/D		173
Converter, 105 MHz to 10.7 MHz		552
Converter, 1-to-5 volt		167
Converter, picoampere-to-voltage		160
Converter, power D/A		174
Converter, RMS-to-DC		160
Converter, SA A/D		166
Converter, serial gray-to-binary		203
Converter, sign-magnitude		160
Converter, temp/freq		169
Converter, 10-bit A/D		167
Converter, 10-bit CMOS A/D		199, 200, 204
Converter, 10-bit D/A		168, 172, 175, 177, 179, 180, 184
Converter, 10-bit D/A converter		204
Converter, 10 kHz F/V		161
Converter, 10-bit D/A		182, 194
Converter, 10-bit multiplying D/A		189, 188
Converter, 10-bit sign multiplying D/A		177, 184
Converter, 10-7 MHz to 45 MHz		528
Converter, 3-digit I²L A/D		190, 191
Converter, 12-bit D/A		171, 172, 183
Converter, 12-bit D/A multiplying		196
Converter, 12-bit multiplying D/A		195, 198
Converter, 12-bit SA A/D		179, 181
Converter, 27 MHz to 45 MHz		546
Converter, unipolar binary D/V		163
Converter, V/F		161, 164, 170, 178, 181, 182, 185
Converter, V/F with 10 kHz FS		193
Cosine function		380
Constant-current motor drive		425
Constant-current source, 100 mA		211
Count-by-5 counter		465
Countdown timer		495
Counter, count-by-5		465
Counter, 8-digit up/down		492
Counter, electronically programmed		498
Counter, 5 MHz frequency		497
Counter, 40 MHz frequency		482
Counter, 9-digit multifunction		489
Counter, 4-decade synchronous		463
Counter, 4-digit unit		458
Counter, 4-stage ring		466
Counter, 9-digit universal		475
Counter, 100 MHz frequency		469, 487, 490
Counter, 100 MHz multifunction		464, 491
Counter, 100 MHz period		487
Counter, ring		476
Counter, 10 MHz universal		480, 483
Counter, 2 MHz period		469
CPU to multiple peripherals interface		251
CPU to cassette interface		252
Cyclic converter		201
Crowbar circuit		419
Crystal oscillator, fundamental		362
Crystal oscillator, 105 MHz		359
Crystal oscillator, overtone		364
Crystal oscillator, 200 MHz		356
Crystal oscillator with AM output		354
Current boost regulator		211, 214
Current boost regulator, –5V 4A		218
Current bypass		210
Current comparator		223
Current limiter, 2-component		217, 220
Current regulator		213
Current source, voltage-controlled		219
Current-to-voltage converter		159
Current transmitter, 4-to-20 mA		601

D

D/A converter		169, 176, 206
D/A converter		
buffered output		201
CMOS		173
8-bit		170, 172, 177, 183, 191
8-bit multiplying		200, 203
8-bit multiplying CMOS		197, 205
low-power 10-bit CMOS		185
multiplying		190
power		174
10-bit		168, 172, 175, 177, 179, 180, 182, 184, 194
10-bit multiplying		184, 188, 189, 204
10-bit sign multiplying		177
12-bit		171, 172, 183
12-bit multiplying		195, 196, 198
Dancing lights, audio		67
Darlingto amplifier		
push-pull		589
push-pull monolithic		597
Data acquisition, 8-channel		440
Data acquisition subsystem, 8-channel		439
Data acquisition system, 3½-digit parallel		446
Data entry system, CMOS keyboard		442
Data selector		
4-channel analog		590
1-out-of-4		591
Data transfer, serial		437
Data transfer system, 3-wire		42
Data transmission system		42
DC meter protection		87
DC motor control, regulated		418
DC power control		428
Decibel measurement		89, 90
Decoder/display, BCD-to-7-segment		481
Decoder		
8-bit BIN to 3-digit decm		441
4-channel SQ logic		134, 137
FM stereo multiplex		130
PLL FM stereo multiplex		132
Derivative-controlled low-pass filter		77
Detector, digitally programmable limit		24
Deflection system, TV vertical		656
Delayed horizontal sawtooth circuit, TV		667
Demodulator		
color TV chroma		622, 623, 658
color TV chroma		643, 653
FM stereo		143, 149, 152, 153
Detector		
FM		150, 151
FM ratio		135
ionization-chamber smoke		45, 48
missing pulse		24
phase-sensitive		25
product		525
SCR gas/smoke		46
signal-level envelope		26
TGS gas/smoke		43, 44
TGS gas/smoke detector		47
Deviation, percent of		378
D/F converter		158
Difference of the squares		378, 387, 395
Differential amplifier		107, 108, 609
Differential amplifier		
balance		591
wide-band		517
Differential-input composite op amp		117
Differential line receiver, 2-wire		41
Differential line system, 2-wire		41
Differential output power amplifier, 3-watt		593
Differentiator		110
Digitally controlled time delay		459
Digitally programmable limit detector		24
Digitally programmable low-pass filter		77
Digitally programmed oscillator		352
Digitally programmed power supply		231
Digitally programmed set point comparator		606
Digital sample-and-hold circuit		479
Digital thermometer, 6-channel scanning		506
Diode-bridge clamping circuit		11
Diode display, real-time 5-digit fluorescent		249
Direct drive LCD, 4-digit		240
Direction control for shunt motor		432
Direction control, series motor		407
Direct-line-operated power supply		231
Discrete level-shifting circuit		599

693

Discriminator, Foster-Seeley 141
Display decoder, 8-bit BIN to
  3-digit decm 441
Display
  8-digit fluorescent triode 248, 249
  5-digit incandescent 250
  real-time 5-digit fluorescent
    diode 249
  12-digit planar gas discharge 242
Divider 372, 379, 383, 384,
  390, 391, 392
Divider
  A/D 205, 389, 395
  low-frequency 388
  two-quadrant 387
Double balanced mixer 527
Doubler
  frequency 53
  low-frequency 54
DPDT switch
  FET 13
  latching 15
DPM
  LCD 98
  3-digit I²L 238, 239, 243, 249
Driver, photo-activated logic 72, 73,
  75
Dual 4-bit storage register 447
Dual-polarity 3½-digit voltmeter 100
Dual-voltage power supply 227
D/V converter
  bipolar binary 162
  unipolar binary 163
DVM
  dual-polarity 3½-digit 100
  LCD 98
  3½-digit 104
  2½-digit 101

E

Eight 10-bit A/D converters 196
EKG input amplifier 57
EKG recorder input circuitry 56
Elapsed time/countdown timer 495
Elapsed time indicator 103
Elapsed time indicator, 5-digit 103
Electronic organ master oscillator 351
Electronically programmed timer 499
Electronic thermometer 509
Electronic thermostat 501
Electronic trombone 64
Electronic tuner control/memory 678
Electro-optical shaft encoder 586
Encoder, electro-optical shaft 586
Envelope detector, signal-level 26
Equalizer AF amplifier 306
Equalizer amplifier, NAB 301
Expander amplifier 121

F

Fast blinker, 1.5-volt 272
Fast-settling op amp 118
Feedforward frequency com-
  pensation 109
Feedforward op amp, unity-gain 120
Fetal heartbeat monitor 55
FIFO memory, high-speed 445
Filter, active 80
Filter, bandpass 80
Filter, digitally programmable
  low-pass 77
Filter, derivative-controlled
  low-pass 77
Filter, notch 78, 81, 82
Filter, voltage-controlled
  low-pass 78
Fire siren 65
First and second CB mixer 579
Flasher, alternating 271
Flasher, battery-operated 267
Flasher, high-voltage 269
Flasher light, automatic 270
Flasher, light target 266
Flasher, low-voltage lamp 265
Flasher, 1.5-volt LED 269

Flasher, 1 kW 272
Flasher, sequential AC 273
Flasher, 6-volt 269
Flasher, trigger switch 265
Flasher, 12-volt 2-wire auto-
  motive 271
Flasher, warning light 266
Floating regulator, 50-volt 223
Floating regulator, 100-volt 226
Flowmeter circuit 96
Fluorescent diode display, real-time
  5-digit 249
Fluorescent light, battery
  operated 69
Fluorescent triode display,
  8-digit 248, 249
FM detector 150, 151
FM IF amplifier 126, 128,
  129, 135, 136, 138,
  141, 142, 143, 149, 155
FM IF amplifier, high-gain 130
FM IF gain block 126, 139
FM limiter 130, 141
FM ratio detector 135
FM sound IF, TV 616, 617
FM sound system, TV 614
FM stereo demodulator 143, 149,
  152, 153
FM stereo multiplex decoder 130
FM stereo multiplex decoder,
  PLL 132
FM stereo processor 124, 125, 130,
  143, 149, 152, 153
FM stereo processor, PLL 132
FM transmitter, 80-watt 175
  MHz 541
FM tuning indicator 126
FM/TV 4.5 MHz sound channel,
  complete 659
FM wireless microphone 574
Follower, unity-gain voltage 113, 114
Follower, voltage 116, 123
Four-stage ring counter 466
Four-quadrant multiplier 382
Foster-Seeley discriminator 141
Free-running multivibrator 50
Frequency and period counter,
  40 MHz 478
Frequency compensation 106
Frequency compensation,
  closed-loop 122
Frequency compensation,
  feedforward 109
Frequency compensation,
  single pole 121
Frequency compensation,
  two pole 118
Frequency counter, 5 MHz 497
Frequency counter,
  40 MHz 478, 482
Frequency counter,
  100 MHz 469, 487, 490
Frequency counter/tach-
  ometer 494, 498
Frequency doubler 53
Frequency doubler, 110 MHz to
  300 MHz 54
Frequency-modulated 52 MHz
  oscillator 360
Frequency-modulated oscillator, 415
  MHz 365
Frequency-sensitive alarm 74
Frequency synthesizer 345, 350
Frequency synthesizer mixer 550
Frequency synthesizer, PLL 369
FSK, self-generating 22
FSK tone generator 22
FSK using the MC1545G 21
FSK with slope and voltage
  detection 21
Full-wave average voltage feed-
  back control 428
Full-wave control, triac 413
Full-wave SCR bridge power
  supply 665

Full-wave trigger circuit 434
Function generator, program-
  mable 346
Function generator, sine 345
Fundamental crystal oscillator 362
F/V converter 164, 186, 187, 188,
  189, 190, 192, 193, 197
F/V converter, 10 kHz 161
FM tuner 157
F/V converter with data transfer
  system 199

G

Gain block, FM IF 126
Gain control automatic 610
Gain control automatic,
  high-performance 603
Gas discharge display,
  4-digit CMOS 246
  12-digit CMOS 247
Gated oscillator 359
General-purpose AF pre-
  amplifier 311
General-purpose
  amplifier 115, 275, 595
Generator, color TV
  subcarrier 637, 662
  power function 344
  precision ramp 348
  programmable function 346
  prog triangle/square-wave 342
  sawtooth 342
  sine function 345
  single pulse 348
  space mark 343
  staircase 346, 347, 349
  voltage-controlled ramp 347
Gray-to-binary converter,
  parallel 200
  serial 203

H

Half-wave average voltage
  control 434
Half-wave synchronous rec-
  tification 610
Half-wave variable motor control 411
Heater control, proportional 430
Heartbeat monitor, fetal 55
Heart rate monitor 59
Heater temperature control,
  triac 422
High CMV analog isolator,
  precision 607
High CMV analog selector 604
High-gain inverting amplifier 106
High-impedance inverting
  amplifier 106
High-intensity lamp dimmer 420
High-low AC switch 413
High-performance gain control 603
High-precision tachometer 91
High-slew-rate power amplifier 105
High-slew-rate op amp, inverting 123
High-speed FIFO memory 445
High-speed integrator 19
High-speed sample-and-hold
  circuit 86
High-speed 10-bit A/D converter 159
High-voltage booster 70
High-voltage flasher 269
Horizontal AFC and oscillator,
  TV 635, 639
Horizontal processor, TV 629
Horizontal self-regulating scan,
  color TV 675
Hysteresis-free power controller 421

I

IF amplifier 129, 143
IF amplifier,
  AM/FM 131, 133, 144, 148
IF amplifier, AM/FM IF 147
IF amplifier, complete TV
  video 679, 683
IF amplifier, FM 126, 128, 135, 136,
  138, 141, 142, 149, 154, 155

IF amplifier, 4.5 MHz color TV sound	613
IF amplifier, linear TV	636
IF amplifier, 10.7 MHz	550, 552
IF amplifier, 30 MHz	518
IF amplifier, 12.5 MHz	577
IF amplifier, two-stage 60 MHz	521
IF amplifier, TV	670, 671
IF amplifier, TV video	617, 621, 630, 633, 652, 685, 686
I/F converter	185, 186
IF gain block, FM	139
IF strip, 30 MHz	518
IF system, color TV video	620
IF system, low-noise TV	669
Ignition, capacitor discharge SCR	587
Incandescent display, 5-digit	250
Indicating one-shot multivibrator	51
Indicator, continuous	270
Indicator, elapsed time	103
Indicator, FM tuning	126
Injection-locked oscillator, 3.58 MHz	612
Input amplifier, EKG	57
Input circuitry, EKG recorder	56
Insertion-loss tester, bandpass filter	79
Instrument amplifier	115
Instrument amplifier, stable	604
Instrumentation amplifier	114, 608
Integrator, high-speed	19
Integrator, long-term precision	19, 20
Integrator, low-drift	18
Integrator, operational	18
Integrator, precision	19
Integrator, simple op amp	20
Intercom, 200 mW babysitter	319
Interface, A/D converter to 8-bit bus	254
Interface, CMOS to 4-digit display	248
Interface, CPU to cassette	252
Interface, 8748 to 8251	261
Interface, 8080A to 8255A interface	257
Interface, 8085 to multiple peripherals	251
Interface, 8085A to 8202	258
Interface, 8085 to RS-232C	258
Interface, 4 to 20 mA	251
Interface, LCD	254
Interface, logic to inductive load	405
Interface, memory mapped device to 8080A	435
Interface, MOS to AC load	420
Interface, M6800 to 115-volt AC load	433
Interface, number cruncher to SC/MP	260
Interface, power down to 80/20	256
Interface, SC/MP to cassette	259
Interface, sample-hold to A/D converter	253
Interface, TTY-to-A/D converter interface	253
Interface, TTY/RS-232C to MC6850	259
Inverting amplifier	109, 122
Inverting amplifier, high-gain	106
Inverting amplifier, high-impedance	106
Inverting-only composite op amp	119
Inverting power amplifier, 1-watt	602
Inverting sample-and-hold circuit	84
Inverting unity-gain op amp	123
Ionization-chamber smoke detector	45, 48
Isolation amplifier	55
Isolator, precision high CMV analog	607
I²L A/D converter, 3-digit	190, 191
I²L DPM, 3-digit	238, 239, 243, 247
Inverter, 5-volt 40-ampere power	232
Inverter, line-operated	236
Inverter, line operated 15 kHz	228
Inverter, 20-watt 3-phase	225

J

JFET chopper	263

K

Kelvin temp/freq converter	166

L

Lamp dimmer, high-intensity	420
Latch	598
Latching DPDT switch	15
LCD, 4-digit direct drive	240
LCD DPM/DVM	98
LCD interface	254
LCD, 3½-digit multiplexed	238
LED, 8-digit multiplexed	243
5-digit	239
16-digit multiplexed	244
LED flasher, 1.5-volt	269
Level control, automatic	609
Level-shifting circuit, discrete	599
Light, automatic night	68
battery operated fluorescent	69
Light density absorbance circuit	501
Light dimmer, 800-watt triac	423, 426
low-cost	422
time-dependent	401
Light-operated relay	73
Light-operated SCR alarm	72
Light-operated 6 kV series switch	76
Light-relay-operated SCR alarm	71
Light-sensitive oscillator, watchdog	357
Light target flasher	266
Light-triggered photoflash slave	65
Lights, audio dancing	67
Limit detector, digitally programmable	24
Limiter, FM	130, 141
2-component current	217, 220
Linear amplifier, 80-watt PEP 1.6 to 30 MHz	531
100-/140-/160-watt	581
160-watt PEP	535
20W 1.6 to 30 MHz SSB	584
Linear amplitude modulator	559
Linear SSB amplifier, 20-watt 1.6 to 30 MHz	561
Linear TV IF amplifier	636
Linear voltage-controlled oscillator	361
Linearization, transducer	593
Line operated 15 kHz inverter	228
Line-operated inverter	236
Line-operated servo motor amplifier	31
Line-voltage compensation	404
Liquid crystal display DPM	98
Liquid flow measurement	96
Liquid-level control, automatic	500
Load-cell amplifier, battery-powered	609
Logarithmic amplifier, op amp	16, 17
Logic driver, photo-activated	72, 73, 75
Logic to inductive load interface	405
Long-term precision integrator	19
Long duration time delay	459
Long-term precision integrator	20
Long time delay power switch	473
Look-ahead carry adder, 4-bit	29
16-bit	27
32-bit	27
Low-cost light dimmer	422
Low-distortion oscillator	353
Low-drift integrator	18
Low-drift low-noise amplifier	112
Low-frequency divider	388
Low-frequency doubler	54
Low-noise amplifier low-drift	112
Low-noise equalizing amplifier	286
Low-noise preamplifier	293
Low-noise TV IF system	669
Low-pass filter, derivative-controlled	77
digitally programmable	77
voltage-controlled	78
Low-power 10-bit CMOS D/A converter	185
Low-voltage-controlled triac switch	429
Low-voltage lamp flasher	265
Low-voltage power supply	233

M

Marine band transmitter, 10-/25-watt	569
Mass spectrometer, programmable	504
Matrix temperature multiplexer	506
Master oscillator, electronic organ	351
MCS-48 based analog processor	436
Measurement, decibel	89, 90
Measurement, liquid flow	96
Measurement, peak-to-peak noise	92
Measurement, ratiometric	90
Measurement, temperature differential	508
Memory core sense amplifier	439
Memory, high-speed FIFO	445
Memory mapped device to 8080A interface	435
Meter, capacitance	97
Meter protection, DC	87
Meter, watt-hour	87
Ministrobe, 3-volt	268
Missing pulse detector	24
Mixer and VCO for CB	585
Mixer, double balanced	527
Mixer, frequency synthesizer	550
Mixer, 900 MHz RF to 45 MHz IF	578
Mixer, 100 MHz	520
Mixer, 105 MHz RF to 10.7 MHz IF	562, 563, 564
Mixer, 105 MHz to 10.7 MHz	553
Mixer, 10.7 MHz to 455 kHz	552
Mixer, 30 MHz RF to 5 MHz IF	523
Mixer, 27 MHz	551
Mixer, 27 MHz AGC-able self-oscillating	562
Mixer, 27 MHz to 10.7 MHz	556
Mixer, 250 MHz RF to 50 MHz IF	516, 519
Model railroading control, SCR	404
Modifier scale factor and offset	394
Modulator, amplitude	524, 542
Modulator, balanced	520, 529
Modulator, linear amplitude	562
Modulator, pulse position	600
Modulator, pulse width	611
Monitor, fetal heartbeat	55
Monitor, heart rate	59
Monostable multivibrator	52
Morse code set	66
MOSFET analog chopper	262
MOSFET chopper	264
MOS at AC load interface	420
Motor control circuit	403
Motor control, pulse-width-modulated	410
Motor control, regulated DC	418
Motor control with timer, universal	486
Motor control, variable AC	429
Motor drive, constant-current	425
M6800 to 115-volt AC load interface	433
Motor speed control, triac	425
Motor-starting switch, triac static	405
Multimeter, 3½-digit autoranging	102
Multicrystal RF oscillator	363, 368
Multifunction counter, 9-digit	489
Multifunction counter, 100 MHz	464, 491
Multiplexed LCD, 3½-digit	239
Multiplexed LED, 8-digit	243

Multiplexed LED, 16-digit 244
Multiplexer, 8-channel analog 510
Multiplexer, 8-channel temperature 509
Multiplexer, 4-channel sequencing 511
Multiplexer matrix temperature 506
Multiplexer, 1-out-of-64 512
Multiplexer, 1-out-of-32 514
Multiplexer, 2-channel analog 510
Multiplexer, 2-level 8-channel 515
Multiplexer, 2-level 16-channel 511
Multiplier 374, 375, 376, 385, 390, 392, 396
Multiplier, analog 377
Multiplier/divider chip 9
Multiplier, four-quadrant 382
Multiplier, one-quadrant 378
Multiplier, two-quadrant 377, 384
Multiplying CMOS D/A converter, 8-bit 197, 205
Multiplying D/A converter 190
Multiplying D/A converter, 8-bit 200, 203
Multiplying D/A converter, 10-bit 184, 204
Multiplying D/A converter, 12-bit 195, 196, 198
Multivibrator, free-running 50
Multivibrator, indicating one-shot 51
Multivibrator, monostable 52
Multivibrator, one-shot 49, 51, 52

N

NAB equalizer amplifier 301
Nicad battery charger, 20 kHz 39
Nicad battery charger, 25 kHz 36
Nicad battery charger, 60-hertz 38, 40
Night light, automatic 68
Noise measurement, peak-to-peak 92
Noninverting amplifier 111, 116
Noninverting amplifier, programmable gain 596
Noninverting op amp 123
Noninverting power amplifier, 1-watt 587, 597
Noninverting sample-and-hold circuit 85
Notch filter 78, 81, 82
Number cruncher to SC/MP interface 260

O

Offset, modified scale factor and 394
One quadrant multiplier 378
One-shot multivibrator 49, 51, 52
One-shot multivibrator, indicating 51
Op amp 110
Op amp, differential-input composite 117
fast-settling 118
high-speed integrator 19
integrator 20
inverting high-slew-rate 123
inverting-only composite 119
inverting unity-gain 123
lagarithmic amplifier 16, 17
noninverting 123
specs tester 99
unity-gain 107, 111, 119
unity-gain feedforward 120
with high bandwidth 114
Operational integrator 18
Organ master oscillator, electronic 351
Oscillator, AF voltage-controlled 366
Oscillator, color TV 3.58 MHz 646, 648, 656
crystal 354
digtally programmed 352
800 kHz experimental RF 355
electronic organ master 351
52 MHz frequency-modulated 360
415 MHz frequency-modulated 365
fundamental crystal 362
gated 359
linear voltage-controlled 361
low-distortion 353
multicrystal RF 363, 368
1 MHz 356
1-second reference 358
100 MHz 358
105 MHz crystal 359
145 MHz RF 370
overtone crystal 364
quadrature 352
16.3 MHz 564
10 MHz 351
3.58 MHz injection-locked 612
200 MHz crystal 356
variable audio 355
voltage-controlled 358, 365
voltage-controlled crystal 367
voltage-controlled sine-wave 357, 370
watchdog light-sensitive 357
Oscilloscope calibrator 88
Output storage register 438
Overtone crystal oscillator 364
Overvoltage protection, triac 421

P

Paper-tape reader 457
Parallel adder, 8-bit 28
Parallel data acquisition system, 3½-digit 446
Peak-to-peak noise measurement 92
Percentage computer 383
Period counter, 2 MHz 469
40 MHz 478
100 MHz 487
Phase control, SCR 590
Phasemeter for sine waves 501
Phase-sensitive detector 25
Photo-activated alarm 74
Photo-activated logic driver 72, 73, 75
Photo-driven pulse streacher 71
Photoflash slave, light-triggered 65
Picoampere-to-voltage converter 160
Planar gas discharge display, 12-digit 242
Plug-in speed control 412
PLL clock 473, 481
PLL FM stereo multiplex decoder 132
PLL frequency synthesizer 369
Power amplifier, high-slew-rate servo motor 105, 30
Power control, AC 432
Power control, DC 428
Power controller, hysteresis-free 421
Power controller, SCR 419
Power D/A converter 174
Power down to SBC 80/20 interface 256
Power function generator 344
Power generation 371, 353
Power inverter, 5-volt 40-ampere 232
Power supply, digitally programmed 231
direct-line-operated 231
dual-voltage 227
full-wave SCR bridge 665
low-voltage 233
programmable 220
servo amplifier 230
12-volt 209
±15-volt 1-ampere 237
24-volt regulated 221
80-watt switching 234
Power switch, long time delay 473
Preamplifier, AF 284
general-purpose AF 311
high-gain AF 295, 309
low-noise 286, 293
servo motor 32, 33, 34, 35
stereo AF 276, 277, 278
Precent of deviation 378
Precision AGC circuit 601
Precision amplifier 605
Precision analog tachometer 91
Precision elapsed time/countdown timer 495
Precision high CMV analog selector 604, 607
Precision integrator 19
Precision integrator, long-term 19, 20
Precision programmable timer 467
Prescaler, fixed divide-by-ten 607
Processor, color TV chroma 641, 657, 664, 688
color TV chroma signal 627
FM stereo 124, 125, 130, 132, 143, 149, 152, 153
TV horizontal 629
TV video 663
video signal 619, 645, 646, 651, 660
Product detector 525
Programmable gain noninverting amplifier 596
Programmable mass spectrometer 504
Programmable power supply 220
Programmable timer, precision 467
Programmable triangle/square-wave gen 342
Programmable 100-hour timer 477
Programmed counter 498
Programmed timer, electronically 499
Programmer, PROM 453, 454, 455
Projection lamp voltage regulator 427
PROM programmer 453, 454, 455
Proportional heater control 430
Pulse amplifier 594
Pulse detector, missing 24
Pulse generator, single 348
Pulse position modulator 600
Pulse power amplifier 602
Pulse streatcher, photo driven 71
Pulse-width-modulated motor control 410
Pulse width modulator 611
Push-pull Darlington amplifier 587
Push-pull monolithic Darlington amplifier 597

Q

Quadrature oscillator 352

R

Ramp generator, precision 348
voltage-controlled 347
Ratio detector, FM 135
Ratiometer, wide-range 89
Ratiometric measurement 90
Reader, paper-tape 457
Read/write memory, CMOS static 449
Real-time 5-digit fluorescent diode display 250
Receiver, AM/FM broadcast 156
AM broadcast 154
remote control for TV 643
2-wire differential line 41
Record/playback amplifier 283
rectification, full-wave synchronous 592
half-wave rectification 610
Reference oscillator, 1-second 358
Reference, -10 volt 217, 218, 230
Register dual 4-bit storage 447
8-bit storage 456

output storage	438	
shift-right/shift-left	451	
Regulated battery charger, automotive	37	
Regulated DC motor control	418	
Regulated power supply, 24-volt	221	
Regulator, current	213	
current boost	211, 214	
fixed negative voltage	222	
precision temperature	508	
projection lamp voltage	427	
RMS	398, 400, 402, 430	
RMS voltage	229	
step-down switching	220	
switching	222	
5-volt	207, 212, 213, 215	
5-volt 5-ampere	208, 211	
5-volt adjustable	219	
5-volt remote shutdown	215	
5 volt shunt	215	
5-volt switching	216	
-5 volt	217	
-5V 4A current boost	218	
6-volt	215	
6-volt 5-ampere	210	
6-volt adjustable	219	
-6 volt	217	
12-volt	215	
12-volt adjustable	219	
-12 volt	217	
15-volt	207, 208, 213, 215	
15-volt 2-ampere	210	
15-volt adjustable	219	
15-volt switching	216	
-15 volts	214, 217	
24-volt	215	
24-volt adjustable	219	
40-volt 100 mA	209	
50-volt floating	223	
100-volt floating	226	
100-volt RMS	226	
Relay-contact protection, triac	407	
Relay, light-operated	73	
Relay-solid-state	427	
Relay-temperature-operated	502	
Relay, time-delayed	493	
triac solid-state	409	
Reminder, car turn signal	274	
Remote control of AC load	426	
Remote control of lamp or appliance	397, 399	
Remote control receiver for TV	643	
Remote shutdown 5-volt regulator	215	
Resolution control, color TV automatic	648, 655	
RF amplifier, 10 MHz	516	
30 MHz	516, 523, 529	
36 MHz	585	
45/58 MHz	519, 575, 576	
100 MHz	529	
105 MHz	563	
105 MHz gate-1-controlled	562	
105 MHz gate-2-controlled	554, 562	
105 MHz low-noise	550	
118 to 136 broadband	553	
144 to 175 MHz	539	
144 MHz to 175 MHz FM	533	
200 MHz	523	
450/432 MHz	578	
900 MHz	555	
30/60/100 MHz	517	
RF oscillator, multicrystal	363, 368	
145 MHz	370	
800 kHz experimental	355	
RF post amplifier, 45 MHz	544	
200 MHz	545	
80-watt	539	
2-state 80-watt	533	
RF scanner/monitor scanning logic	571, 573	
RF scanner/monitor, scanning logic for	558	

RGB video output, complete	612, 618	
Ring counter	476	
four-state	466	
RMS circuit, true	380, 387	
RMS regulator	398, 400, 402, 430	
100-volt	226	
RMS-to-DC converter	160	
high-crest-factor	165	
RMS voltage regulator	229	

S

SA A/D converter, 8-bit	202	
SA A/D converter, 12-bit	166, 173, 179, 181	
Sample-and-hold circuit	83, 84, 86	
Sample-and-hold circuit, digital	479	
Sample-and-hold circuit, high-speed	86	
Sample-and-hold circuit, inverting	84	
Sample-and-hold circuit, non-inverting	85	
Sample-and-hold, unity-gain	85	
Sawtooth generator	342	
Scanning digital thermometer, 6-channel	506	
Scanning logic for RF scanner/monitor	571, 573	
Scanning logic, 10-channel	558	
SCR alarm, light-operated	72	
SCR alarm, light-relay-operated	71	
SCR crowbar life tester	94	
SCR gas/smoke detector	46	
SCR model railroading control	404	
SCR phase control	590	
SCR power controller	419	
Selector, 4-channel analog data	590	
Selector, high CMV analog	604	
Selector, 1-out-of-4 data	591	
Self-generating FSK	22	
Sense amplifier, memory core	439	
Sequencer, time	467	
Sequencing multiplexer, 4-channel	511	
Sequential AC flasher	273	
Sequential UJT-SCR timer	472	
Series chopper	262, 263	
Serial data transfer circuit	437	
Series motor direction control	407	
Series motor speed control	407	
Series-shunt analog switch	14	
Series-shunt chopper	262, 263	
Series switch, light-operated 6 kV	76	
Servo amplifier power supply	230	
Servo motor amplifier, line-operated	31	
Servo motor preamplifier	33, 34, 35	
Servo motor power amplifier	30	
Servo motor preamplifier	32	
Set point comparator, digitally programmed	606	
Shaft encoder, electro-optical	586	
Shift-right/shift-left register	451	
Shunt motor, direction control for	432	
Shunt motor, speed control for	432	
Sign-magnitude converter, 14-bit	192	
Sign-magnitude D/A converter	160	
Signal-level envelope detector	26	
Signal processor, color TV chroma	627	
Signal processor, TV video	645, 646, 651, 660	
Signal processor, video	619	
Single pole frequency compensation	121	
Sine function	373	
Sine function generator	345	
Sine-wave oscillator, voltage-controlled	357	
Siren, fire	65	
Siren, whooper	64	
Slope and voltage detection, FSK	21	

Solid-state relay, triac	427	
Solid-state-relay, triac	409	
Sorter, component	95	
Sound channel, 4.5 MHz TV	647	
Sound channel, TV	620, 654	
Sound channel, TV 4.5 MHz	649, 659	
Sound IF amplifier, 4.5 MHz color TV	613	
Sound system, complete TV	623	
Source, 100 mA constant-current	211	
Space mark generator	343	
Speed control for shunt motor	432	
Speed control, half-wave variable motor	411	
Speed control, plug-in	412	
Speed control, pulse-width-modulated	410	
Speed control, series motor	407	
Speed control, triac motor	425	
Speed control, variable	415, 417	
SQ logic decoder, 4-channel	134, 137	
Squarer	389	
Squares, difference of	378	
Square root of the sum of the squares	386	
Square rooter	372, 385, 388, 391, 394	
SSB linear amplifier, 20-watt 1.6 to 30 MHz	561, 584	
Stable instrument amplifier	604	
Staircase generator	346, 347, 349	
Stereo AF amplifier, 2-watts-per-channel	284	
Stereo AF preamplifier	276, 277, 278	
Stopwatch/timer, 4-digit 7 function	485	
Storage register, dual 4-bit	447	
Storage register, 8-bit	456	
Storage register, output	438	
Subcarrier regenerator, color TV	637, 662	
Summing amplifier	120	
Summing/scaling amplifier	594	
Switch, AC-controlled triac	424	
Switch, bistable	14	
Switch, DPDT FET	13	
Switch, high-low AC	413	
Switch, light-operated 6 kV series	76	
Switch, long time delay power	473	
Switch, low-voltage-controlled	429	
Switch, 3-position static	424	
Switch, triac zero-point	411	
Switch, touch	406	
Switch, video	62	
Switching 5-volt regulator	216	
Switching 15-volt regulator	216	
Switching power supply, 80-watt	235	
Switching regulator	222	
Switching regulator, step-down	220	
Sync delay, TV vertical	666	
Synchronous counter, 4-decade	463	
Synchronous rectification, half-wave	610	
Synchronous rectification, full-wave	592	
Synthesizer, frequency	345, 350	
Synthesizer, PLL frequency	369	

T

Tachometer frequency counter	494, 498	
high-precision	91	
precision analog	91	
Tape recorder AF amplifier	292	
Tape recorder position controller	456	
Transmitter, 10-/25-watt marine band	569	
Temperature alarm, thermistor-operated	504	
Temperature differential measurement	506	

Temperature, matrix temperature	506
Temperature multiplexer, 8-channel	509
Temperature-operated relay	502
Temperature regulator, precision	508
Tester	
buzz box continuity	88
coil	88
1 MHz bandpass filter	79
op amp specs	99
SCR crowbar life	94
TTL SOA	93
TGS gas/smoke detector	43, 44, 47
Thermister-operated temperature alarm	504
Thermister thermometer	507
Thermometer	
acoustic	503
electronic	509
6-channel scanning digital	506
thermister	507
Thermostat, electronic	501
Three-position static switch	424
Thyristor feedback control	419
Thyristor half-wave control	412
Time delay	471
Time delay, digitally controlled	459
Time-delayed relay	493
Time delay, log duration	459
Time delay power switch, long	473
Time delay, ultralong	465, 479
Time delay with constant-current charging	461
Time-dependent light dimmer	401
Timer	
electronically programmed	499
4-digit 7-function	485
precision, programmable	467
programmable 100-hour	477
sequential UJT-SCR	472
20-minute	461
universal motor control with	486
Time sequencer	467
Tone generator, FSK	23
Touch switch	406
Tracking voltage reference	212
Transducer linearization	593
Transmitter	
80-watt 175 MHz FM	541
4-to-20 mA current	601
118 MHz to 136 MHz AM	522
7-watt 118 to 136 MHz AM	549
13-watt 188 to 136 MHz AM	549
27 MHz AM	565
2.5-watt AM	522
Triac AC static contactor	424
Triac control	423
Triac control circuit	397, 399
Triac control, 240-volt	431
Triac full-wave control	413
Triac heater temperature control	422
Triac light dimmer, 800-watt	423, 426
Triac motor speed control	425
Triac overvoltage protection	421
Triac relay-contact protection	407
Triac solid-state relay	409
Triac static motor-starting switch	405
Triac zero-point switch	411
Trigger circuit, full-wave	434
Trigger switch flasher	265
Trigger, triac	403
Triode display, 8-digit fluorescent	249, 250
Trombone, electronic	64
True RMS circuit	380, 387
TTL SOA tester	93
TTY/RS-232C to MC6850 interface	259
TTY-to-A/D converter interface	253
Tuner	
AM broadcast	127, 128, 131, 132, 133, 139, 146
AM/FM stereo	140, 145

Tuner control/memory, electronic	678
Tuner	
FM	157
TV VHF	684
TV VHF varactor	673, 674
27 MHz autodyne	551, 577
TV AFT	681
TV AFT/AFC	615, 680
TV AFT system	619
TV chroma IF amplifier	687
TV delayed horizontal sawtooth circuit	667
TV FM sound IF	616, 617
TV FM sound system	614
TV 4.5 MHz sound channel	649
TV horizontal AFC and oscillator	639
TV horizontal AFC and oscillator	635
TV horizontal processor	629
TV IF amplifier	670, 671
TV IF amplifier, linear	636
TV IF system, low-noise	669
TV sound channel	616, 620, 654
TV sound channel, 4.5 MHz	647
TV sound system	
complete	623
4.5 MHz	615
TV UHF varactor tuner	674
TV vertical circuit, complete	631
TV vertical deflection system	656
TV vertical sync delay	666
TV VHF tuner	684
TV VHF varactor tuner	673
TV video IF amplifier	617, 621, 630, 633, 652, 685, 686
TV video IF amplifier, complete	679, 683
TV video processor	663
TV video signal processor	645, 646, 651, 660
Two pole frequency compensation	118
Two-quadrant multiplier	377, 384
Two-stage 60 MHz IF amplifier	521
Two-wire differential line receiver	41
Two-wire differential line system	41

U

UHF amplifier	
10-watt 450 to 470 MHz	542
13-watt	537
25-watt	526
25-watt 450 to 470 MHz	543
220 to 225 MHz	537
450 MHz to 512 MHz	526
UHF varactor tuner, TV	674
Ultralong time delay	465, 479
Unit counter, 4-digit	458
Unity-gain feedforward op amp	120
Unity-gain op amp	107, 111, 119
Unity-gain op amp, inverting	123
Unity-gain sample-and-hold circuit	85
Unity-gain voltage follower	113, 114
Universal counter	
9-digit	475
10 MHz	480, 483
Universal motor control with timer	486
Up/down counter, 8-digit	492

V

V/F converter	161, 164, 170, 178, 181, 182, 185
V/F converter, differential-input	165
V/F converter with data transfer system	199
V/F converter with 10 kHz FS	193
Varactor tuner, TV UHF	674
Varactor tuner, TV VHF	673
Variable AC motor control	429
Variable audio oscillator	355
Variable gain amplifier	387

Variable speed control	415, 417
Variable voltage regulator	221
VCO	358, 365
VCO, AF	366
VCO and mixer for CB	585
VCO, crystal	367
VCO, linear	361
VCO, sine-wave	370
Vector computer	379, 386, 389
Vector magnitude function	381
Vertical circuit, complete TV	631
Vertical deflection system, TV	656
Vertical sync delay, TV	666
VHF tuner, TV	684
VHF varactor tuner, TV	673
Video amplifier	60, 61
Video amplifier, complete color TV	677
Video amplifier, 30 MHz	63
Video amplifier, wide-band	62
Video IF amplifier, complete TV	679, 683
Video IF amplifier, TV	617, 621, 630, 633, 652, 685, 686
Video IF system, color TV	620
Video output, complete RGB	612, 618
Video processor, TV	663
Video signal processor	619
Video signal processor, TV	645, 646, 651, 660
Video switch	62
Voltage control, half-wave average	434
Voltage-controlled amplifier	598
Voltage-controlled crystal oscillator	367
Voltage-controlled current source	219
Voltage-controlled low-pass filter	78
Voltage-controlled oscillator	358, 365
Voltage-controlled oscillator, AF	366
Voltage-controlled oscillator, linear	361
Voltage-controlled ramp generator	347
Voltage-controlled sine-wave oscillator	357, 370
Voltage follower	116, 123
Voltage follower, high-current output	105
Voltage follower, high-impedance input	105
Voltage follower, unity-gain	113, 114
Voltage offset null	595
Voltage reference, tracking	212
Voltage regulator, fixed negative	222
Voltage regulator, projection lamp	427
Voltage regulator, RMS	229
Voltage regulator, variable	221
Voltmeter, dual-polarity 3½-digit	100
Voltmeter, 3½-digit	104
Voltmeter, 2½-digit	101

W

Warning light flasher	266
Water seepage alarm	505
Watt-hour meter	87
Wattmeter, AF	89
Weighing system, 199 lb 15 oz max	505
Weighted averaging amplifier	108
Wein bridge oscillator, stabilized	365
Wide-band differential amplifier	517
Wide-band video amplifier	62
Wide-range ratiometer	89
Wireless microphone, FM	574
Whooper siren	64

Z

Zero-point switch, triac	411